# Crowdfunding for Sustainable Entrepreneurship and Innovation

Walter Vassallo
*MC Shareable, Monaco*

A volume in the Advances in Business Strategy and Competitive Advantage (ABSCA) Book Series

IGI GLOBAL
DISSEMINATOR OF KNOWLEDGE
www.igi-global.com

Published in the United States of America by
Business Science Reference (an imprint of IGI Global)
701 E. Chocolate Avenue
Hershey PA, USA 17033
Tel: 717-533-8845
Fax: 717-533-8661
E-mail: cust@igi-global.com
Web site: http://www.igi-global.com



Library of Congress Cataloging-in-Publication Data

Names: Vassallo, Walter, 1975- editor.
Title: Crowdfunding for sustainable entrepreneurship and innovation / Walter Vassallo, editor.
Description: Hershey, PA : Business Science Reference, [2017] | Includes bibliographical references and index.
Identifiers: LCCN 2016017494| ISBN 9781522505686 (hardcover) | ISBN 9781522505693 (ebook)
Subjects: LCSH: Crowdfunding. | Venture capital. | New products. | Technological innovations. | Entrepreneurship.
Classification: LCC HG4751 .C786 2017 | DDC 658.15/224--dc23 LC record available at https://lccn.loc.gov/2016017494

This book is published in the IGI Global book series Advances in Business Strategy and Competitive Advantage (ABSCA) (ISSN: 2327-3429; eISSN: 2327-3437)

British Cataloguing in Publication Data
A Cataloguing in Publication record for this book is available from the British Library.



For electronic access to this publication, please contact: eresources@igi-global.com.

# Advances in Business Strategy and Competitive Advantage (ABSCA) Book Series

Patricia Ordóñez de Pablos
*Universidad de Oviedo, Spain*

ISSN: 2327-3429
EISSN: 2327-3437

Mission

Business entities are constantly seeking new ways through which to gain advantage over their competitors and strengthen their position within the business environment. With competition at an all-time high due to technological advancements allowing for competition on a global scale, firms continue to seek new ways through which to improve and strengthen their business processes, procedures, and profitability.

The **Advances in Business Strategy and Competitive Advantage (ABSCA) Book Series** is a timely series responding to the high demand for state-of-the-art research on how business strategies are created, implemented and re-designed to meet the demands of globalized competitive markets. With a focus on local and global challenges, business opportunities and the needs of society, the **ABSCA** encourages scientific discourse on doing business and managing information technologies for the creation of sustainable competitive advantage.

Coverage

- Tacit Knowledge
- Strategic management
- Strategy Performance Management
- Value Creation
- Economies of Scale
- Adaptive Enterprise
- Competitive Strategy
- Foreign Investment Decision Process
- Cost Leadership Strategy
- Customer-Orientation Strategy

IGI Global is currently accepting manuscripts for publication within this series. To submit a proposal for a volume in this series, please contact our Acquisition Editors at Acquisitions@igi-global.com or visit: http://www.igi-global.com/publish/.

The Advances in Business Strategy and Competitive Advantage (ABSCA) Book Series (ISSN 2327-3429) is published by IGI Global, 701 E. Chocolate Avenue, Hershey, PA 17033-1240, USA, www.igi-global.com. This series is composed of titles available for purchase individually; each title is edited to be contextually exclusive from any other title within the series. For pricing and ordering information please visit http://www.igi-global.com/book-series/advances-business-strategy-competitive-advantage/73672. Postmaster: Send all address changes to above address. 

## Titles in this Series

*For a list of additional titles in this series, please visit: www.igi-global.com*

*Analyzing the Relationship between Corporate Social Responsibility and Foreign Direct Investment*
Marianne Ojo (North West University, South Africa)
Business Science Reference • copyright 2016 • 253pp • H/C (ISBN: 9781522503057) • US $185.00 (our price)

*Handbook of Research on Driving Competitive Advantage through Sustainable, Lean, and Disruptive Innovation*
Latif Al-Hakim (University of Southern Queensland, Australia) Xiaobo Wu (Zhejiang University, China) Andy Koronios (University of South Australia, Australia) and Yongyi Shou (Zhejiang University, China)
Business Science Reference • copyright 2016 • 743pp • H/C (ISBN: 9781522501350) • US $365.00 (our price)

*Ethical and Social Perspectives on Global Business Interaction in Emerging Markets*
Minwir Al-Shammari (University of Bahrain, Bahrain) and Hatem Masri (University of Bahrain, Bahrain)
Business Science Reference • copyright 2016 • 372pp • H/C (ISBN: 9781466698642) • US $185.00 (our price)

*Strategic Approaches to Successful Crowdfunding*
Djamchid Assadi (Burgundy School of Business, France)
Business Science Reference • copyright 2016 • 378pp • H/C (ISBN: 9781466696044) • US $205.00 (our price)

*Handbook of Research on Entrepreneurship in the Contemporary Knowledge-Based Global Economy*
Neeta Baporikar (University of Pune, India)
Business Science Reference • copyright 2016 • 607pp • H/C (ISBN: 9781466687981) • US $345.00 (our price)

*Competitive Strategies for Academic Entrepreneurship Commercialization of Research-Based Products*
Anna Szopa (Jagiellonian University, Poland) Waldemar Karwowski (University of Central Florida, USA) and David Barbe (University of Maryland, College Park, USA)
Business Science Reference • copyright 2015 • 317pp • H/C (ISBN: 9781466684874) • US $195.00 (our price)

*Incorporating Business Models and Strategies into Social Entrepreneurship*
Ziska Fields (University of KwaZulu-Natal, South Africa)
Business Science Reference • copyright 2016 • 362pp • H/C (ISBN: 9781466687486) • US $190.00 (our price)

*Technological Solutions for Sustainable Business Practice in Asia*
Patricia Ordóñez de Pablos (Universidad de Oviedo, Spain)
Business Science Reference • copyright 2015 • 375pp • H/C (ISBN: 9781466684621) • US $200.00 (our price)

# Editorial Advisory Board

# Table of Contents

**Foreword** .......... xiv

**Preface** .......... xvi

**Chapter 1**
The Crowd Economy: From the Crowd to Businesses to Public Administrations and Multinational Companies .......... 1
*Epi Ludvik Nekaj, Crowdsourcing Week, USA*

**Chapter 2**
Towards a New Economy: Co-Creation and Open Innovation in a Trustworthy Europe .......... 20
*Domenico Rossetti di Valdalbero, European Commission, Belgium*
*Bogdan Birnbaum, European Commission, Belgium*

**Chapter 3**
Crowdsourcing New Tools to Start Lean and Succeed in Entrepreneurship: Entrepreneurship in the Crowd Economy .......... 37
*Priti Ambani, The Next Billion, New Zealand*

**Chapter 4**
Crowdsourcing Social Innovation: Towards a Collaborative Social Capitalism .......... 54
*Emanuele Musa, Babele, Romania*

**Chapter 5**
The World's Oldest Profession - Now and Then: Disruption of the Commercial Banking Model .......... 78
*Steven A. Cinelli, REMARQ, USA*

**Chapter 6**
The Crowdfunding Market, Models, Platforms, and Projects .......... 90
*Melek Demiray, Istanbul Technical University, Turkey*
*Sebnem Burnaz, Istanbul Technical University, Turkey*
*Yonca Aslanbay, Istanbul Bilgi University, Turkey*

**Chapter 7**
Horizontal Crowdfunding Platforms: Socio-Economic Impact and Exogenous/Endogenous System of Rules .......... 127
*Tomaso Greco, Università degli Studi dell'Insubria, Italy*

**Chapter 8**
Vertical and Community-Based Crowdfunding Platforms ........ 140
*Tomaso Greco, Università degli Studi dell'Insubria, Italy*

**Chapter 9**
The Crowdfunding Communities and the Value of Identification for Sustainability of Co-Creation ........ 155
*Melek Demiray, Istanbul Technical University, Turkey*
*Yonca Aslanbay, Istanbul Bilgi University, Turkey*

**Chapter 10**
Principles in Crowdfunding Benefits and Validation Tools: How to Ensure Campaign Success ........ 175
*Paul Morrell Niederer, Crowdsourcing Week, Australia*

**Chapter 11**
Crowdfunding as a Marketing Tool ........ 187
*Fabio Allegreni, Crowd Advisors, Italy*

**Chapter 12**
Legal Aspects and Regulation in Crowdfunding: Comparisons across Countries ........ 204
*Alessandro Maria Lerro, Associazione Italiana Equity Crowdfunding, Italy*

**Chapter 13**
How to Launch a Successful Crowdfunding Campaign ........ 224
*Ronald Kleverlaan, CrowdfundingHub, The Netherlands*

**Chapter 14**
Real Estate Crowdfunding: 2015 and Beyond ........ 238
*Steven A. Cinelli, REMARQ, USA*

**Chapter 15**
The Application of Crowdfunding to the Energy Sector ........ 266
*Chiara Candelise, Bocconi University, Italy*

**Chapter 16**
A Crowd-Funder Value (CFV) Framework for Crowd-Investment: A Roadmap for Entrepreneurial Success in the Contemporary Society ........ 288
*Payal Jain, University of Northampton, UK*

**Compilation of References** ........ 310

**About the Contributors** ........ 329

**Index** ........ 333

# Detailed Table of Contents

**Foreword** ............................................................................................................................ xiv

**Preface** ............................................................................................................................ xvi

**Chapter 1**
The Crowd Economy: From the Crowd to Businesses to Public Administrations and Multinational Companies ............................................................................................................................ 1
*Epi Ludvik Nekaj, Crowdsourcing Week, USA*

A digital transformation is underway. One that is redefining the essence of human interaction and with ideas, share unused resources and create new on-demand services that are customisable and unique. These are only a few examples of real productivity that when layered on the Internet creates an abundance of resources and opportunity. This people-powered abundance is called the crowd economy. It is the way the society lives, works and plays. There is a new paradigm shift that challenges traditional notions of the "norm" while expanding possibilities. The hallmark of the digital age is social connections that are boosted by the web and mobile networks. These technological advances have taken collaboration and cooperation to a level never seen before. Social connections through the web have gone beyond social media likes and shares and has evolved into social productivity - a phenomenon that arises when networked crowds collaborate to solve problems, raise funds, and come up with innovative ideas and solutions.

**Chapter 2**
Towards a New Economy: Co-Creation and Open Innovation in a Trustworthy Europe .................... 20
*Domenico Rossetti di Valdalbero, European Commission, Belgium*
*Bogdan Birnbaum, European Commission, Belgium*

This chapter focuses on the transition towards a new economy, where Europe can find untapped sources of growth and employment, renewing the legitimacy of public policy-making, especially through greater citizens' involvement and by delivering better public services for all. The chapter analyses the perspectives of Europe in 2050 following three dimensions: Global and European demographic changes; Economics, energy, environment and technological prospects; Geopolitics and governance. The concepts of trust, co-creation, sharing and sustainable economy, innovation, and social innovation are presented as the key ingredients of the "new" economy.

**Chapter 3**
Crowdsourcing New Tools to Start Lean and Succeed in Entrepreneurship: Entrepreneurship in the Crowd Economy ........ 37
*Priti Ambani, The Next Billion, New Zealand*

The very essence of the new entrepreneur is shattering tradition. On the heels of the new social internet, we are seeing the rise of the solo-entrepreneur or intrapreneur who embraces globalisation, failure and successes and collaboration. Powerful networks brought together by crowdsourcing are supplying the tools to start lean, innovate and solve complex problems. This chapter by Priti Ambani explores the changing ecosystem and the effect of networked crowds on starting lean and succeeding with entrepreneurship.

**Chapter 4**
Crowdsourcing Social Innovation: Towards a Collaborative Social Capitalism ........ 54
*Emanuele Musa, Babele, Romania*

The growing awareness of humanity's finite resources and recognition of the limitations of one-off projects are prompting step changes in development planning. Sustainable development addresses the limitations of current practices; its aim is to achieve the triple bottom line of economic prosperity, environmental quality and social equity, meeting the needs of present society without compromising resources for future generations. Collective intelligence is considered by both Charles Leadbeater (former advisor to Tony Blair) and MIT as one of the most powerful ways to tackle complex problems, like climate change. This chapter explores the principles of crowdsourcing, its applications and the main trends. It presents theories, practices and examples of the use of crowdsourcing to innovate in the area of sustainable development for the common good. It announces the rise of collective brain-power to the challenge of creating better and more effective forms of civic and social engagement to solve problems on a world scale.

**Chapter 5**
The World's Oldest Profession - Now and Then: Disruption of the Commercial Banking Model ........ 78
*Steven A. Cinelli, REMARQ, USA*

Modern banking found its roots during the Renaissance period casted by the European merchant banks. Their success was due in large part to their aggregation, absorption and deployment of information about borrowers, structures and markets. In the 21st century, banking again is being advanced due to insights developed by vast amounts of information and data, this time gathered and managed through new technologies and models, in quest of efficiency, improved risk management and improved portfolio performance. New entrants into the business of banking operate outside of existing regulatory structures, and may enjoy a level of competitive flexibility compared to existing commercial banks. Might this portend the end of the so-called modern commercial banking model, or might it serve as a strategic imperative for the banks to adapt to innovation?

**Chapter 6**
The Crowdfunding Market, Models, Platforms, and Projects ............................................................ 90
*Melek Demiray, Istanbul Technical University, Turkey*
*Sebnem Burnaz, Istanbul Technical University, Turkey*
*Yonca Aslanbay, Istanbul Bilgi University, Turkey*

The changing structure and intensity of competition on one hand, the ongoing pressure on creating differentiated offerings lead to seek out new ways to improve technology as well as new ways to perform business. Technological developments enable market participants or other interest parties to engage in the process of both innovation and production, resulting with an online social network market model of co-creation. This chapter focuses on describing different aspects of recently developing crowdfunding market models where new projects are offered and funded by large number of individuals from general public. The current extent and dynamics of the growing crowdfunding market, various models of crowdfunding, crowdfunding platforms and the projects that are crowdfunded are analyzed through the chapter by giving examples and comparative comments. The analysis of the market is concluded by the foreseen future trends.

**Chapter 7**
Horizontal Crowdfunding Platforms: Socio-Economic Impact and Exogenous/Endogenous System of Rules ............................................................ 127
*Tomaso Greco, Università degli Studi dell'Insubria, Italy*

In recent years arose a multitude of crowdfunding platforms of different nature and with different goals. In this chapter some peculiar aspects of crowdfunding platforms will be examined, both in terms of their socio-economic impact and in terms of the exogenous and endogenous system of rules with which they interact. Crowdfunding is not governed exclusively by the rules of law and the regulations of the platforms that host it. Every campaign involves the construction of a community, with its specific rules, that are in part pre-established by the campaign creator and in part the result of the interaction of the community. However, these rules do not live in a closed set, building a case history to which the future campaigns will not necessarily have to conform, but with high probability they will use as a parameter. Although it is not possible to identify the rules that are common to all the crowdfunding campaigns, could be observed an evolution mostly harmonic between different campaigns with respect to the methods, the mechanisms of involvement and, of course, the rules.

**Chapter 8**
Vertical and Community-Based Crowdfunding Platforms ............................................................ 140
*Tomaso Greco, Università degli Studi dell'Insubria, Italy*

If the inherent nature of the horizontal crowdfunding platforms makes the relationship between campaign creator and backers basically without intermediation of the platform, in the case of vertical platforms there are different and sometimes complex levels of intermediation. The vertical crowdfunding platforms can be considered more than a subset of the horizontal crowdfunding, as an independent system of relationships and rules which has in common with the horizontal crowdfunding a large part of the fundraising methods, but that is also profoundly different with respect to the aggregative procedures and the system of rules.

**Chapter 9**
The Crowdfunding Communities and the Value of Identification for Sustainability of Co-Creation ..... 155
*Melek Demiray, Istanbul Technical University, Turkey*
*Yonca Aslanbay, Istanbul Bilgi University, Turkey*

For a sustainable future, the actors of the market need a change in spirit that will elevate the life-chances of people by being an "empowered part" of the economic system. A participative, open knowledge economic and market system that will ensure the material well-being, the social visibility, the happiness and the consciousness of the individual is essential. Crowdfunding, as a recent online social community network market model is to be a new socio-technical system of co-creation through self "making and funding". The aim of this study is to discuss the role of identification in co-creation for sustainability of newly rising crowdfunding communities. In line with this objective, the following two aspects are clarified: the key characteristics of crowdfunding platforms as online communities and the role of identification for co-creation in online crowdfunding communities.

**Chapter 10**
Principles in Crowdfunding Benefits and Validation Tools: How to Ensure Campaign Success ..... 175
*Paul Morrell Niederer, Crowdsourcing Week, Australia*

This chapter provides the key benefits of a crowdfunding campaign together with a rich variety of validation tools provided to enhance the chance of a crowdfunding campaigns success. Benefits cannot be achieved in a crowdfunding campaign if it is not successful. Thorough validation of the campaign upfront can heighten success. The author in an internationally respected crowdfunding authority who has managed over 300 crowdfunding campaigns through the platforms he has been involved with. The author lists 11 benefits and 11 validation tools together with his own road tested processes plus a plethora of links to tools to assist both the novice and experienced crowdfunder.

**Chapter 11**
Crowdfunding as a Marketing Tool ..... 187
*Fabio Allegreni, Crowd Advisors, Italy*

Most people, at a first sight, perceive Crowdfunding as a specific and exciting process to raise money. Of course, this is correct, but it is not the whole truth. A successful crowdfunding campaign must be addressed to a community of people, rather than to an indiscriminate crowd, which is also typical for any marketing strategy. Leveraging on this key element, crowdfunding can amplify almost any kind of communication activity, provided it is able to meet the needs of a community and to provide the benefits that the community expects. Therefore, if well exploited, crowdfunding is a powerful tool in the hands of marketers, whose potential goes far beyond the evidence of pre-selling a new product, to reach new marketing horizons such as branding, business development, CRM, credit scoring and even HR management.

**Chapter 12**
Legal Aspects and Regulation in Crowdfunding: Comparisons across Countries ............................ 204
*Alessandro Maria Lerro, Associazione Italiana Equity Crowdfunding, Italy*

Crowdfunding started with the donation-based and reward-based models, but soon its disruptive capacity to innovate the traditional approach to finance made it evolve into an alternative finance instrument, suitable to fund new business. Peer to peer lending and equity crowdfunding are investment instrument that can truly sustain the growth of businesses and generate profits. As any investment instrument, lending and equity crowdfunding are heavy regulated in order to protect investors and maintain a proper trust in the system. This chapter will go through the American and European main regulatory principles, offering a picture of the overall approach to this innovative and disruptive financing tool.

**Chapter 13**
How to Launch a Successful Crowdfunding Campaign ..................................................................... 224
*Ronald Kleverlaan, CrowdfundingHub, The Netherlands*

This chapter provides insights into the important characteristics of a successful crowdfunding campaign. It starts with providing the fundamentals of crowdfunding and shows the reasons why it is important to find the right target groups and communities for a campaign. It describes in details the Return on Involvement model, what the funders want in return for their support. In the second part the additional results of crowdfunding are discussed, such as market research, marketing, brand loyalty and the possibility to use crowdfunding to raise additional funding. The last part of this chapter will focus on the background of setting up and running a successful campaign.

**Chapter 14**
Real Estate Crowdfunding: 2015 and Beyond ........................................................................... 238
*Steven A. Cinelli, REMARQ, USA*

Like a juggernaut, "crowdfunding" has hit the media, the financial markets and the common narrative by storm. Incipient in many ways, the intermediation of financing transactions online has become a billion-dollar industry. As technology has advanced, even recreated, industries, there seems none more primed than "finance", an inherently information business. By creating improved efficiency, both art and science of finance are enriched. A 21st century vestige of the 1980's syndication business, real estate seems to be enjoying the fruits of the crowd, with $1 billion of property financings conducted online in 2014, with an expected $2.5 billion this year. For sponsors and investors, there appears legitimacy to the online approach, underscored by the level of venture capital now finding home in this burgeoning sector. Yet, like its progenitor, might real estate crowdfinance find legislative, regulatory, and practical headwinds, stunting its progress? Still early, with business models, scalability and sustainability still suspect, the current momentum seems promising.

**Chapter 15**
The Application of Crowdfunding to the Energy Sector ........ 266
*Chiara Candelise, Bocconi University, Italy*

Crowdfunding in energy begins as a response to reduced investments (both governmental and private) into the transition to decarbonized energy systems and to the spread of innovative business models and approaches conductive of greater participation of citizens and communities in distributed renewable energy projects. This chapter presents results of a worldwide overview of the use of crowdfunding in the energy industry. Evidence gathered from available energy crowdfunding platforms highlight a very new, but quite dynamic sector. The crowdfunding tool has been applied in most of its forms, ranging from peer-to-business lending to pure donation, with strong environmental and social mission and the explicit aim of increasing participation of citizens in sustainable energy investment. Evidence also shows that, despite maintaining their environmental and clean energy focus, some energy platforms have begun to move from niche, grass root initiatives into larger projects and collaboration with energy private sector and institutional finance.

**Chapter 16**
A Crowd-Funder Value (CFV) Framework for Crowd-Investment: A Roadmap for Entrepreneurial Success in the Contemporary Society ........ 288
*Payal Jain, University of Northampton, UK*

Crowd-funding is used by business and social entrepreneurs to fund projects that impact society in many ways. Although crowd-funders fund and support entrepreneurial projects, stakeholders are less aware of crowd-funder motivations and behaviours towards products/services, which impact on the success of these projects. The purpose of this research, therefore, is two-fold: a) development of crowd-funder value framework for understanding crowd-funder motivations; and b) provision of robust theoretical basis to the construct of 'crowd-funding'. Different social constructs, such as social identity, and symbolic, crowd and relational social capital related to crowd-funding are discussed based on symbolic convergence theory of communication and social identity theory. This framework will act as a roadmap to understand how crowd-funder motivations impact and create value for individuals, organisations and society, and inform how social, environmental and economic value and impact can be maximised through crowd-funding initiatives.

**Compilation of References** ........ 310

**About the Contributors** ........ 329

**Index** ........ 333

# Foreword

The financial industry is being disrupted. Fintech solutions are breaking up the traditional banking system and community based finance solutions are challenging the current way of investing in businesses. New financing models such as crowdfunding and community empower are democratizing access to credit and are offering individuals and communities the possibility to support, co-create, contribute and invest in public and private initiatives they think are important. Basically, we are re-inventing finance.

The elements making crowdfunding successful are not new. In fact, collecting funds from a group of people for a specific project is traditionally the most common way to support innovation and entrepreneurship. Because funding comes from strong connections in a network, it provides much more than just money. It creates a strong community of ambassadors who will help to realize and support the project with more than just their capital contributions. They will commit their time, energy and network to make it successful. This creates a much more sustainable model and will also focus on the social return next to their financial return.

Society has changed in the last century from a decentralized, local economy with strong social connections between investors and projects, to a more centralized investment and banking model where the connection between investors and projects disappeared. These models were based on the assumption that individuals are only interested in financial returns. History shows that people are also very much interested on the social impact and social return. By using web-based collaboration tools, online payment solutions and blockchain solutions a lot of small and large initiatives are currently experimenting with new financial and social models to bring these social connections back in sustainable business models.

I am very proud we found the most experienced international academics, lawyers, entrepreneurs and practitioners in crowdsourcing and crowdfunding to contribute to this publication and give their expert views on these trends. A lot of great examples of crowdsourcing and crowdfunding initiatives are globally showing good results. These examples were researched and documented in separate publications, but never before such a wide range of experts joined forces to share their knowledge in a combined publication. This book is a standard reference for future research and academic publications, but also a very interesting read for all the different stakeholders in our new, but fast growing industry.

The authors covered all areas of this emerging industry. From the basic definitions of crowdsourcing and crowdfunding to detailed analysis on crowdsourcing for co-creating innovation and new businesses. The book embraces a vast number of analyses, which include, among others, the different crowdfunding models, the use of crowdsourcing and crowdfunding in different industries, the legal and regulatory aspects, the market trends, the added value of crowdfunding for marketing and market research.

This book is very insightful for readers who are interested to get an overview of this new, fast growing industry, but even the most experienced practitioner will learn something new.

I will encourage everyone not just to read these great chapters, but also to start experimenting and implementing the solutions these experts describe in their articles. These crowd and community-based finance solutions offer a lot of opportunities to create sustainable, more innovative businesses that will focus on a combination of financial and social returns.

As written in the Preface,

*innovation means to anticipate future needs, create new business models, without losing revenue. There are tremendous unexplored opportunities to catch thanks to crowdsourcing and crowdfunding, two sides of the same coin that lead to a revolution of social and economic models. The reading of this book will help to understand the changes taking place to grasp opportunities and advantages.*

Together we can create a more balanced financial system that is not focused only on financial returns, but will also incorporate the social returns, resulting in more sustainable and inclusive business models.

*Ronald Kleverlaan*
*CrowdfundingHub, The Netherlands*
*April 2016*

**Ronald Kleverlaan** *is the Founder of CrowdfundingHub – European Expertise Centre for Alternative and Community, as well as Finance and Special Advisor Alternative Finance European Commission.*

# Preface

Trying to predict the future has little value unless you first seek to understand the forces that drive the changes, the subtle nuances that can create future opportunities. Innovation means to anticipate future needs, create new business models, without losing revenue.

We always talk about vision, by reading this book anyone can become a "visionary thinker", one who knows how to translate trends and changes into unique opportunities.

The book is not limited to innovation. Innovation is a driver which results in a positive change, that makes life better. The book provides a precise view of the World to come, a broad view of the Knowledge Era in which we live, in order to understand the changes taking place to grasp opportunities and advantages.

I have been always fascinated by looking at the evolution of economy and society. When I graduated in 1999 I prepared a thesis on how internet would have changed the World, disrupting old fashion schemes, imagining new business models, new opportunities, new C2C, P2C and P2P environments. After many years I realised that the timing or place to present that research was wrong, but still there is a need to discover the potentials offered by the analyses of our disruptive evolution and the characteristics of the Era in which we live, the so called Knowledge Era.

There are tremendous unexplored opportunities to catch thanks to crowdsourcing and crowdfunding, two sides of the same coin that lead to a revolution of social and economic models.

By definition, crowdsourcing is the process of obtaining needed services, ideas, or content by soliciting contributions from a large group of people, and especially from an online community, while crowdfunding refers to open calls to the public, generally via internet, to finance a project through either a donation, a monetary contribution in exchange for a reward, product pre-ordering, lending, or investment. It is wrong and misleading to see crowdfunding as an emerging alternative source of financing only. It represents a winning opportunity for the entire system, the World in which we are living, with great social-economic implications, impacts on new business models, on new life styles, on how to communicate science, how to stimulate innovation and participation, how to support entrepreneurship and research while promoting a renewed attention to today's social challenges.

This work is aimed at presenting how new models that break the mold in certain sectors can be applied in any sector. The paradigm shift. Open innovation and social innovation as tools for sustainable development, even in the evolution of learning. Crowdfunding and crowdsourcing as tool to sustain innovation and development. How the evolution of post-Fordism brings to a new capitalism. How the crowd economy drive the success. The third Industrial Revolution that leads us to live in what is called the Era of knowledge understood as a new way to live and work that revolves around the human capital, identified as a aggregating and propulsive force of know-how. As stated in the book by James Surowiecki "The Wisdom of Crowds: Why the Many Are Smarter Than the Few and How Collective Wisdom Shapes Business, Economies, Societies and Nations".

Crowdsourcing and crowdfunding are the beginning of a new version of fair democratic capitalism where each individual can take part and contribute.

Global populations are interacting and creating value on collaborative platforms creating value, providing access and sharing resources. This is crowd capital. The crowd economy is a dynamic ecosystem of productive people who participate through a platform with a purpose to achieve mutually beneficial goals. The meeting between the "long-tail" demand and "blue oceans" of opportunities creates crowd capital that boosts economies.

It is therefore reductive categorizing the phenomenon of crowdfunding as a mere alternative channel to access credit. Because it impacts the entire economic and social system.

Today, millions of people are bakers, in 2020 there will be billions as also stated by Jeremy Rifkin in "Third Industrial Revolution".

Starting from the questions 'why', 'how', 'what if' I have structured this work, with an interdisciplinary approach that counts numerous research contributions from a wide variety of disciplines including applied sciences, sociology, marketing, economics, law, policy and regulatory frameworks.

Observing the changing society we can project ourselves into the future. It is not true that the future is unpredictable and uncontrollable. The future creates the present. We can control the future and seize the opportunities it offers us to improve our lives. If any new project is not aligned with the problems, the needs and requirements of the future, the future won't allow the project to be implemented. This book will give an opportunity to everyone to make sure that their projects will be successfully implemented. Today anyone can be a protagonist of a sustainable project, even those you don't have a project can be proactively involved and benefit from the successful implementation of positive changes.

We project ourselves in a sustainable way in the future thanks to being protagonists of right choices today. We can control the future less than what we might want, but a lot more than what we can imagine.

I believe that we need not genes, but people again inspired and driven by passion. It is time to turn problems into opportunities, to ask ourselves the question "what if", to see with different eyes what is around us, to create something new, to be able to say at the end of the day "I did something great." The real progress that we would incur in the coming years requires fresh, creative and innovative ideas. I bet these years of deep crisis and deep recession will be born the decade that will inspire new services, products, methods, lifestyles, ideas and more. Looking back, many of the great discoveries that have revolutionized entire systems such as television, radio, xerography were invented during the years of deep recession.

When I decided to write this book I clear my objective: to share with others a projection of the future because the story and the solutions are created together. Past, present and future are part of the life of each of us.

Observing the changing society and project it into the future is the mission of this book. By doing so we can improve our future, we can not only understand, but to be the key players of the changes, to be the protagonists of the opportunities to be seized and implemented.

Jobs are over and the future is income generation, says innovation strategist. The old lifecycle path of childhood dependence, followed by a learning phase directed to gaining employment, leading to "productive" years of jobs and career building and the ultimate capping of the cycle with retirement and fall back on social security is fading. Instead today's life path follows more of a wave pattern that oscillates between phases of actively learning new skills, being productive in society through a range of short term assignments and realistically even unemployed for periods of time. What is being witnessed today is a rise of a new life path that is peppered with life-long learning opportunities, short and medium term monetizing engagements that builds social capital. So new-age productive workers are building a diverse portfolio of skills, experiences, network tools, excess capacity and physical assets that aid in passive and active income generation, and last a lifetime.

The alternative finance industry is a vibrant industry that develops very quickly. Community shares, civic crowdfunding, real estate crowdfunding and marketplace lending are examples of the developments that take place in the industry.

The growing awareness of humanity's finite resources and recognition of the limitations of one-off projects are prompting step changes in development planning. Sustainable development addresses the limitations of current practices and features at the centre of almost every political, environmental, scientific and economic discussion held today. Its aim is to achieve the triple bottom line of economic prosperity, environmental quality and social equity, meeting the needs of present society without compromising resources for future generations.

The 1992 Rio Earth Summit was the moment when sustainable development captured worldwide attention. It established why sustainable development is necessary; the project now is how to execute it. The 21st century is the era of mass innovation - more ideas, knowledge and information being shared by more people than ever before. Can we harness collective brainpower to innovate in the area of sustainable development for the common good? Google, Wikipedia and Linux are already using collective intelligence to develop new solutions through online communities. Could a collaborative online community do the same for sustainable development?

The global economy is in flux, in a new place full of possibilities and also full of unknowns. The power shifts from a handful of people to a democratic group of many is creating building innovation while tearing down archaic systems that do not work. As platforms mature and the crowd becomes more empowered and knowledgeable, there will be an organic restructuring and platforms that offer periphery services for true well-being of all stakeholders will distinguish themselves from the rest of the pack. From a political and civic point of view, governments should also prepare for the rise of this new economy. The crowd economy is not simply a buzzword but pathways to infinite opportunities.

A transition from ownership towards accessibility can be observed over the last decade in the economic exchanges. In the traditional economy, customers buy products and become owners (*proprietas*). In a sharing economy, consumers pay for temporary access-rights to a product or service (*usus*).

The world is moving into a world of open innovation, user innovation and co-creation. A world where the digital and physical are coming together, where new knowledge is created through global collaborations involving thousands of people from across the world and from all walks of life.

Co-creation is the active flow of information and ideas among different sectors of society (academia, government at all levels, business, third sector and citizens) which allows for a better understanding of participation, engagement and empowerment when developing policy and programmes, creating and improving services and tackling systemic change within each dimension of society. Through engaging citizens to redesign and remake their cities and communities together with businesses, public authorities, universities and all stakeholders, co-creation can lead to citizen empowerment, jobs creation through innovation, more efficient and effective public administration and strengthening democracy through open government. Co-creation can increase the innovation capacity of all sectors, by increasing the sources for new ideas, and facilitating cross-fertilisation through knowledge and experience sharing. Through the coproduction of services with citizens and business, co-creation can reduce costs and build on design driven ideas and testing of new solutions.

The traditional business model based on *proprietas* is rapidly changing due to two main drivers: one the one hand, technological advancement allows the model of sharing economy to spread different markets and become convenient and flexible. On the other hand there is a shift towards peer-to-peer accessibility. According to this model based on *usus*, companies are providing access for consumers to company owned property, or in peer-to-peer models, online platforms are facilitating access for consumers to consumer owned property.

Another key element is the community engagement, the worldwide review indicate that the social and ethical advocacy of projects are key factors to drive the crowd.

The power of the crowd which is the key in the sourcing but also in the funding. More generally crowdfunding, despite beginning as a grass root phenomenon, it has quickly scaled up and moved into financing innovation and providing seed capital for innovators and start-ups, raising money in the billion euros range and being increasingly used also by established businesses of any dimension and ranging from PA to Multinationals to finance growth and expansion activities.

By 2016 the crowdfunding industry is on track to account for more funding than venture capital, according to a recent report by Massolution[1].

Just five years ago there was a relatively small market of early adopters crowdfunding online to the tune of a reported $880 million in 2010.

Fast forward to today and we saw $16 billion crowdfunded in 2014, with 2015 estimated to grow to over $34 billion.

In comparison, the VC industry invests an average of $30 billion each year.

The crowdfunding market is expected to grow across all models over the next decade: the World Bank predicts that the total crowdinvesting market potential will reach EUR 70 billion per year by 2025, excluding North America. This simply shows that crowdfunding models can be used easily to support start-ups and established businesses and can become an alternative solution for early stage financing for several projects.

On the other hand, very few startups are maximizing the opportunity of the internet today through the power of crowdsourcing. Most go as far as crowdfunding and if that is extent of the grasp of crowd applications, entrepreneurs are missing the big picture.

But crowdfunding is much more, as we will see in the following chapters. It has emerged as effective way to raise money with the added power of social media. Crowdfunding, beside its primary role of alternative financing has a strong marketing and communication function, which can favour the use of crowdfunding by institutional actors, corporate and large companies for public advocacy and marketing activities. Indeed, crowd instruments can drive a positive change and help citizens, research, public administrations and companies of any size in improving their image among their customers as well as in the wider public by communicating and marketing their support to clean and participatory investments or simply by being supporters and users of a tool generally perceived to be a more democratic form of financing.

The book will analyse in detail the crowdfunding models and how they serve specific purposes beyond just money, like meet the first customers of the business, presell, test and finetune the product/service, receive feedback and procure market validation. We will see how crowdsorcing and crowdfunding are powerful tools to sustain growth, as a response to the ongoing transition.

Crowdfunding will not only draw individual entrepreneurs but also business firms, NGOs, universities, local and central governments. Crowdfunding activities inside the company have potential to become widespread because of their advantages. This kind of approach expands the level of creation inside the company and fosters technology-related improvement.

Expanding on the idea of crowdsourcing, crowdfunding implies mobilizing the crowd to finance projects which are posted by their proponents either on personal pages or on dedicated websites, known as crowdfunding platforms. Funding is provided by web users in exchange for some monetary claims on the project revenues, for an economic reward, or simply for donation. Suggestions and feedbacks for improving the projects are also welcome by proponents. Scholars and practitioners agree that crowdfunding holds the potential of changing traditional funding practices in general, but – more interestingly – especially for entrepreneurial projects and startups by complementing financing provided by professional investors like venture capitalists, business angels, or equity providers (Giudici et al., 2012).

The first part of the book illustrates the transitional age in which we live, so as to explain the various aspects, providing scientific, economic, social and dynamic basis for growth. A path in which we are the actors, regardless of spatial and cultural constraints, with repercussions in the spheres of education, work, society, organization, participation and information seen also in relation to accepting the role of "active" citizen in the protection, defense and enhancement of public affairs.

Readers will discover how the sharing economy can produce dynamic changes in the workplace and in the fields of analysis and development, understanding what factors contribute to change and what to expect in the future.

All of our markets, systems and technologies have become incredibly fluid and interconnected and are leading to unknown areas. For most of us, the chaotic nature of interconnected trends may appear like a needle of a compass in permanent rotation. This disorder calls for answers, that the text aims to provide, bearing in mind that uncertainties create great opportunities for those who can read and translate them into a radical change.

The resto of the book is devoted to crowdfunding, which is analysed in all its aspects and projections towards the future.

The book offers a clear understanding of the knowledge Era and what it offers, while crowdfunding represents the *deus ex machina* to implement it in a sustainable manner.

The book explores the potential of crowsourcing and crowdfunding, which has wider implications on society, labour, finance, social inclusion, sustainable development and growth, project management, public investments, creativity, culture, innovation, prosperity, social entrepreneurship and entrepreneurship. It will be discovered by reading the book that any idea and project can be successfully realised and there are no limits in terms of area of business, budget needed and typology of the proponent, meaning that anyone can launch a successful crowdfunding initiative.

## OBJECTIVE OF THE BOOK

This comprehensive and timely publication aims to be an essential reference source, building on the available literature in the field of crowdsourcing and crowdfunding while providing for further research opportunities in this dynamic field.

Designing crowd actions involves the rethinking on how to communicate science, how to stimulate innovation and participation, while promoting a renewed attention to today's social challenges. Crowdsourcing and crowdfunding support entrepreneurship and research; is a winning opportunity for the entire system.

It is envisaged that this text will provide the resources necessary for anyone, policy makers, entrepreneurs, citizens, NGOs, traders, cooperatives, associations, researchers and students to adopt and implement crowdfunding solutions in achieving their goals.

The book contributes to understand the World in which we live in order to maximise the opportunities people can get. It doesn't merely discuss technologies, but models and opportunities to be taken from the correct 'reading' of the Knowledge Era.

The reading of the text intends to give anyone the possibility of transforming problems into opportunities, asking the question " what if", seeing with new eyes all that surrounds us, creating something new.

It will give free rein to our potential in order to meet our needs. Reading the book will contribute to the art of exploration, to the perception and understanding of where we are, of our needs and of the forces that cast us into the future.

Understanding the knowledge Era will let us see the future that belongs to us, or one in which we can be successful protagonists, finding a career, being the architects of a solution, creating new professional figures, conceive new business, consumption and lifestyle models, turn our ideas and dreams into reality.

There are no books which cover the topic in such way. By reading the book it will be possible to understand the essence of the changes, the interlinked elements which moves our society forward. Crowdsourcing and crowdfunding is the beginning of a more democratic and equitable version capitalism where now every individual has an opportunity to take part.

## TARGET AUDIENCE

The book is addressed to a wide audience which encompass: students, researchers, citizens and general public, entrepreneurs, startups, associations, cooperatives, public institutions and policy makers.

## OVERVIEW OF THE CHAPTERS

### Chapter 1

### The Crowd Economy: From the Crowd to Businesses to Public Administrations and Multinational Companies

The chapter is devoted to the power of the crowd. A digital transformation is underway. One that is redefining the essence of human interaction and with ideas, share unused resources and create new on-demand services that are customisable and unique. These are only a few examples of real productivity that when layered on the Internet creates an abundance of resources and opportunity. This people-powered abundance is called the crowd economy: it the way the society lives, works and plays. There is a new paradigm shift that challenges traditional notions of the "norm" while expanding possibilities.

The hallmark of the digital age is social connections that are boosted by the web and mobile networks. These technological advances have taken collaboration and cooperation to a level never seen before. Social connections through the web have gone beyond social media likes and shares and has evolved into social productivity - a phenomenon that arises when networked crowds collaborate to solve problems, raise funds, come up with innovative sustainable solutions.

## Chapter 2

### Towards a New Economy: Co-Creation and Open Innovation in a Trustful Europe

This chapter focuses on the transition towards a new economy, where Europe can find untapped sources of growth and employment, renewing the legitimacy of public policy-making, especially through greater citizens' involvement and by delivering better public services for all. The chapter analyses the perspectives of Europe in 2050 following three dimensions: Global and European demographic changes; Economics, energy, environment and technological prospects; Geopolitics and governance. The concepts of trust, co–creation, sharing and sustainable economy, innovation, and social innovation are presented as the key ingredients of the "new" economy.

## Chapter 4

### Crowdsourcing New Tools to Start Lean and Succeed in Entrepreneurship: Entrepreneurship in the Crowd Economy

The very essence of the new entrepreneur is shattering tradition. On the heels of the new social internet, we are seeing the rise of the solo-entrepreneur or intrapreneur who embraces globalisation, failure and successes and collaboration. Powerful networks brought together by crowdsourcing are supplying the tools to start lean, innovate and solve complex problems. This chapter explores the changing ecosystem and the effect of networked crowds on starting lean and succeeding with entrepreneurship.

## Chapter 5

### Crowdsourcing Social Innovation: Towards a Collaborative Social Capitalism

The growing awareness of humanity's finite resources and recognition of the limitations of one-off projects are prompting step changes in development planning. Sustainable development addresses the limitations of current practices; its aim is to achieve the triple bottom line of economic prosperity, environmental quality and social equity, meeting the needs of present society without compromising resources for future generations. Collective intelligence is considered by both Charles Leadbeater (former advisor to Tony Blair) and MIT as one of the most powerful ways to tackle complex problems, like climate change. This chapter explores the principles of crowdsourcing, its applications and the main trends. It presents theories, practices and examples of the use of crowdsourcing to innovate in the area of sustainable development for the common good. It announces the rise of collective brain-power to the challenge of creating better and more effective forms of civic and social engagement to solve problems on a world scale.

## Chapter 6

### The World's Oldest Profession: Now and Then – Disruption of the Commercial Banking Model

Modern banking found its roots during the Renaissance period casted by the European merchant banks. Their success was due in large part to their aggregation, absorption and deployment of information about borrowers, structures and markets. In the 21st century, banking again is being advanced due to insights developed by vast amounts of information and data, this time gathered and managed through new technologies and models, in quest of efficiency, improved risk management and improved portfolio performance. New entrants into the business of banking operate outside of existing regulatory structures, and may enjoy a level of competitive flexibility compared to existing commercial banks. Might this portend the end of the so-called modern commercial banking model, or might it serve as a strategic imperative for the banks to adapt to innovation?

## Chapter 7

### The Crowdfunding Market, Models, Platforms, and Projects

The changing structure and intensity of competition on one hand, the ongoing pressure on creating differentiated offerings lead to seek out new ways to improve technology as well as new ways to perform business. Technological developments enable market participants or other interest parties to engage in the process of both innovation and production, resulting with an online social network market model of co-creation. This chapter focuses on describing different aspects of recently developing crowdfunding market models where new projects are offered and funded by large number of individuals from general public. The current extent and dynamics of the growing crowdfunding market, various models of crowdfunding, crowdfunding platforms and the projects that are crowdfunded are analyzed through the chapter by giving examples and comparative comments. The analysis of the market is concluded by the foreseen future trends.

## Chapter 8

## Chapter 3

### The History and Evolution of Crowdsourcing

In this chapter a detailed history of Crowdsourcing phenomena is given, in both the research and practice, focusing on the private sector, to illustrate the diverse and rapidly emerging phenomena that comprise this body of research. In terms of actual organizational practice, Crowdsourcing is generally implemented by organizations through two modalities; in-house, and through intermediaries. For example, organizations can launch Crowdsourcing initiatives on their own branded platforms, like Dell did with IdeaStorm, or they can commission Crowdsourcing intermediaries to host contests on their web properties for a fee.

## Horizontal Crowdfunding Platforms: Socio-Economic Impact and Exogenous/Endogenous System of Rules

In recent years arose a multitude of crowdfunding platforms of different nature and with different goals. In this chapter some peculiar aspects of crowdfunding platforms will be examined, both in terms of their socio-economic impact and in terms of the exogenous and endogenous system of rules with which they interact. Crowdfunding is not governed exclusively by the rules of law and the regulations of the platforms that host it. Every campaign involves the construction of a community, with its specific rules, that are in part pre-established by the campaign creator and in part the result of the interaction of the community. However, these rules do not live in a closed set, building a case history to which the future campaigns will not necessarily have to conform, but with high probability they will use as a parameter. Although it is not possible to identify the rules that are common to all the crowdfunding campaigns, could be observed an evolution mostly harmonic between different campaigns with respect to the methods, the mechanisms of involvement and, of course, the rules.

# Chapter 9

## Vertical and Community-Based Crowdfunding Platforms

If the inherent nature of the horizontal crowdfunding platforms makes the relationship between campaign creator and backers basically without intermediation of the platform, in the case of vertical platforms there are different and sometimes complex levels of intermediation. The vertical crowdfunding platforms can be considered more than a subset of the horizontal crowdfunding, as an independent system of relationships and rules which has in common with the horizontal crowdfunding a large part of the fundraising methods, but that is also profoundly different with respect to the aggregative procedures and the system of rules.

# Chapter 10

## The Crowdfunding Communities and the Value of Identification for Sustainability of Co-Creation

For a sustainable future, the actors of the market need a change in spirit that will elevate the life-chances of people by being an "empowered part" of the economic system. A participative, open knowledge economic and market system that will ensure the material well-being, the social visibility, the happiness and the consciousness of the individual is essential. Crowdfunding, as a recent online social community network market model is to be a new socio-technical system of co-creation through self "making and funding". The aim of this study is to discuss the role of identification in co-creation for sustainability of newly rising crowdfunding communities. In line with this objective, the following two aspects are clarified: the key characteristics of crowdfunding platforms as online communities and the role of identification for co-creation in online crowdfunding communities.

## Chapter 11

### Principles in Crowdfunding Benefits and Validation Tools: How to Ensure Campaign Success

This chapter provides the key benefits of a crowdfunding campaign together with a rich variety of validation tools provided to enhance the chance of a crowdfunding campaign's success. Benefits cannot be achieved in a crowdfunding campaign if it is not successful. Thorough validation of the campaign upfront can heighten success. The author in an internationally respected crowdfunding authority who has managed over 300 crowdfunding campaigns through the platforms he has been involved with. The author lists 11 benefits and 11 validation tools together with a plethora of links to tools to assist both the novice and experienced crowdfunder.

## Chapter 12

### Crowdfunding as a Marketing Tool

Most people, at a first sight, perceive Crowdfunding as a specific and exciting process to raise money. Of course, this is correct, but it is not the whole truth. A successful crowdfunding campaign must be addressed to a community of people, rather than to an indiscriminate crowd, which is also typical for any marketing strategy. Leveraging on this key element, crowdfunding can amplify almost any kind of communication activity, provided it is able to meet the needs of a community and to provide the benefits that the community expects. Therefore, if well exploited, crowdfunding is a powerful tool in the hands of marketers, whose potential goes far beyond the evidence of pre-selling a new product, to reach new marketing horizons such as branding, business development, CRM, credit scoring and even HR management.

## Chapter 13

### Legal Aspects and Regulation in Crowdfunding: Comparisons across Countries

Crowdfunding started with the donation-based and reward-based models, but soon its disruptive capacity to innovate the traditional approach to finance made it evolve into an alternative finance instrument, suitable to fund new business. Peer to peer lending and equity crowdfunding are investment instrument that can truly sustain the growth of businesses and generate profits. As any investment instrument, lending and equity crowdfunding are heavy regulated in order to protect investors and maintain a proper trust in the system. This chapter will go through the American and European main regulatory principles, offering a picture of the overall approach to this innovative and disruptive financing tool.

## Chapter 14

### How to Launch a Successful Crowdfunding Campaign

This chapter provides insights into the important characteristics of a successful crowdfunding campaign. It starts with providing the fundamentals of crowdfunding and shows the reasons why it is important to find the right target groups and communities for a campaign. It describes in details the Return on Involvement model, what the funders want in return for their support. In the second part the additional results of crowdfunding are discussed, such as market research, marketing, brand loyalty and the possibility to use crowdfunding to raise additional funding. The last part of this chapter will focus on the background of setting up and running a successful campaign.

## Chapter 15

### Real Estate Crowdfunding: 2015 and Beyond

Like a juggernaut, "crowdfunding" has hit the media, the financial markets and the common narrative by storm. Incipient in many ways, the intermediation of financing transactions online has become a billion-dollar industry. As technology has advanced, even recreated, industries, there seems none more primed than "finance", an inherently information business. By creating improved efficiency, both art and science of finance are enriched. A 21st century vestige of the 1980's syndication business, real estate seems to be enjoying the fruits of the crowd, with $1 billion of property financings conducted online in 2014, with an expected $2.5 billion this year. For sponsors and investors, there appears legitimacy to the online approach, underscored by the level of venture capital now finding home in this burgeoning sector. Yet, like its progenitor, might real estate crowdfinance find legislative, regulatory, and practical headwinds, stunting its progress? Still early, with business models, scalability and sustainability still suspect, the current momentum seems promising.

## Chapter 16

### The Application of Crowdfunding to the Energy Sector

Crowdfunding in energy begins as a response to reduced investments (both governmental and private) into the transition to decarbonized energy systems and to the spread of innovative business models and approaches conductive of greater participation of citizens and communities in distributed renewable energy projects. This chapter presents results of a worldwide overview of the use of crowdfunding in the energy industry. Evidence gathered from available energy crowdfunding platforms highlight a very new, but quite dynamic sector. The crowdfunding tool has been applied in most of its forms, ranging from peer-to-business lending to pure donation, with strong environmental and social mission and the explicit aim of increasing participation of citizens in sustainable energy investment. Evidence also shows that, despite maintaining their environmental and clean energy focus, some energy platforms have begun to move from niche, grass root initiatives into larger projects and collaboration with energy private sector and institutional finance.

## Chapter 17

### A Crowd-Funder Value Framework for Crowd-Investment: A Roadmap for Entrepreneurial Success in the Contemporary Society

Crowd-funding is used by business and social entrepreneurs to fund projects that impact society in many ways. Although crowd-funders fund and support entrepreneurial projects, stakeholders are less aware of crowd-funder motivations and behaviours towards products/services, which impact on the success of these projects. The purpose of this research, therefore, is two-fold: a) development of crowd-funder value framework for understanding crowd-funder motivations; and b) provision of robust theoretical basis to the construct of 'crowd-funding'. Different social constructs, such as social identity, and symbolic, crowd and relational social capital related to crowd-funding are discussed based on symbolic convergence theory of communication and social identity theory. This framework will act as a roadmap to understand how crowd-funder motivations impact and create value for individuals, organisations and society, and inform how social, environmental and economic value and impact can be maximised through crowd-funding initiatives.

*Walter Vassallo*
*MC Shareable, Monaco*

## ENDNOTE

1 Massolution report 2015.

# Chapter 1
# The Crowd Economy:
## From the Crowd to Businesses to Public Administrations and Multinational Companies

**Epi Ludvik Nekaj**
*Crowdsourcing Week, USA*

## ABSTRACT

*A digital transformation is underway. One that is redefining the essence of human interaction and with ideas, share unused resources and create new on-demand services that are customisable and unique. These are only a few examples of real productivity that when layered on the Internet creates an abundance of resources and opportunity. This people-powered abundance is called the crowd economy. It is the way the society lives, works and plays. There is a new paradigm shift that challenges traditional notions of the "norm" while expanding possibilities. The hallmark of the digital age is social connections that are boosted by the web and mobile networks. These technological advances have taken collaboration and cooperation to a level never seen before. Social connections through the web have gone beyond social media likes and shares and has evolved into social productivity - a phenomenon that arises when networked crowds collaborate to solve problems, raise funds, and come up with innovative ideas and solutions.*

## BACKGROUND ON THE CROWD ECONOMY AND WHY THE DEFINITION MATTERS

If one follows the emergence of these new solutions that rely on networks and their interdependencies, you will find they thrive on collective action. And this collective action is more than a bunch of people coming together. We are seeing the maturing of the sharing economy to save resources, open innovation for idea enhancements, P2P funding for startups to scale or crowdsourcing for a user generated video or design. The author is making the argument that all these interactions are not just about sharing or collaboration but fall under a more powerful human-powered umbrella called the crowd economy. The new democratic economics has evolved with a number of different definitions that Rachel Botsman (2013), Founder of Collaborative Lab has described in a post on *Fast Coexist*.

DOI: 10.4018/978-1-5225-0568-6.ch001

- **Sharing Economy:** Economic model to share underutilised resources.
- **Peer-to-Peer Economy:** Person-to-person marketplaces that facilitate the sharing and direct trade of assets built on peer trust.
- **Collaborative Consumption:** An economic model based on sharing, swapping, trading, or renting products and services, enabling access over ownership.
- **Collaborative Economy:** An economy built on distributed networks of connected individuals and communities versus centralized institutions, transforming how we can produce, consume, finance, and learn.

These definitions surely fit many of the new models of economics that have become popular over the last 5-8 years but a more powerful term that encapsulates the democratic nature of thinking is presented in the word the 'crowd economy'. The fundamentals of the crowd economy is anchored in crowdsourcing. It is a synergy of elements like, networked crowds, a democratic mind set, self-awareness, exponential problems that need collective action and technological advances that morph the crowd economy into a more purpose-driven and valuable ecosystem. It is very important to anchor oneself on the true fundamentals of crowdsourcing to maximize the opportunities of crowd economy. The broad crowd economy is characterized by five fundamentals - people, purpose, platform, participation and productivity.

The crowd economy is a dynamic ecosystem of productive people who participate through a platform with a purpose to achieve mutually beneficial goals.

## The 5 P's of the Crowd Economy

Inspired by 4 P's of the marketing; Product, Price, Place & Promotion, the 5 P's bring a deeper understanding about the complexities of crowd interactions and the interplay between people, purpose, participation and productivity (see Figure 1).

*Figure 1. CrowdEconomy book*

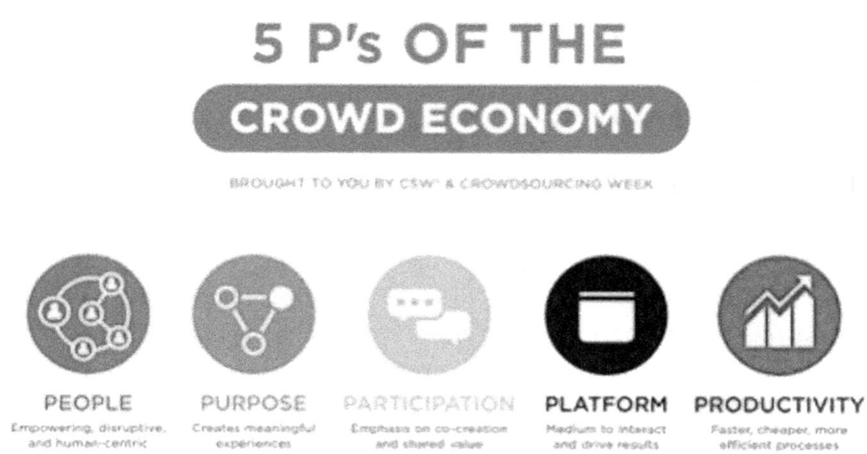

1. **The People:** The crowd economy is empowering, inclusive, disruptive and human centric.
Human-centric values need to be embedded in applications geared towards the crowd economy where the community is the starting point. The crowd economy or collective action is not about mob behavior but very targeted cooperative solutions that help communities better their lives. People-powered platforms are forging these interconnections between users that are breaking down the barriers between creators, producers and end users. By empowering people, organizations are finding new, previously unimagined pathways and solutions to complex problems as in the case with NASA. Jennifer Gustetic, Assistant Director of Open Innovation at the US Government Office of Science and Technology attests that highly technical submissions from aerospace enthusiasts and amateurs matched the quality and accuracy of experienced space scientists.

2. **The Purpose:** The crowd economy creates meaningful experiences and shared value.
The crowd economy embodies a culture of shared value creation and social responsibility that distinguishes itself from the traditional one-dimensional thinking and practices of the old economy. People driven initiatives often embody a larger mission to create solutions that work for, and, with all stakeholders. There is more than one channel of communication and the notion that everyone can further his or her purpose is life changing. Patient Innovation is an impressive platform that helps patents drive innovation to find treatment options for their health problems. Patients often develop valuable solutions to help them cope with their disease. In some cases, they saved their own lives (Oliveira, 2014).

3. **The Participation:** Co-creation and participation are emphasized in the crowd economy and communities take an active stake in crafting positive futures.
The power of participation to accelerate innovation is best seen through crowdfunding, that has enabled early ideas get a jumpstart. Crowd verdict is critical to validate business plans and ideas and working with them only bring financial support but also value product input and iteration. Participation from 60K+ supporters made the Coolest Cooler the most funded project on Kickstarter, exceeding its modest goal of $50,000 to raise over $13 million (Grepper, 2014). Crowdfunded companies are reaching new goals and successes – Jon Medved's OurCrowd, the leading hybrid venture capital equity crowdfunding platform for accredited investors, announced that its portfolio company ReWalk Robotics has completed a successful IPO. When speaking of crowd participation, it is hard not to mention LEGO, a company that turned its fanatic following into an active design community.

4. **The Platform:** Crowds need a medium to interact and produce results.
This pillar of the crowd economy has manifested in the form of technology, connectivity and mobile networks. Soon the Internet of Things will contribute to this medium, amplifying human interactions with powerful data. Platforms like Airbnb and Uber have become synonymous with peer marketplaces and have led to new business paradigms taking shape.

5. **The Productivity:** Crowd economy fosters faster, cheaper, better and resource efficient processes.
Digital crowd applications for civic activities, disaster relief and humanitarian work are creating widespread impact. Helping and participation comes naturally to us and the networked web has fitted this mind set with wings. When Malaysian Airlines Flight 370 suddenly disappeared in March

this year, a massive digital search ensued that mobilized eight million volunteers, amounting to 20,000 years of volunteer time! Participation in the crowd economy is not always at this scale but imagine the possibilities (Meier, 2014).

The crowd economy is the infrastructure for creating crowd capital. It is one of the fundamental changes seen this century in terms of opportunities that previously did not exist, the distribution that takes place in seconds, and the decentralized powers and abundance. For long people-centric approaches to problem solving have been associated with a narrow scope that focus on either a marketing gimmick or a cost cutting measure or the worst, a fad. But the spectrum of possibilities with the real crowd economy is limitless. Collaborative consumption, sharing economy, crowdsourcing - whatever the name a powerful crowd economy is the foundation that is breeding the mind set to tackle critical challenges of today.

## How Crowd Capital Is Generated

We are living in an age of abundance - wealth, information and opportunity. The problem is that there is uneven distribution and hence the economy is inherently unstable. How can we sustain economic growth in such a climate? Through crowd capital.

The Oxford dictionary defines capital as: "wealth in the form of money or other assets owned by a person or organization." To most, crowd capital sounds just like crowdfunding. But it is not just about crowdfunding. Crowds and communities provide more than funds which are turning out to become the building blocks of the new economy.

Analyst and strategy consultant Barbara Gray explains "blue ocean" of opportunity meets the "long-tail" demand to create sharing economy giants like Airbnb and Uber - and economic abundance (Gray, 2014). I call this crowd capital, where value for a more robust economy and outlook is created. Let's take a look at the current scenario.

## What Is the "Blue Ocean" of Opportunity?

In her article, Gray (2014) references authors W. Chan Kim and Renee Mauborgne and their book 'The Blue Ocean Strategy: How to Create Uncontested Marketplaces and Make Competition Irrelevant', that calls for a different business strategy using technology to create vast "blue oceans" of opportunity in untapped markets. How do companies create blue oceans of opportunity?

Not through tightly regulated corporate structures and regulation. This is possible through technology of course but also through a democratic approach to business that hands-off more power to users and consumers - a key feature found in the collaborative platforms like Airbnb or Uber that are easily expanding into regions projecting huge markets for growth.

## What Is the "Long Tail" Demand?

The long tail of demand refers to the unique, uncommon but numerous instances of specific consumer demand that emerges when consumers are offered almost unlimited choice and personalization.

With the new collaborative and crowd economy, business platforms are catering to individual needs, choices and demand and focussing on this long-tail demand, previously restricted because using limited retail space to cater to short instances of unique demand did not make economic sense. This long tail demand detailed in Chris Anderson's book 'The Long Tail', is fueled on the supply side by democratically availability of assets, time and finances.

Here's how crowds are building capital and laying the foundation for a different but more democratic economy.

## Crowd Capital through Crowdfunding

As I've written many times in my publications, the crowd has more money that Carlos Slim or Apple. The crowdfunding initiative on Greece bailout is a sneak preview of what a concerned citizen can do to empower the crowd to participate in solving a political-economic problem.

Crowdfunding isn't just a trend or movement or initiative. It is a disruptor that's empowering even the average citizen to manage and seek finances without banks for credit, angels for seed funding or VCs. According to the latest report on global crowdfunding, Asia has slightly surpassed Europe in terms of crowdfunding volume. And Europe's alternative finance market is on it's way to reach €7 billion in 2015 (Crowdsourcing.org, 2015). Incumbent finance and banking sectors and policymakers across Europe are being prodded to take action. Once again, innovative business and people in need of credit (long-tail of demand) that were not previously on the target list for banks and VCs are meeting the "blue ocean" of opportunity through crowds.

## Crowd Capital through Blockchain

Quoting Stacey Higginbotham on Fortune (2015), "…as our economy becomes more digital and distributed, we need tools that can scale and still offer accountability. Blockchain does that." Bitcoin is powered by a recordkeeping system, known as the blockchain, a distributed ledger that can be used to track transactions.

Crowd currencies, powered by democratic protocols like Blockchain will become part of Europe's financial systems in the future because of the value they bring in terms of transparency and innovation. The Euro Banking Association (EBA) says blockchain technology has the potential to reduce costs, improve product offerings and decrease transaction time for banks, even saving them up to $20 billion a year (Perez, 2015). In the case of blockchain, again, the "long-tail" demand of banking access and almost real-time transactions is powered and verified by distributed entities, a "blue ocean" of resources (Perez, 2015).

## Crowd Capital through Asset Sharing

Global populations are interacting and creating value on collaborative platforms creating value, providing access and sharing resources. *This* is crowd capital. The crowd economy is a dynamic ecosystem of productive people who participate through a platform with a purpose to achieve mutually beneficial goals. Barbara Gray (2014) points out that hugely successful companies on the collaborative economy are bringing together the "long-tail" demand and "blue oceans" of opportunities - and this again creates crowd capital that boosts economies. Local economies do not need an Uber or Airbnb to create crowd capital. Home-grown solutions can create impact, with the right support from innovators and buy-in from government.

Another representation of the crowd economy and the resulting crowd capital can be visualised by the 14 broad applications of networked crowds, jointly developed by Crowdsourcing Week, CSW2 and Wikibrands. By applying the 5P's of crowdsourcing to different verticals, crowd capital can manifest in the following ways (see Figure 2).

Here is a breakdown of the different sectors of the crowd economy:

1. **Crowd Currencies:** Democratic currencies, backed by no- central authority but by self governing crowds through technology frameworks like the blockchain. e.g. Bitcoin.
2. **Peer-to-Peer Lending/Commerce:** Lending instruments and loans available through peers and direct social lending and socialized commerce between people without a financial or retail intermediary e.g. Lending Club, Etsy.
3. **Equity-Based Crowdfunding:** Members of the crowd become part-owners of the company which is raising funds in the form of shares and dividends e.g. Crowdcube.
4. **Non-Equity-Based Crowdfunding:** Members of the crowd give money to support a worthy cause or as an exchange for a "reward," typically the product or service that the crowdfunding company produces or provides e.g Kickstarter.
5. **Sharing Economy:** A collaborative economic model built around the sharing of human and physical resources incl. shared creation, production, distribution, trade and consumption of goods, space, travel and services by different people and organizations e.g. Airbnb.
6. **Customer Co-Creation:** A partnership between an organization and group of customers, to jointly produce a mutually valued outcome while encouraging a more active involvement from the customer to create a value rich experience e.g. Lego Mindstorms, Threadless.
7. **Social Business:** Companies who encourage genuine listening, sharing, engagement and commerce through open, social channels to create value across the corporation or within various functions e.g. Dell, Betabrand.

*Figure 2.*

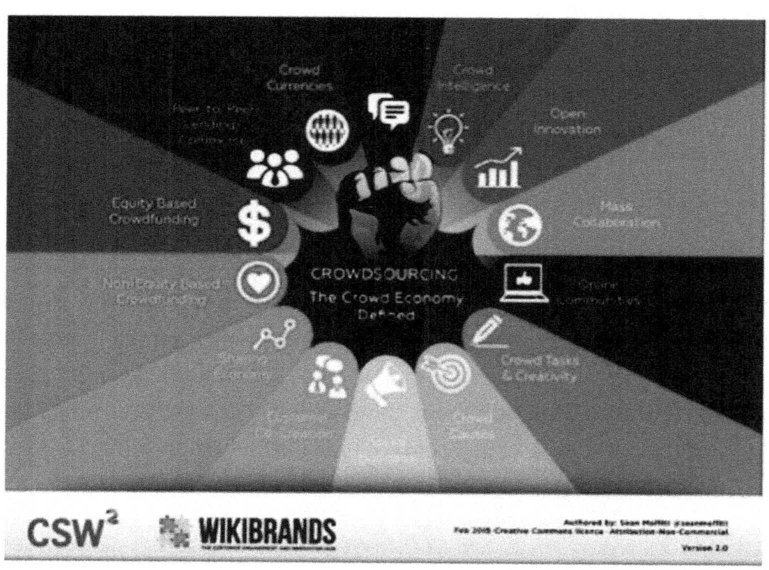

8. **Crowd Causes:** Individuals, not-for-profits or purpose-driven organizations, who engage collectively with passionate and interested stakeholders to raise the profile, impact and fundraising potential of their causes e.g. Charity Water.
9. **Crowd Tasks and Creativity:** The effort of obtaining requested labour, services, solutions, ideas, or content by soliciting contributions from groups of people, typically online and geographically dispersed, rather than via traditional employees or suppliers, e.g. Amazon Mechanical Turk, Crowdspring, Freelancer.
10. **Online Communities:** Groups of individuals gathering and participating on web-based platform(s) and interacting with one another on an ongoing basis driven by a common, tribal attraction for a belief, idea, product, brand, idea, cause or business, e.g. Reddit.
11. **Mass Collaboration:** A form of collective action that occurs when large numbers of people work independently on a single project, often modular in nature, that produce shared understandings and outputs, e.g. Wikipedia.
12. **Open Innovation:** Use of inflows and outflows of knowledge to accelerate innovation inside organizations, and expand markets for external applications, through more permeable, open organizational boundaries and combining the organization's resources with external co-operators, e.g. Nine Sigma, OpenIDEO.
13. **Crowd Intelligence:** The shared or group intelligence and insight that emerges from the collaboration, collective efforts, and/or competition of many individuals, e.g. Netflix, Quora.
14. **Citizen Engagement:** Digitally-connected platforms and activities to help inform, organize, empower and mobilize citizens, communities, national, regional and global movements, e.g. Code for America, Challenge.gov.

### Crowd Capital Needs to Be Considered as an Economic Stabilizer

2.5 Billion people will connect to the Internet in the next five years. Connectivity can help fuel crowd capital as more people get online. This a no-brainer. But are we ready? Discussions on the crowd or collaborative economy that include innovators, policymakers and regulators must perceive crowd capital emerging from the on demand, digital platforms and adopt a long-term view of the benefits.

Crowd capital is democratic capital that helps billions take ownership at the macro level. Rather than relying on the top one per cent for economic stability, we are looking at billions helping to create a more robust financial ecosystem through collaboration.

## CROWD CAPITAL AMONG YOUTH AND SILVER GENERATION

With the proliferation of the digital economy, the paradigms for education and learning job skills have changed. With technology, traditional jobs and employment pathways are being disrupted. The new workforce comprising of the millennial generation and younger populations are defining the future of work. In this new ecosystem barriers between vacation and work, office and home are blurring. Another trend is emerging that is relevant to the older, silver generation. The new ways of creating wealth through crowd capital is challenging the notion of retirement with the new ecosystem presenting numerous opportunity for baby boomers to activate their time, resources and contributions for economic benefit. This new economy is also making changes to social net infrastructure. The opportunities are vast, but they also require a change in thinking. Helping global youth and the silver generation make a smooth transition is critical to tackling rising unemployment and job skills gap. Crowd consultancy CSW2 has formulated a crowd-economy based approach called C.A.B.L.E. for youth aged 17-23 years old and silver generation which stands to connect, activate, build, learn and employ. Each step is designed to bring these age groups closer to crowd economy while empowering them at the same time.

- Connect (and participate in crowdsourcing platforms).
- Activate (the passion, skills, talents and needs).
- Build (social currency across the platforms recommended by CSW).
- Learn (mechanics and apply).
- Employ (build crowd capital).

## IMPORTANCE OF CROWD CAPITAL

The first 3 billion people connected to the Internet at the time of reasonably well functioning markets was utilised with a mindset of receiving information that can be described as an "Internet of download." These permeated attitudes on how content, information, entertainment, consumer products were consumed in the old framework. It is a very passive mind set in which it does not immerse itself into the full potential of Internet. Whereas, the new "Internet of upload" mind set brings people closer to the most advanced opportunities that could be experienced. This type of mind set enables and empowers individuals around the world to work and finance projects, co-create products, buy and sell goods instantly. The "Internet of upload" mind set brings crowds together for a purpose. What is being witnessed today is the emergence of a new era - a world of abundance - taking shape in different forms and sizes that will have a ripple effect on the different sectors and pillars of the society.

Looking back, it is worth noting why crowd economy evolved during the 2008 world economic recession. During this time, there was a shortage of jobs and financing but an abundance of technology, talent and information. New projects were not being financed by banks, but by the people. New products were introduced from ground up and idle goods being shared among peers. This dull economic period saw the birth of hundreds of startups in the crowd economy that brought people together based on their passions, skills, talents and tangible resources. It is worth noting that during this time the world's internet population passed 1.5 billion and the access to internet was on exponential growth- more data has been created from 2008 - 2015 than ever before as this diagram represent (see Figure 3).

So when we talk about crowds, it is important to talk about data. Just like land was the basis of the agriculture age, iron was the basis of the industrial age, in this case data is the basis of the information age. And unlike the land or iron which is limited in resources, data is unlimited and infinitely reproduced - by anyone, anywhere.

It is worth noting that the world is in real transition. Baby boomers are retiring. The new arising workforce taking over which is much more dynamic, flexible and even cheaper. Generation of wealth is being transmitted from old business models to new business models and in some cases disrupting their own business models. Advent of technologies and robots replacing and automating most boring jobs for humans can lead to very unpredictive territories in the future.

Now what happens when new internet users immerse themselves in the *"internet of upload"* mind set and become part of crowd economy. Although these 4 billion people who remain disconnected from

*Figure 3. Digital information*

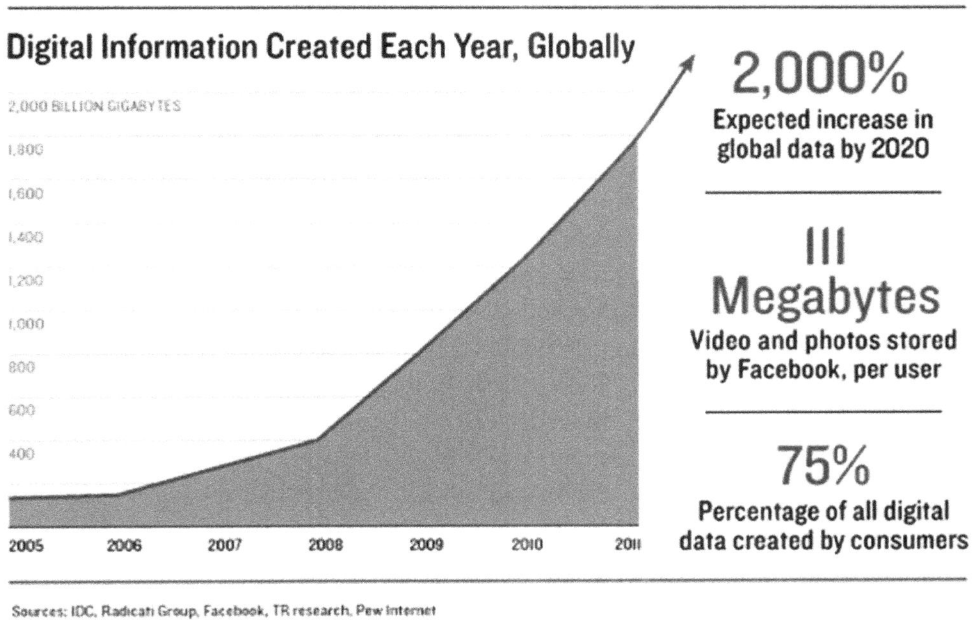


the internet are overwhelmingly concentrated in developing world economies, studies show that new users of the internet prove to demonstrate a quick adaptation of internet-based tools irrespective of socio-economic disadvantages.

The cloud removes physical limitations that hold us back and the crowd provides exponential innovation. When the cloud meets crowd, social productivity is enhanced and infinite opportunities are inevitable.

A slew of statistics tells us that the cloud would be worth $107 billion by 2017; by that time, the world's population would reach approximately 7.4 billion people and the internet population would reach 3.6 billion, nearly half will be connected to the internet, setting high expectations for what we call the crowd economy - a new paradigm defined by connected individuals cooperating over the Internet (Butler, 2014).

Now we are talking about social productivity where networks of people driven by their passion, skills, talents and tangible resources are LinkedIn nodes to create a powerful change. This powerful change that is unstoppable and very disruptive to centralization will have large effects how large organizations including governments are embracing it.

## SOCIAL REPUTATION AND TRUST IN THE CROWD ECONOMY

And when it comes to social productivity reputation, it is much different than just a social media reputation. What is emerging with the crowd economy is a new reputation paradigm that is a blend of digital and physical worlds that is no longer based on superficial descriptions but instead driven by actions. People are getting empowered by simply participating in these platforms while at the same time creating social reputation that is no longer based only on social media but is based on their actions across these platforms. An example of this can be seen with Upwork, Freelancer, Airbnb, Uber, 99designs, Peerby, BlaBlaCar, Kaggle, Kickstarter, Crowdcube, Lending Club, Quora, and many more platforms who are enabling people by simply participating.

As an early adopter of a platform like Airbnb, that aggregates homes available for short-term rentals, the author was able to live in different homes globally while traveling to evangelize Crowdsourcing Week. This was not only a money saving proposition but also an opportunity to meet some talented people who also doubled up to support Crowdsourcing Week events in their home city, through logistics, design and marketing services. At the same time the author was building his social reputation which is the key to get priority preference on future bookings on a social platform like Airbnb.

This is far greater and more accurate than social media driven parameters of the past. With an advent of data mining and science, algorithm and artificial intelligence sophisticated platforms will emerge that will accurately describe people and what they can do in the instant. Two important examples of social productivity based reputation is 'Trust Man', created by the founders of BlaBlaCar (the ride sharing app) and Traity created by Juan Cartagena, both of which are setting the basis for an Internet protocol for trust and reputation which is an important aspect of the crowd economy.

Trust Man is based on the D.R.E.A.M.S. framework that is based on the six pillars of trust which is able to harness declared information of the user, real life ratings, engagement, activity, moderation and yes social.

- **D for Declared:** Information is the foundation of a trusted online profile, it is the information that is volunteered by the user, telling the community a bit more about themselves. No one trusts a stranger, so this is the first, essential, step in moving away from anonymity towards online trust. For example, users can declare a name, age, their preferences or even give a quick description of themselves in their own words.
- **R for Rated:** Have long been used online because users trust content that has been created by a third party. However, unlike older online services like ebay or tripadvisor, for example, collaborative services ask users to rate one another, after having met "In Real Life", enabling people to build valuable peer-reviewed reputations and to create inter-personal trust in a community.
- **E for Engaged:** In order to feel completely comfortable transacting with a fellow user of a sharing service, you need to believe that the other party will respect their engagement. That's why a sharing service must always allow its members to financially commit to their transaction, via a pre-payment service.
- **A for Activity:** The D.R.E.A.M.S. framework gains its full value within a specific context. For example, a positive rating about a person's ironing skills does not indicate that they will be a good driver to ride share with, therefore, a successful sharing service must be activity-based.
- **M for Moderated:** All information transferred by users of a sharing service must be third-party verified, whether this is the verification of contact or bank details or the approval of User Generated Content. Users need to know that everything they see online meets a required level of goodwill and authenticity, as ensured by the third-party providing the sharing platform.
- **S for Social:** Networks allow users to connect their online identity with their real world identity, be it socially, via Facebook, or professionally, via LinkedIn. Indeed, collaborative services are one facet of a person's online presence: their personal responsibility and good reputation are essential to their interactions on sharing services.

On the other hand, Traity aims to become the standard for people´s reputation. Unfortunately today digital reputation is scattered all around the place. But is it fair or productive that the reputation one earns on eBay can only be used inside those walls? Who owns a reputation? the individual or the platform? When one signups with Traity, they build their reputation profile, verify themselves on new sites and use the *ReputationAPI* to verify one is a trusted user, bringing control and privacy back to the individual.

The above examples are at the beginning of what social reputation space looks like at the moment but expect to see real advancement on social reputation as crowd economy grows and becomes embedded in everyday lives.

## CROWD CAPITAL AND ENTREPRENEURS

### Past Models

When one looks at the opportunities for entrepreneurs to capitalise on the crowd economy it is imperative to look at the year 2002. During this time the dot com bubble collapsed and tech companies were on a speculative mode (see Figure 4).

Figure 5 points out that from 1997-2002 as an internet startup, the cost to entry or founding a startup was on the millions of dollars mark. This means, most of that money raised from VCs went into operation costs and not so much into the revenue generation. Instead, very few had a business model that supported the business in the first place.

Today, there is a completely different scenario. Many of the entry products and services for a startup to go cash flow positive have been demonetized and democratized. This means today one can literally found an internet startup business with five thousand dollars or less and a website. And there are opportunities for a startup to produce a MVP (minimum viable product) that can be tested on the market (see Figure 5).

*Figure 4. Rise and fall Nasdaq by WSJ*

*Figure 5. Cost to launch an internet startup*

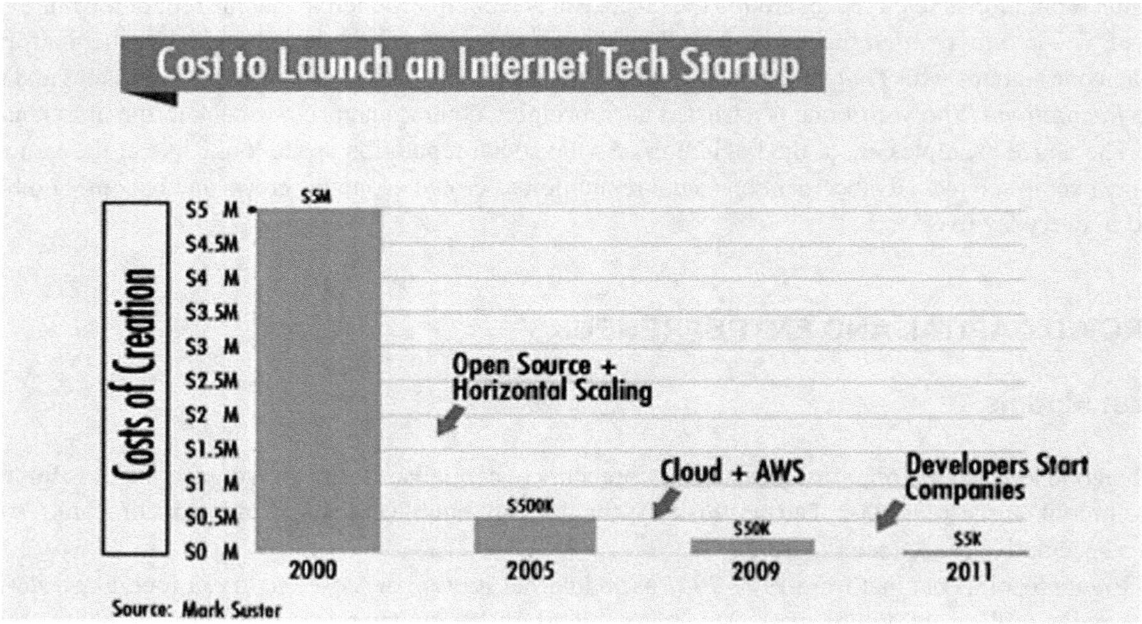


This is why it's important for every internet entrepreneur to have an intrinsic understanding of the crowd economy. Today, an entrepreneur doesn't have to invest a crazy amount of time and money just to find out the initial assumptions of the market were wrong or misplaced. Market development and product fine tuning happens simultaneously with a crowdsourcing approach. Understanding the techniques and knowhow of different networks that can accelerate the development of the MVP and testing the market is very important in order to navigate through hyper competitive world of startups.

This close connection with the market players is helping startups, globally, come up with innovate business models that are working to solve everyday problems. On the other hand, very few startups are maximizing the opportunity of the internet today through the power of crowdsourcing. Most go as far as crowdfunding and if that is extent of the grasp of crowd applications, entrepreneurs are missing the big picture. The synergies between the crowd and a startup are many and can be inculcated right through the lifecycle of the startup:

- **Branding:** Crowdsourced platforms like 99designs, designcrowd, crowdSPRING, jovoto and many more lets entrepreneurs get a "crowd" perspective on logo and branding designs that cannot be afforded by a design studio.
- **Website Design and Development:** Experienced freelancers on platforms like Upwork or Freelancer can provide services from technical coding to website development and content development.
- Quora for crowdsourced answers to questions.
- Platforms like Usertesting and Utest for software and user testing.

When securing talent through platforms, take a look at the following factors on the provider's profile:

- Reviews,
- Portfolio,
- Overall dollar amount for the jobs taken,
- Number of jobs taken,
- Country of origin.

Look at this formula very carefully in terms of who can deliver the project based on the specifics listed including the time of delivery. R(Reviews) + P(Portfolio) = $/N(C). Combining the number of reviews + type of portfolio can clarify the type of work they can deliver. Looking at the overall dollar amount earned with jobs divided by number of jobs can determine the talent and expertise level of the freelancer. Let's say the overall dollar amount earned on jobs taken is $15K and number of jobs are 3. Dividing $15K by 3 denotes an average of $5K per job. Considering the country of operation of the freelancer can one a good estimate of the talent, I am considering. To know more about each country and its GDP per capita one can use the most up to date portal from CIA factbook.

Crowdfunding is very popular among startup founders, especially reward -based crowdfunding, but in many instances founders of startups look at crowdfunding only as means of finance. But crowdfunding is much more. It has emerged as effective way to raise money with the added power of social media. There are four types of crowdfunding platforms: donation, reward, equity and lending. Each of these serve specific purposes beyond just money. Here is a chance to meet the first customers of the business, presell, test and finetune the product, receive feedback and procure market validation. If crowdfunding startups don't have a strategy for the above by-products of crowdfunding, they are not utilising the platform and opportunity to its maximum potential.

Let's look at donation driven crowdfunding. Causes that work with platforms like Razoo and End-Crowd typically also get backers that are willing to donate their time too. In addition to giving time backers also become social media evangelists of the organisation or brand and they become your biggest promoters of your cause.

Now, when it comes to pre selling products and services reward driven crowdfunding is the avenue for entrepreneurs to test drive their product and get the feedback in the instant. Many times entrepreneurs are overwhelmed by million dollar raises they hear on the news and forget realistic goals. This also goes without saying that a failed crowdfunding project does not necessarily veto an idea or product. What it signifies that the product may need to be improved or the target market may be wrong. Running a crowdfunding campaign requires a high level of time and commitment. Crowd tools lower the barrier to entry and reduce resource use, but they necessitate a deep level of understanding of the business model and product /service being promoted.

Answering the question "why" helps one recognize the real benefits of using crowd tools. For many entrepreneurs the success is just to raise the funds and this is where more than 50% of campaigns fail. They fail because the focus is so much on the raise and they fail to deliver the product which was promised at the first place. Never over promise the client otherwise it will backfire sooner or later. The same applies here, lots of entrepreneurs over promise, they raise more funds than initially expected and compromise on the most critical part of their campaign which is to deliver the product.

Now, equity crowdfunding is a very different approach on raising funds. An approach that is intertwined with a country's legal frameworks. In equity crowdfunding, UK is the frontrunner not just with number of platforms leading the charts worldwide like CrowdCube, Seeders, Syndicate Room and others but also through government tax incentives given to micro-investors. In the US market, clearly the JOBS Act is not really working out well on the federal level as many would have loved to see so what one ends up seeing in US is numbers of states bypassing federal law and make it legal to do equity crowdfunding.

Knowing which country to crowdfund in and which platform to use is only the first step. You need to follow the logistics of equity crowdfunding in order to do a successful launch. Preparing for a thorough biz plan, favorable company articles for shareholders, realistic 3 year financial forecasts and great video pitch is important when doing an equity crowdfunding. In the case of the author's company, the team spent a year or more creating momentum and traction on the business case while at the same time building a community around the world. This approach leads to a favourable valuation in terms of revenues but also in terms of market share and potential. Idea here is to bring the best opportunity to your community and build crowd capital and not the other way around.

## Future Evolution

Just as the author of "Third Industrial Revolution" Jeremy Rifkin said that now millions of people are becoming bankers, it is not too bold to envision a world by 2020 where billions of people will become bankers. Crowdsourcing and crowdfunding is the beginning of a more democratic and equitable version capitalism where now every individual has an opportunity to take part.

## Crowds and Government

### Past Models

The first time we saw governments harnessing the power of crowds was an offering by the British government called the The Longitude Rewards. This was a prize competition for the invention of a simple and practical method to determine the exact location of the ship's longitude at sea. The rewards, established through an Act of Parliament (the Longitude Act) in 1714, were administered by the Board of Longitude.

For decades, many governments had full grip on media, broadcast, telecommunication, transportation, shipping etc. This is still existent in many countries around the world, however empowering technologies are changing the pace of human behaviours where now governments are catching up to this rapid adaptation by its own citizens. First wave came during the social media days. During the early days of social media, we saw the rise of human voice, a voice where every individual has an opportunity to express and connect. This enabled individuals to raise their concerns to the highest level and bypassing the gatekeepers. Something fundamentally changed here and yet many governments, even democratic ones failed to understand the capacity and impact of such tools.

Although this first wave caught many governments unprepared, yet we saw resistance in one side and resilience on the other side. This makes us look at some of the scenarios that occurred since the birth of social media and why it happened:

- Slow adaptation.
- Failing to monitor.
- Failing to participate.

Exponential innovation powered by social media and crowds is not only disrupting business but also governance. Not only are old companies, not ready to evolve going out of business but also traditional, entire industries are becoming irrelevant if unable to adapt fast enough.

It's important and worth noting that very few governments have identified this new trend and some have reacted in such way by just going against it without really understanding the real value. Although, for many governments the term innovation has been key focus to grow their economies yet they fail to innovate their policies that are directly clashing with the new crowd innovation happening everywhere. In the current dynamic, individuals are solving major problems where in the past only big governments had the capability of doing. What this highlights is that governments must understand and become a part of the ecosystem in order to master the art of participation. Job creation today does not start by stopping innovation which is empowering citizens just because it is unsettling, new and seems foreign while disrupting old models. It starts first by looking at the policy and adapting them to current socio-economic needs. Governments also need to look at the taxations of individuals in a more pragmatic way and come up with solutions and not impose regulations and special taxes for such innovations. Uber's case is a perfect example of this fightback. The on-demand ride hailing platform is in trouble worldwide from Paris to Delhi. Uber has been called the "nastiest, most morally corrupt company ever", Bloomberg reports, and the reason is because many old businesses are too vested into their old model monopolised markets. Uber's model may not perfect but the company defines disruption and governments should see an opportunity to work with companies like Uber to foretell the future and be prepared for it.

The United States government is one example which is engaging American citizens as strategic partners in solving, creating, and innovating. The Presidential Innovation Fellows program is an initiative that pairs talented, diverse individuals from outside government with top Federal innovators to implement game-changing projects that make the Federal Government work better for the American people. And this is one of many other initiatives that has taken place across many government agencies in US resulting in new innovative ways of collaboration beyond the "usual suspects" to expand and diversify the talent pool tackling problems.

In 2012 White House created a platform called We the People as a right to petition the government as guaranteed by the First Amendment of the United States Constitution. If a petition gets enough support, White House staff will review it, ensure it is sent to the appropriate policy experts, and issue an official response. The We the People API currently provides read-only access to data within We the People, the White House petitions system. This API provides access to data on all petitions that passed the 150 signature threshold required to become publicly-available on the site.

In the event of the launch of We the People platform, President Barack Obama said "My administration is committed to creating an unprecedented level of openness in government. We will work together to ensure the public trust and establish a system of transparency, public participation and collaboration. Openness will strengthen our democracy and promote efficiency and effectiveness in government."

This is indeed one forward thinking government making such initiatives in order to make a more inclusive transparent government for the people by the people.

Governments can use crowd-powered technologies to "leap-frog" to the latest technological offerings.

One great example can be seen with "FireChat," a product of Open Garden the brainchild of Micha Benoliel. Micha's goal is to become a big telco, a full mobile operator, but 100 per cent built on software and this is how he's making it work.

FireChat sends encrypted private messages through the network. In three or four years the world will have four to five billion smartphones on the market, 80 per cent of them in emerging markets, 80 per cent of those in big cities where people don't necessarily pay for data.

FireChat's goal is to create a network out of smartphones without capital expenditure. Firechat becomes a carrier, a mobile operator but with no infrastructure. It lets users create live chatrooms which tens of thousands of people can access simultaneously to share real-time news -- one reason the app has been popular among activists in Hong Kong, Iraq and Russia to stay in touch even when governments were blocking phone signals. The service works by creating a "daisy chain of smartphones" -- using Airdrop, Bluetooth and Bluetooth Low Energy on an iPhone, and Wi-Fi Direct, Bluetooth and Bluetooth Low Energy on Android. When phones are in proximity they detect each other and establish a link. With offline private messaging, messages are forwarded through the P2P (peer-to-peer) network until the first person in the group connects to the internet to forward them to non-proximate recipients. At five percent urban penetration, it can guarantee message delivery to anyone on the network within 20 minutes, even with no signal.

In the beginning we had SMS almost replaced by IM; now we have what is called OM -- off-the-grid messaging. The real opportunity for this is in the emerging markets to give people off the grid, the first opportunity to connect. The big opportunity for government with a technology like Firechat is during crisis communication. For example, residents in the Philippines were encouraged to install FireChat as part of an earthquake awareness campaign.

### Future Evolution

We will see many more governments opening up - they must. Opening up different data stacks so solutions to the problems can be sourced and developed by its own citizens. Now for example Singapore is one of the leaders in open government and open data innovation as they opened up stacks of data open for people to participate.

## Crowds and Multinationals

### Past Models

Since the industrial age, the world's economy operated based on a unidirectional, dig and burn mentality. This has lasted over two and a half centuries and just now we are starting to recognize the effects and dangers of mass production that has affected every vertical even food and agriculture. Now, mass production is inherently not evil and full credit must be given to human potential that was unveiled during industrial age but where the path has gone astray is an industrial age with too much power in the hands of the few and unbridled capitalism. The human centric economy changes everything with a reliance on circular feedback loops and democratic participation. The combined factors of the information age, social media and crowd-powered participation has also forced multinationals to present authentic stories and become transparent.

The Circular Economy project by the Ellen MacArthur foundation strives to bring collaborative economy principles to mass production and industrialization. Companies like GE and IBM are continuously innovating moving beyond the industrial age to the age of the industrial Internet, connecting behaviours, devices and cars to maximize efficiency and reduce resource usage. This new mind set for multinationals is not a "nice to have" but a "must-have" that helps these companies stay relevant. Successful companies from Coke to Unilever are using crowdsourcing in innovative ways to stay ahead of the innovation curve and compete with startups.

The planet currently houses 7+ billion people. And if you look further, we have billions of planets on the universe for humans to explore and bring the age of asteroid mining. Till recently, planetary exploration was exclusively for governments but over the last decade we started to see the advancements for commercial purposes and a major part of the credit for this work goes to Peter Diamandis and the XPRIZE Foundation. Through XPRIZE, Peter Diamandis is working with corporations and foundations to bring forth solutions that aim to solve problems on a scale never seen before. The organisation uses incentivised prize competitions to crowdsource cutting edge solutions in space exploration, learning energy and more.

Another key area for disruption is multinational banks. With crowdfunding, peer-to-peer loans and fintech companies coming into the picture, banks have no choice but to revisit their century-old business models. New age workers want more from banks and innovative businesses like Transpay (that allows businesses to transfer cross-border payment to global freelancers via direct bank transfer with much lower fees) are all rewriting the ways banks provide service. On their toes, major banks are adopting the technology behind Bitcoin, called blockchain that uses a democratic record-keeping system to secure transactions and shorten transaction times.

Multinationals are adopting crowd economy principles to stay relevant and manage risk, but also they see a major economic advantage in harnessing collaboration. Being active players in the crowd -digital economy leads to expanding brand value, new business ideas, loyal customers and fast-rack innovation.

### Future Evolution

Multinationals want to innovate like startups and it's possible through a full scale adoption of the crowd economy. The next few years will see the principles of the crowd deeply embedded into the core DNA of organisations that want to survive and thrive. The crowd economy will also help build resilience into these companies that will see some old models and businesses fade away while new revenue streams open.

## The Challenges in the Crowd Economy and Future Work

As the crowd economy matures, and there is a greater understanding of long term repercussions, expect to see the following key points surface in discussions:

- **Future of Work:** With temporary freelance work and project-based work gaining in popularity because of crowd platforms, questions on the social safety net and standard of living emerge. The problem with this thinking is that people are trying to fit the new economy paradigm into the old way of life. If millennials are choosing freelance work with less than secure income one hand, they are also not laying emphasis on buying houses or cars.
- **Changing Mind Sets:** The crowd economy is a drastic change in mind set that will require time for full-scale adoption. Most people are still grappling with some elements of the crowd economy like crowdfunding or platform services like Uber, but do not see the full picture. More collaboration between digital disruptors and government is needed to hasten adoption and reap the full benefits of a collaborative society.

- **Digital Divide and Information Inequality:** The crowd economy could bring about information inequality between the rich with access to technology and the poor with no connectivity especially in developing countries. Ironically, the crowdsourcing could benefit these very people the most. The crowd ecosystem doesn't get diluted with more participation and in fact improves outcomes for everybody. A full functioning crowd economy is for everyone.

## CONCLUSION

The global economy is in flux, in a new place full of possibilities and also full of unknowns. The power shift from a handful of people to a democratic group of many is creating building innovation while tearing down archaic systems that do not work. As platforms mature and the crowd becomes more empowered and knowledgeable, there will be an organic restructuring and platforms that offer periphery services for true well-being of all stakeholders will distinguish themselves from the rest of the pack. From a political and civic point of view, governments should also prepare for the rise of this new economy. The crowd economy is not simply a buzzword but pathways to infinite opportunities.

## REFERENCES

*About the D.R.E.A.M.S Framework from TrustMan by BlaBlaCar.* (n.d.). Retrieved from www.betrustman.com/#about-framework

Botsman, R. (2013). *The Sharing Economy Lacks A Shared Definition.* Retrieved from: www.fastcoexist.com/3022028/the-sharing-economy-lacks-a-shared-definition

Butler, B. (2014). *IDC: Cloud will be $107B industry by 2017.* Retrieved from: www.networkworld.com/article/2175333/cloud-computing/idc--cloud-will-be--107b-industry-by-2017.html

Crowdsourcing.org (2015). *Global Crowdfunding Market to Reach $34.4B in 2015, Predicts Massolution's 2015CF Industry Report.* Retrieved from: www.crowdsourcing.org/editorial/global-crowdfunding-market-to-reach-344b-in-2015-predicts-massolutions-2015cf-industry-report/45376

Gray, B. (2014). *Brady Capital Research Nov 24, 2014 II. Economic Abundance: Where the Long Tail Meets the Blue Ocean.* Retrieved from: http://www.linkedin.com/pulse/20141124150237-10448698-ii-where-the-long-tail-meets-the-blue-ocean?trk=mp-reader-card

Grepper, R. (2014). *COOLEST COOLER: 21st Century Cooler that's Actually Cooler.* Retrieved from: https://www.kickstarter.com/projects/ryangrepper/coolest-cooler-21st-century-cooler-thats-actually

Higginbottom, S. (2015). Everyone's Betting on Blockchain because We Need Technology That Can Scale. *Fortune.* Retrieved from: http://fortune.com/2015/05/29/blockchain-big-because-scale/?iid=sr-link1

Meier, P. (2014). *iRevolutions. Results of the Crowdsourced Search for Malaysia Flight 370 (Updated).* Retrieved from: https://irevolutions.org/2014/03/15/results-of-the-crowdsourced-flight-370-search/

Oliveira, P. (2014). *Patient Innovation: Crowdsourcing Healthcare Solutions - Crowdsourcing Week Global 2014.* Retrieved from: https://www.youtube.com/watch?v=sblJKVFv3_s

Perez, Y. B. (2015). *Santander: Blockchain Tech Can Save Banks $20 Billion a Year*. Retrieved from: www.coindesk.com/santander-blockchain-tech-can-save-banks-20-billion-a-year/

Rowan, D. (2015). *FireChat will become a totally free mobile carrier*. Retrieved from www.wired.co.uk/news/archive/2015-08/05/firechat-full-mobile-carrier-free

The Network. (2013, May 29). *Cisco's Visual Networking Index Forecast Projects Nearly Half the World's Population Will Be Connected to the Internet by 2017*. Retrieved from https://newsroom.cisco.com/press-release-content?type=webcontent&articleId=1197391

*The White House Presidential Innovation Fellows*. (n.d.). Retrieved from https://www.whitehouse.gov/innovationfellows

*United States Census Bureau*. (n.d.). Retrieved from www.census.gov/population/international/data/worldpop/table_population.php

Van Zuylen-Wood, S. (2015). *Bloomberg Businessweek The Struggles of New York City's Taxi King*. Retrieved from www.bloomberg.com/features/2015-taxi-medallion-king/

# Chapter 2
# Towards a New Economy: Co-Creation and Open Innovation in a Trustworthy Europe

**Domenico Rossetti di Valdalbero**
*European Commission, Belgium*

**Bogdan Birnbaum**
*European Commission, Belgium*

## ABSTRACT

*This chapter focuses on the transition towards a new economy, where Europe can find untapped sources of growth and employment, renewing the legitimacy of public policy-making, especially through greater citizens' involvement and by delivering better public services for all. The chapter analyses the perspectives of Europe in 2050 following three dimensions: Global and European demographic changes; Economics, energy, environment and technological prospects; Geopolitics and governance. The concepts of trust, co-creation, sharing and sustainable economy, innovation, and social innovation are presented as the key ingredients of the "new" economy.*

## INTRODUCTION

Two strong trends are emerging in the European Union that, at first glance, appear contradictory: on the one hand, individual empowerment, and on the other, the collaborative economy.

Individuals in Europe have rarely, through history, been masters of their destiny to the extent they are today. A person can live, travel and work throughout the whole European Union. They can contact anyone across the world from a PC, tablet or smart phone. Using a 3D printer, one can make virtually anything from the comfort of its shelter. Thanks to the "selfie", not only a great painter or a famous aristocrat can have its own portrait. By merely speaking its *own* language, one can communicate with half the globe with the help of automatic translation programme. Legislation on euthanasia and abortion place the rights of the individual above any superior norm. In short, the 21st century is that of the most extreme form of "egolatry".

DOI: 10.4018/978-1-5225-0568-6.ch002

On the other hand, the collaborative economy has never been as fashionable. Driven by population growth (from 2.5 billion people in 1950 to 9 billion in 2050), the expansion of urban areas (from two megacities in 1950 to around 30 in 2050), the knowledge society (see search engines and worldwide ICT companies) and the social networks (from Facebook to Twitter), this new "collaborative economy" seems to be taking off in sectors such as tourism, transport and urban farming. Bottom-up shared consumption is gaining in popularity. In the same time there is an ever increasing demand for top-down sustainable environment, high-quality public spaces, public transport and well-planned city centres with properly maintained green areas.

Europe has to propose solutions to the current issues with and effective short, medium and long term action as suggested by the report Global Europe 2050 and the work performed by the European Strategy and Policy Analysis System (European Commission [EC], 2013a; European Union [EU], 2014). Ageing population, securing sustainable resources, developing clean energy supplies, migration, instability in the neighbourhood represent the main issues at stake. The EU has to resist the charm of fanatical individualism/nationalism and keep in mind the founding principles of the European construction: de facto solidarity (the 1951 Schuman Declaration), social and economic cohesion (the counterpart to the internal market) as held dear by Jacques Delors, and intelligent, inclusive and sustainable growth. The words "fairness" and "European social resilience" ought to begin to outweigh short-term national self-interest and the "I want my money back" attitude proclaimed by an ever growing chorus – with 28 voices standing together on this point. In this context the EU should take further steps to build resilient and sustainable economy (Rossetti di Valdalbero, 2015).

Europe should reap the benefits of individual empowerment and the collaborative economy by rediscovering the concept of trust and stimulating the concept of co-creation (Moedas, 2015a). While the giants of information and communication technology across the Atlantic may be dominating the global marketplace with the Koreans, Chinese and Japanese close behind, Europe has a huge capacity for social, marketing, technological and organisational innovation to be exploited. But it needs to be given the financial and legal means to be able to grow and thrive. The book by Mariana Mazzucato is worth absorbing, and its title "*The Entrepreneurial State*", speaks for itself (Mazzucato, 2013).

Over centuries, Europe has developed a particular relationship with innovation and nature. Precaution and progress go hand in hand. Man works the earth and uses it for his advantage while respecting and nurturing it. The environmental question which is well described by Michel Serres in *"Le contrat naturel"* permeates every part of European society (Serres, 1992). Proof of this can be seen in European regulations on the environment, agriculture and health, animal welfare, biodiversity and the fight against climate change, as well as the promotion of research, energy efficiency and renewable energy sources. When Europe presents itself as the pioneer of the "circular economy", it has in mind the interests of the individual as well as that of the collective and the environment (EC, 2014a). When Europe brings together the strands of individualism and the collaborative economy, it encourages the social entrepreneurs, the marriage between profit and volunteering, between the economic viability and the social purpose of the companies.

## PERSPECTIVES ON 2050: EUROPE IN A CHANGING WORLD

A number of European forward-looking studies (EC, 2014b) point a number of currently unfolding trends at the global level in each of the following three dimensions: Global and European demographic changes; Economics, energy, environment and technological prospects; Geopolitics and governance (EC, 2013a; EU, 2014).

### Global and European Demographic Changes

In the global demographic and societal dimension three main trends can be observed: A continued imbalance in population growth, the changing reproduction and family lifestyle and a generational shift coupled with high youth unemployment. 60% of young in Greece and 50% in Spain and Croatia are unemployed; 14 million are "NEETs", *Not in Employment, Education or Training*, i.e. in no man's land of inaction, abandonment and despair (Eurostat, 2015).

In the EU there were 96 million 15-24 year olds in 2010. In 2030 this generation will be 35-44 years. In 2050, the cohort from 55 to 64 will still be in working life. The education of EU young people is high: more than half of all students are in upper secondary education attending vocational programs. More than one third of 15-24 year olds are NEETs. Every fourth of unemployed 15-24 year olds and third unemployed 25-29 year olds have been unemployed for more than 12 months, so in difficult transition to work in the future. This will be also true for that half of employed 15-24 year olds who is in a low skilled or elementary occupation. When more than half (65%) of employed 15-24 year olds have used to work on a temporary contracts or in a part-time jobs their work attitudes have already changed.

A continued imbalance in population growth will represent the main trend over the next decades. Improvements in health, education and living standards will continue to drive population growth. In the absence of major pandemics, global natural disasters or multi-regional wars, organizations such as the UN and OECD estimate that the world population will reach around 9 billion people in 2050. With 2 billion more people on earth will mean a more crowded world with ever more competition on natural resources. By 2050, UN statistics suggest that pensioners in the 'rich world' will comprise one-third of the population and one-tenth will be over 80.

The concept of 'old age' will have to be redefined. In many developed economies, 55 is already the new middle age and one can see a future where what used to be the average age of retirement becomes the mid-point of the average adult life. Longer life expectancy (for e.g. in Europe the life expectancy was 70 years old in 1960 and more than 80 today) will also means that there will be an ageing population. By 2050 in the EU, the oldest old (85 and more) will represent a population equivalent to 25 million people, i.e. the population of The Netherlands and Belgium together. The share of health expenditure in the EU is expected to increase from 8% today to 12% by 2050 (EC, 2014c). Regional differences in growth rates and life expectancy at the world level will result in a larger proportion of people living in Asia and, after 2050, in Africa.

Taking the average fertility rate into account, the natural replacement rate - by which a population stays level is 2.1 children per woman - will be reached globally by around 2020.

There are several reasons for the decline in overall fertility. There is, for example, clear evidence to suggest that as people get more affluent they want smaller families. The transition in fertility rate from five children to two, which took place over 130 years in Britain (1800-1930), happened in just twenty years (1965-1985) in South Korea. Some countries are experiencing even more profound change largely due to the movement from countryside to urban areas where the cost of feeding and housing a larger family is often prohibitive. Fertility is also falling because more women are better educated and work outside.

The young generation in Europe is searching for a job but also for more attractive jobs to better fit with life aspirations. Flexibility is a requirement of many employers but also of many young people. The young generation is also hyper-connected, so virtual and collaborative networks will grow in importance. Workforces will become more balanced with a greater spread of ages, more ethnic diversity and more women in the workforce.

## Economics, Energy, Environment, and Technological Prospects

Smart growth (fostering knowledge, innovation, education and digital society), sustainable growth (making production more resource-efficient while boosting competitiveness) and inclusive growth (raising participation in the labour market, acquisition of skills and fight against poverty) remain until 2050 the leitmotifs of EU policies.

The shift to Asia in economic terms is a fact since the beginning of the century. It is difficult to make generalizations for a - still poor - continent of 3 billion people but Asian consumers are growing richer, with average earnings in many countries tripling over the past ten years. On economic development, history is accelerating: it took 155 years for Britain to double its GDP per capita, 50 years for the US and just 15 years for China (Schinas, 2012). Over the last ten years, Asia has accounted for half of the world's GDP growth. Forecasts indicate that Asian growth will continue to outpace Europe and the US. A paradoxical future trend is towards "less poverty but with more new poor".

With economic growth coupled with resource consumption, one can expect more constraints on key resources. As people become wealthier they use more energy (e.g. for air conditioning, heating, computing, mobility), more water, eat and waste more food. Efficiency gains have been largely offset by the 'rebound effect', where technological improvements have ultimately led to greater consumption and not less consumption.

Transport is a particular challenging sector as concerns the long term (2050) targets of $CO_2$ emission reduction. It is also a sector where an important research effort should be concentrated to produce technological, social and public sector innovations. Climate change, urbanization and relative resource shortages may fuel a shift away from large petrol-engine cars to small electric and hybrid vehicles. There will be also a boom in cheap cars and bikes in emerging countries. Tax rates, license charges, road charging and parking fees will increasingly be linked to vehicle type. Anti-car and anti-driver sentiment and regulation will probably increase.

The industrialized world needed over a century to develop the infrastructure to supply natural resources to the point of use (pipelines, sewage farms, ports, rail and road links, electricity grids). Making significant changes to this infrastructure may require decades. Delays in investment in infrastructure to cope with changing demand will result in supply and demand gaps. The question is not only about energy security but also about resource security – land, food, water and raw materials with these latter that are geographically concentrated.

The recent progresses and new-found prosperity in China, India, Brazil and South Africa have knock-on effects on the demand and availability for global resources. Global weather may become more volatile and severe (hurricanes, disastrous flooding, extreme heat and shortage of water) while rising sea levels could devastate low-lying cities. Systemic governance of climate change and sustainability challenges have been discussed at the recent Conference of the Parties of the United Nations Framework Convention on Climate Change (COP-21) in Paris from 30 November to 11 December 2015. But there is a need of a paradigm shift in the global economy, starting with manufacturing efficiencies and energy saving solutions, new and clean options for the residential and transport sectors.

Global connectivity, brought about by technology, globalization, low-cost travel and migration, is changing how people live, how they work and how they think. ICT, the Internet and new mobile devices will continue to play a pervasive and transformational role. Mobile devices are becoming a key technology to access the internet and promise to continue heavily transform our way of living. In 2000, there were around 700 million mobile devices, most of which did not connect to the World Wide Web, 370 million people had access to the internet, and the likes of Google (incorporated 1998) and Facebook (2004) were unknown or barely gathering momentum. In 2015, there are 7.6 billion mobile devices worldwide, many of which are smart and connected to Internet (GSMA Intelligence, 2015).

Machines and technology will be a dominant feature of the future. Computers and artificial intelligence may become more performant than human brain, at which point humanity will be faced with a dilemma. If machines are more intelligent than their makers, what is to stop them taking over? However, machines do not think. Even if they can be said to 'think' they only think about what they are doing, which leaves the gates wide open for humans to have empathy, imagination, creativity and ideas. Developing smarter tools and infrastructure, embedding intelligent technology, will need to be balanced by greater awareness and people's cognitive and creative abilities to find out intelligent uses of the intelligent technologies.

## Geopolitics and Governance

Political developments are inherently unpredictable, given the wide scope for human agency in shaping the course of events. However, drawing from the development of structural trends and from the evolving strategic outlook of key international actors, some insights into the future international political system emerge.

In the absence of a clear and established international political system, like the one based on the East-West confrontation during the Cold War, globalization itself will be the most influential factor shaping international politics. The ways in which the new economic powers will be willing and able to translate their weight into political influence will be key to the future of international relations. Real and perceived risks from terrorism (cf. ISIS) may also play a key role in defining new alliances worldwide.

As indicated in the European ESPAS report, one can expect the emergence of a multipolar world with fewer hegemonic powers or "superstates" and many more middle powers claiming their share of influence in global affairs. It is more likely that there will be an evolution from today's G7/G20 to tomorrow's G50 framework rather than a consolidation towards a G2 arrangement (EC, 2013a; EU, 2014).

People can bank online, bet online, watch television online, so why not vote online once digital access is almost universal and e-illiteracy eradicated? E-voting can expand in the future to cover a wide set of policy issues, allowing citizens to check policy actions and express their opinion more frequently. This will have to be backed by adequate information and awareness raising activities to enable citizens to provide their judgment on an increasing range of topics. Besides e-voting, global e-action groups and virtual protests are already of common use and will thrive in the future, making politics more connected to real-time for the best and the worst. Representative democracy will radically change as the ever-increasing global demand of empowered citizens to shape major decisions and to control government will continue.

As a form of government, democracy is about the way in which government is appointed, by popular vote, and not the way power is exercised. The constitutional feature of Western democracies is the distinctive set of principles and institutions of liberalism, with an emphasis on check and balances (cf. Montesquieu) and respect for individual rights. Political regimes can, however, be democratic while not being liberal. Conversely, some non-democratic regimes can guarantee the rule of law, proper services and a degree of respect for human and civil rights, while not permitting electoral competition challenging the ruling elite (hybrid regimes).

The control exerted by private interests on policy making (cf. lobby) and the corresponding loss of prestige and influence of traditional political parties, undermine an open public debate and the definition of collective preferences.

## TRUST: SUBSTRATUM OF THE SHARING ECONOMY AND CO-CREATION

Trust is at the roots of the European economic system. In finance, one speaks about "trust in the markets". Without trust between producers and consumers, there will be no free market. In the emerging sharing economy, trustful relationships are a pre-requisite.

A transition from ownership towards accessibility can be observed over the last decade in the economic exchanges. In the traditional economy, customers buy products and become owners (*proprietas*). In a sharing economy, consumers pay for temporary access-rights to a product or service (*usus*). In this context, a new pact based on trust between Man and Man, between Man and Technology is needed. This would allow the co-creation process to be efficiently implemented.

### Trust as the Essential Ingredient for Economic and Social Relations

Trust is at the core of our European values. It is at the basis of all human relations. If one believes, as Aristotle, that "Man is a social animal", then trust is the food of such animal. Trust is also at the roots of the economic system. In the emerging sharing economy "C2C" - Consumer to Consumer, trustful relationships are a pre-requisite.

As Jean-Claude Juncker (2015) said in his State of the Union Speech before the European Parliament: "There is not enough Europe in our Union. There is not enough union in our Union". One key problem is certainly the lack of trust: among Member States, among institutions and among actors.

Trust is the keyword in current debates on migration, radicalisation and inclusive societies. Trust and trustworthiness are the "glue" that holds societies together, that allow business to flourish and personal interactions to be meaningful. In politics, trust is the twin sister of legitimacy. If you do not trust the elections and the voting system, you cannot have a democracy.

The current headlines call for a large debate on Trust: millions of young people are unemployed in Europe and do not believe in the capacity of the socio-economic system to help them. In the millennium to come, Europe has to make efforts to switch from an inward-looking to an outward-looking perspective. Commissioner Carlos Moedas (2015a) is making a plea for research and innovation to be more "Open to society" and more "Open to the world". The words of Italo Calvino (1957) have to be repeated: "If you raise a wall, you should think at what you leave outside, what you lose". Research cooperation and science diplomacy can also contribute to conflict prevention and resolution. Scientists cooperate across borders, even when there is little trust and political dialogue between their respective governments (Moedas, 2015b).

The world is moving into a world of open innovation, user innovation and co-creation. A world where the digital and physical are coming together, where new knowledge is created through global collaborations involving thousands of people from across the world and from all walks of life.

Co-creation is the active flow of information and ideas among different sectors of society (academia, government at all levels, business, third sector and citizens) which allows for a better understanding of participation, engagement and empowerment when developing policy and programmes, creating and improving services and tackling systemic change within each dimension of society. Through engaging citizens to redesign and remake their cities and communities together with businesses, public authorities, universities and all stakeholders, co-creation can lead to citizen empowerment, jobs creation through innovation, more efficient and effective public administration and strengthening democracy through open government. Co-creation can increase the innovation capacity of all sectors, by increasing the sources for new ideas, and facilitating cross-fertilisation through knowledge and experience sharing. Through the coproduction of services with citizens and business, co-creation can reduce costs and build on design driven ideas and testing of new solutions (EC, 2014d).

## Progress and the New Pact between Man and Technology

Europe is facing different threats. First, an excessive place to the individual leads to selfies, narcissism and *Egolaïtrie*. When the individual is the ultimate end, it opens the doors to intolerance, narrow-mindedness, fanatic movement and xenophobia. The Nazis and the Soviet communists did not have a God. The second threat deals with the purely materialistic view of man and the world. Reason without faith does not give happiness. *Science sans conscience n'est que ruine de l'âme* as taught by François Rabelais. Consumerism provides a false happiness. Europe should avoid to copy the *American way of life*. Egoism should be balanced by altruism; Personal motivation by social goals; private profits by common goods; The Ltd by non-profit and volunteer activities. With the corporate social responsibility and the emergence of social entrepreneurs, business is going in the right direction. The third European threat is the fear of neo-colonialism and non-interventionism: Ultra-pacific movements and indifference are two sides of the same coin. Promoting a better world needs strengths, courage and willingness.

Glasses, prostheses and dental crowns are all elements which help and relieve the human being. But in the case of implants or an organ transplant - though vital for the patient - they raise concerns, fear and rejection. DNA and GMOs, nuclear energy, pesticides and antibiotics are some Science and Technology issues that raise concerns of society. The relationship between man and nature has always been balanced between respect and the need for its exploitation. Gardeners of the earth and global sharks have been fighting since ever. Today there is a heated debate between "precaution" and "innovation", pressure to move quickly (aircraft, highways, high-speed trains) and slowly, quiet and calm regeneration (residential areas, parks, green areas). Naomi Oreskes and Erik Conway (2010) have written about how a handful of scientists obscured the truth on issues from tobacco smoke to global warming.

Do technologies have the aim to serve human being or to enslave it? Our smart phones clearly illustrate this: they provide the most complete freedom to work, to connect socially or inform where and what you want to. But they are also emblematic of a new stress addiction to the reign of immediacy and complete isolation.

Dozens of classical writers - Rabelais, Voltaire or Teilhard de Chardin - stressed the need to raise awareness among the public on the effects of technology on the society (Rossetti di Valdalbero, 2014). In the 19th century the German geographer Alexander von Humboldt highlighted the interactions between the forces of nature, the environment, animals and plant life. In the first half of last century, Russian-American author Isaac Asimov coined the term "Robot" ("hassle" in Slavic).

Progress in different domains BANG (*bits, atoms, neurones genes*) and NBIC *(nanotechnology, biotechnology, informatics, cognitive sciences*), has been accompanied by different societal challenges. Kasparov is probably not playing chess any more against the unbeatable IBM computer. Asimo, a small Honda robot, has started to learn by itself on correcting errors. In less than one million years the human being has developed from oral animal basic sounds to multilingual interconnectivity almost global and permanent. According to a European foresight study, the time devoted to information for both business and private purposes, will continue to increase by the end of the century (Chateau & Rossetti di Valdalbero, 2011). Social media used on smartphones were used to mobilise people up and bring minds during the Arab Spring. But at the same time images of the Libyan dictator mutilated and ISIS atrocities are on the same screen. Middle ages and dark hyper-modernity coexist on line. Technology may serve as the best and the worst of the mankind. Some authors refer to the German philosopher Hans Jonas sentence 'The Prometheus, now completely unbound, to whom science has granted unprecedented powers (...) calls for an ethics which prevents human power to become a curse for him' (Landes, 1969).

## Sharing Economy or Moving from *Proprietas* to *Usus*

A transition from ownership towards accessibility can be observed over the last decade in the economic exchanges. Whereas in the traditional markets customers would buy products and become owners (*proprietas*), in a sharing economy consumers pay for temporary access-rights to a product or service (*usus*).

The traditional business model based on *proprietas* is rapidly changing due to two main drivers: one the one hand, technological advancement allows the model of sharing economy to spread different markets and become convenient and flexible. On the other hand, there is a shift towards peer-to-peer accessibility. According to this model based on *usus*, companies are providing access for consumers to company owned property, or in peer-to-peer models, online platforms are facilitating access for consumers to consumer owned property.

As Tom Goodwin said in a recent declaration to social media: something interesting is happening: Uber, the world's largest taxi company, owns no vehicles. Facebook, the world's most popular media owner, creates no content. Alibaba, the most valuable retailer, has no inventory. And Airbnb, the world's largest accommodation provider, owns no real estate. And many other examples can be quoted from the Vellib/Villo of shared bikes in Paris and Brussels to the Spotify music streaming service.

This shared economy business model – that acts as a broker between consumers - is becoming particularly disruptive to conventional solutions as it is able to serve the same needs at a significantly lower price. Differentiation strategies are based on the mechanism that drives matchmaking, which can be either demand-driven, supply-driven or a combination of both.

A big challenge from a client's perspective is the lack of trust in online activities and transactions. Companies in the sharing economy are therefore trying to come up with measures to boost confidence like peer-to-peer rating systems and ID checks (EC, 2013b)

A main worry is also regulatory uncertainty. Will Uber or room-renters be subject to taxation? The policy-makers should find the right balance between the need to protect the consumers from harm, support competition but at the same time assure a fair taxation system. Indeed, this new Consumer-to-Consumer market requires new type of measures and regulations.

## EUROPE: HUB OF INNOVATION AND GARDENER OF THE PLANET

Innovation is Europe's best option to get the economy back on track. Europe has many competitive strengths: the talent and creativity of its people, a strong industrial base, a vibrant services sector, a performing education system, its position as the world's biggest trading bloc and with the largest internal market in the world it is home to many of the world's leading innovative companies. Moreover, it has a leading position in many fields of knowledge and key technologies such as health, food, renewable energies, environmental technologies and transport (EC, 2013c). Europe has all the assets to find new growth opportunities thanks to education, research and innovation. Europe has and can develop its strengths in circular economy and social innovation. The Europe 2020 strategy and the Innovation Union flagship initiative emphasise the need to sustain and where possible promote growth enhancing expenditures and the need of an innovation-friendly environment. The knowledge base in Europe is being strengthened through the of Horizon 2020 - EU Framework Programme for Research and Innovation, the focusing of European Regional Development Funds on smart specialisation strategies, progress towards the European Research Area, and support for skills development (EC, 2010). Europe can reap the benefits from these advantages in terms of economic prosperity, sustainability and quality of life.

Economic growth, research and innovation cannot overlook the environmental and biophysical limits of the planet. Therefore, a new pact between man, technology and its environment is needed. Man should become the "Gardner of the planet". In Europe, a person consumes 4 tons of oil equivalent per year. Only at consumption-level, 100 kg of food is wasted per person per year (around 150 kg at retail-level). Responsible decisions are needed for 7 billion people today and even more responsible decisions are needed for 2050 and the 9 expected billion people. The new relationship between Man and its environment should tackle the issues of food, water, air and biodiversity. Rather than pure exploitation, Europe should be ahead in thinking about responsible use of the resources of our planet. Man is the host and the guardian of the earth (Rossetti di Valdalbero, 2013).

### Research and Innovation as a Key Source of Growth

Innovative solutions need to be safe, accepted by society and sustainable. These challenges provide a significant opportunity to create new shared value through innovation and social innovation. Society's challenges may well reflect the transition to innovative solutions, and today's challenges are perhaps best seen as examples of Joseph Schumpeter's creative destruction model where the failure of old approaches fuels the motivation for change and shapes the future (Schumpeter, 1942).

Evidence shows that much of recent productivity gains come from innovation and that, on average, those countries that invested more in research and innovation (R&I) before and during the crisis have been the most resilient during the economic downturn (Conte, 2014).

Innovation-friendly framework conditions and the quality of public policies are essential ingredients for successful innovation. In recent years, the Commission has made a concerted effort to reduce internal market fragmentation and restore economic confidence. It fostered the functioning of the Single Market, took steps to complete the Banking Union and the Euro zone and took measures to facilitate and diversify access to finance (cf. business angels, technology transfer vehicles or more traditional bank lending) and to streamline legislation and reduce regulatory burdens, and is committed to fostering the long term financing of the European economy. The Commission has also taken initiatives to improve across

Europe the public procurement and demand-side instruments, to dismantle barriers to science-business cooperation and mobility, and to promote efficient intellectual property rights system (EC, 2015a).

*The public sector plays a decisive role in promoting growth. By adopting a more entrepreneurial approach it can raise productivity, efficiency and the quality of the public service and stimulate the private sector. Public procurement represents around one fifth of GDP across the Union and can have an important leverage effects on innovative solutions. However, to fully exploit the benefits, the procurement authorities should coordinate effort across Europe in order to avoid a fragmentation of demand.*

The EU budget for the multiannual financial framework 2014-2020 clearly puts the emphasis on smart and inclusive growth with an increased budget for the research and innovation programme (Horizon 2020). However, this important investment from the EU budget should be complemented with Member States investment both from public and private sources.

Member States have agreed to a Europe 2020 R&D investment target of 3% of their GDP, but most of them are far from their national R&D target under Europe 2020. In order to meet this target, Member States need to step up their investments in research and innovation; ensuring its leverage over the private investment as well as setting up favourable framework conditions in line with industrial policy orientations.

The current budgetary restrictions call for a maximisation of the impact of the public spending by improving its quality. This can stimulate a virtuous cycle by leveraging higher investments levels from private sector and generating increasing economic return. Three axes have been identified by the European Commission where the Member states should step up their efforts: improving the quality of strategy development and the policy making process; improving the quality of programmes, focusing of resources and funding mechanism; and optimising the quality of the public institutions performing research and innovation.

Effective public policies need to be accompanied by innovation-friendly framework conditions in order to stimulate Europe's companies to innovate. The European Commission has increased its effort to reduce internal market fragmentation and restore economic confidence. Deepening and completing the single market is a key priority. The Single Market is still fragmented and high tech areas such as ICT, digital networks, content and services, and healthcare are characterized by inefficiencies business investments in R&I.

## Towards a Circular Economy

The Commission is presenting a new, more ambitious circular economy strategy, in order to transform Europe into a more competitive resource-efficient economy, addressing a range of economic sectors (Katainen, 2015). The circular economy roadmap envisages revised legislation on waste, fertilizers, and water reuse, enhanced commitments on Eco-design, strategies for handling plastics and chemicals, and funding for innovation. The proposal will be fully aligned with the priorities of the Commission in particular with the first one on "A New Boost for Jobs, Growth and Investment". The new package will include legislative proposal on waste targets (EC, 2015b).

The economy has mainly operated on a 'take-make-dispose' model – a linear model where every product is bound to reach its 'end of life'. This traditional pattern of growth developed since the industrial revolution in the market economies is based on the assumption that resources are abundant, available, easy to source and cheap to dispose of. However, every year in the EU, nearly 15 tonnes of materials are used per person, while each EU citizen generates, on average, more than 4.5 tonnes of waste annually, almost half of which is disposed of in landfill sites. The linear economy, which relies exclusively on resource extraction, is no longer a viable option. A transition to a circular economy shifts the focus to reusing, repairing, refurbishing and recycling existing materials and products. What used to be regarded as 'waste' can be turned into a resource (EC, 2014a; EC, 2015b). Circular economy aims at keeping the added value in products for its entire lifecycle and reduce waste by reusing parts that can be still create further value. The objective is to move from "take-make-dispose" model to "take-make-recycle".

The implementation of the circular economy requires changes throughout value chains, from product design to new business and market models, from new ways of turning waste into a resource to new modes of consumer behaviour. This implies systemic change and innovation in technologies but also in organisation, society, finance methods and policies. Specific areas of intervention may include materials production and use, product design, distribution, consumption phase, public procurement, labelling and product information, waste management, development of markets for secondary raw materials (e.g. organic fertilizers), improving framework conditions in sectors such as sustainable chemical production, bio-economy, food, construction, plastics, critical raw materials, water use, and improving cross-sectoral cooperation, e.g. by the promotion of industrial symbiosis, repair and re-use and enabling the development of new business models.

Even in a highly circular economy there will remain some element of linearity as virgin resources are required and residual waste is disposed of. The draft circular economy package mentions "long-term recycling targets on waste for paper, glass, plastics, metal and wood packaging, which will reinforce municipal waste targets" (EC, 2014a). And the action plan sees a major role for research and innovation funding under the EU's Horizon 2020 programme. This is expected to drive the investment in innovation needed to steer the transition to a new circular economy.

As indicated in the Communication, industry already recognises the strong business case for improving resource productivity. It is estimated that resource efficiency improvements all along the value chains could reduce material inputs needs by 17%-24% by 2030 and a better use of resources could represent an overall savings potential of €630 billion per year for European industry (Meyer et al., 2011). Business driven studies based on product-level modelling demonstrate significant material cost saving opportunities for EU industry from circular economy approaches and a potential to boost EU GDP by up to 3.9% by creating new markets and new products and creating value for business (Ellen MacArthur Foundation, 2012).

## Europe Leader in Social Innovation and Open to the Word

The EU is promoting a new path for European research and innovation policy. This strategy is fit for purpose for a world that is open, digital and global. The notion of an Innovation Union is meant to convey a shift from a conceptual idea to action, focused on creating knowledge and also creating jobs. Innovation from this perspective is meant to go far beyond traditional industrial innovation to include both technological and social innovation.

### Open Innovation

Open Innovation, as coined by Henry Chesbrough, can be defined as the use of purposive inflows and outflows of knowledge to accelerate internal innovation (Chesbrough, 2003). The original notion of

Open Innovation was largely based on transferring knowledge, expertise and even resources from one company or research institution to another. The basic premise of Open Innovation is to open up the innovation process (Huizingh, 2011). This assumes that firms can and should use external ideas as well as internal ideas, and internal and external paths to market, as they look to advance their performance.

However, the concept of Open Innovation is constantly changing and is moving from bilateral transactions and collaborations towards networked, multi-collaborative innovation ecosystems. This means that a specific innovation can no longer be seen as the result of predefined and isolated innovation activities but rather as the outcome of a complex, intertwined and less-deterministic co-creation process involving knowledge flows across the entire economic and social environment.

Open Innovation is a new paradigm based on a Quadruple Helix Model where government, industry, academia and civil participants work together to co-create the future and drive structural changes far beyond the scope of what any one organization or person could do alone. This model encompasses also user-oriented innovation models to take full advantage of ideas' cross-fertilisation leading to experimentation and prototyping in real world setting.

There are five key elements in the Open Innovation process (EC, 2016):

1. Networking;
2. Collaboration (involving partners, competitors, universities, and users);
3. Corporate Entrepreneurship (cf. corporate venturing, start-ups and spin-offs);
4. Proactive Intellectual Property Management (new markets for technology); and
5. Research and Development (for achieving competitive advantages in the market).

The Commission has launched several European Fund and new actions to get more innovation impact out of Horizon 2020. One important way is also to create synergies with the Structural Funds. Horizon 2020 has made a huge step forward in supporting innovation. Europe does not yet have a world class scheme to support the very best innovations in the way that the European Research Council is the global reference for supporting excellent science and that is why Commissioner Moedas is working on establishing a European Innovation Council (Moedas, 2015b).

## Social Innovation

Jean-Claude Juncker (2014) declared in November 2014 upon receiving the European Parliament's support for his list of Commissioners: 'I want Europe to be dedicated to being triple-A on social issues, as much as it is to being triple-A in the financial and economic sense'. The EU with its strong legacy in social democracy, solidarity, civic participation, justice and fairness, Europe constitutes especially fertile grounds when it comes to sustainably enabling and growing social innovation. In this context, social innovation represents one of the avenues to explore. Innovation partnership, social innovation figures prominently appear in the flagship initiatives Innovation Union, European Platform against poverty, A Digital Agenda for Europe and the Active and healthy ageing. Social innovation is also included in the HORIZON 2020 framework programme for research and innovation as well as in European cohesion policy.

In the past, societal challenges such as the ageing of Europe, migration waves, social exclusion or sustainability were primarily perceived and addressed through a traditional non-profit philanthropic models. These activities have often been highly dependent on government subsidies or private donations and faced the difficulty of realising a long-lasting, sustainable difference. Today, these societal trends are increasingly perceived as opportunities for innovation. New trends in demography, community and social media, poverty, the environment, health and wellbeing, or ethical goods and services are more and more analysed not only through the "philanthropy" lenses but also as growth markets opportunities.

Social innovation can be defined as process where civil society actors develop new technologies, strategies, ideas and/or organisations to meet social needs or solve social problems (EC, 2013d). Social innovation describes the entire process by which new responses to social needs are developed in order to deliver better social outcomes. It represents new responses to pressing social demands, which affect the process of social interactions and they are not only good for society but also enhance society's capacity to act. It is aimed at improving human well-being. Social innovations are innovations that are social in both their ends and their means. They rely on the inventiveness of citizens, civil society organisations, local communities, businesses and public servants and services.

Social innovations typically go through stages. They start as ideas, which may then be piloted or prototyped. If successful there is a process of sustaining the new model in the implementation stage – perhaps as a new venture or as a new policy within an existing institution. The final stage is to scale up so that the new approach makes a real impact and becomes part of the norm. The challenge for policy makers is to identify which ideas are the most promising to take to the pilot stage, and to identify which pilots are best able to improve on existing models of practice. Then selecting from among those pilots, the projects that should be implemented to become sustainable ventures and the ventures that should be scaled up to achieve systemic changes. It is important that regional authorities design programmes that stimulate a pipeline of projects at each stage which can then be promoted to the next.

Social innovation is inextricably linked to the concepts of solidarity and subsidiarity. The subsidiarity principle is a dynamic concept that allows decisions to be taken on the most appropriate level of government to manage a policy and on the actors that should take that decision. Whereas solidarity, cohesion and equality of opportunity represents key values of the European Union. Social innovation can be further strengthened by the value of solidarity and can be promoted within the subsidiarity principle.

A lot of business model experimentation is ongoing and there is an emergence of hybrid organisational models, horizontal business models designed to create at once economic and social value. There is also a great need and potential for social innovation in the public sector.

## Open to the World

Challenges in areas like energy, health, food and water are global challenges. And Europe should be leading the way in developing global research partnerships to address these challenges. This means speaking with one voice, and close coordination with Member States' international collaborations. Europe is a global leader in science, and this should translate into a leading voice in global debates. In order to remain relevant and competitive, the EU needs to engage more in science diplomacy and global scientific collaboration. It is not sufficient to only support collaborative projects but also to enable partnerships between regions and countries. Embed science in EU diplomacy is a key priority. The UK Royal Society – quoted by Commissioner Moedas, identifies three dimensions of science diplomacy (Moedas, 2015c):

- **Science in Diplomacy:** Informing foreign policy objectives with scientific advice. For example, calling on scientific advice to decide whether citizens are at risk from a natural disaster in a foreign country.

- **Diplomacy for Science:** Facilitating international science cooperation. For example the signature of the Ukraine's Association to Horizon 2020. Ukraine is now able to participate in the largest EU research funding programme, under the same terms and conditions as any EU member state. This major step, both for Europe and for Ukraine, is a clear expression of diplomacy for science: making every effort to support Ukraine's sustainable and self-determined scientific development, based on a relationship of mutual respect and reciprocal commitment.
- **Science for Diplomacy:** Using science cooperation to improve international relations between countries. An example of such science for diplomacy dating back to the Cold War was the IIASA, the International Institute for Applied Systems Analysis located in Austria.

Research and innovation are increasingly interlinked internationally, aided by rapidly developing information and communication technologies. The number of internationally co-authored scientific publications and the mobility of researchers are increasing. Research organisations are establishing offices abroad and companies are investing outside their home countries, in particular in the emerging economies. Global research and innovation were, until recently, dominated by the European Union, the USA and Japan. As the emerging economies continue to strengthen their research and innovation systems, a multipolar system is developing in which countries such as Brazil, China, India and South-Korea exert increasing influence. The share of the BRICS in global expenditure on R&D has more than doubled since 2000. The Union also has a clear interest in its neighbouring countries developing their research and innovation capacity.

Global challenges are important drivers for research and innovation. Our planet has finite resources which need to be cared for sustainably; climate change and infectious diseases do not stop at national borders, food security needs to be ensured across the globe. The Union needs to strengthen its dialogues with international partners to build critical mass for tackling these challenges. To remain a major global player, the Union must promote itself as an attractive location for carrying out research and innovation and be successful in the global competition for talent, while at the same time preserving its economic interests, for instance as regards the protection of intellectual property (EC, 2012).

The scientific values of rationality, transparency and universality foster common understanding, build trust and promote cooperation between peoples, regardless of cultural, national or religious background.

## CONCLUSION

Europe should find untapped sources of growth and employment, renewing the legitimacy of public policy-making, especially through greater citizens' involvement and by delivering better public services for all.

A new European Renaissance scenario is needed. In such a scenario, the EU will continue to enlarge and become stronger. It will consolidate its political, fiscal and military integration. Innovation systems will become more efficient with an increased role given to users. Investment in technological and services innovations will have a direct impact on economic and social development. Member States will work together to make the European Research Area fully functional with research agendas being decided in common across Europe. Through Horizon 2020, the European Research Area and the Innovation Union, the first steps have been taken towards a European Renaissance.

Europe's research, science and innovation ecosystem is more open, more collaborative and more participatory than most of the others in the world. The European history is long: its first universities date back to the twelfth century and its Scientific Academies were established during the Renaissance. The biggest Research Centres were built almost a century ago.

Since 2000 years, Europe has been at the centre of the world: it has given the foundations of philosophy and the pillars of the law. The fathers of theology and the first encyclopaedists were Europeans. The most influential economists and sociologists were coming from European continent. Europe should build on this legacy and kick-start a new Renaissance era.

This optimistic but realistic scenario should be based on three intertwined principles: co-creation, trust and resilience.

Co-creation in all sectors of society provides for manifold opportunities to fully exploit it. Co-creation builds on new forms of interaction among key actors in modern societies, involving citizens, the market and the state. Collaboration between these actors should be organized to share information, knowledge and resources. Open data help to co-create user-friendly, personalised and proactive services that lower the administrative burden and improve the user experience.

Trust is at the basis of all human relations and at the core of European values. Trust and trustworthiness hold societies together. Trust is at the roots of the economic system. In the collaborative economy, trustful relationships are a pre-requisite. In politics, trust is the twin sister of legitimacy. Without trust in the elections, there is no democracy.

Resilience is the capacity to successfully deal with difficult events and to adapt and overcome adversity. In a rapidly changing world, resilience creates stability, which in turn promotes job creation, economic growth and environmental sustainability. To change the world, a resilient Union is needed.

Co-creation, trust and resilience should be at the core of future Research and Innovation policies. These will support in the same time European competitiveness, European sustainability and European citizen's quality of life. But these policies should enable individuals to engage in innovation as co-creators, promote social innovation, and allows firms to test and roll-out solutions in the new changing world.

If Europe, a continent so proud of its roots and shared values, can innovate sustainably, it will become the beating heart of a new and positive union between individual empowerment and the collaborative economy.

## REFERENCES

Calvino, I. (1957). *The Baron in the Trees*. Turin: Einaudi.

Chateau, B., & Rossetti di Valdalbero, D. (Eds.). (2011). *World and European Energy and Environment Transition Outlook*. Luxembourg: European Commission.

Chesbrough, H. (2003). *Open Innovation: The new imperative for creating and profiting from technology*. Boston: Harvard Business School Press.

Conte, A. (2014). *Efficiency of R&D Spending at national and regional level*. Luxembourg: EC, Joint Research Centre.

EC-European Commission. (2010). Europe 2020: A strategy for smart, sustainable and inclusive growth. *COM*, *2020*, 3.

EC-European Commission. (2013a). *Global Europe 2050*. Luxembourg: European Union.

EC-European Commission. (2013b). *Accessibility Based Business Models for Peer-to-Peer Markets*. Brussels: European Commission.

EC-European Commission. (2013c). Innovation Union Competitiveness Report 2013[December.]. *SWD*, *505*, 2.

EC-European Commission. (2013d). *Guide to social innovation*. Brussels: European Commission.

*EC-European Commission* (2014a), *Communication, Towards a circular economy: A zero waste programme for Europe*. COM(2014)398, 2 July.

EC-European Commission. (2014b). *EU research in Foresight*. Brussels: European Commission.

EC-European Commission. (2014c). *The 2015 Ageing Report*. Brussels: European Commission.

EC-European Commission. (2014d). *Scoping Paper of Horizon 2020, Societal Challenge 'Europe in a changing world - inclusive, innovative and reflective societies*. Brussels: European Commission.

EC-European Commission. (2014e). *Communication Enhancing and focusing EU international cooperation in research and innovation: A strategic approach*. COM(2012)497, 14 September.

EC-European Commission. (2015a). Communication on steps towards completing Economic and Monetary Union. *COM*, *600*, 21.

EC-European Commission. (2015b). Communication, Closing the loop - An EU action plan for the Circular Economy. *COM*, *614*, 2.

EC-European Commission. (2016). *Open Innovation 2.0*. Retrieved from https://ec.europa.eu/digital-single-market/en/open-innovation-20

Ellen MacArthur Foundation. (2012), Towards the Circular Economy: Economic and business rationale for an accelerated transition. Isle of Weight, UK: Author.

European Union. (2014). *Global trends to 2030 – Can the EU meet the challenges ahead?* ESPAS report. Brussels: European Union.

Eurostat. (2015). *Statistics by theme*. Eurostat website.

Huizingh, E. K. R. (2011). Open innovation: State of the art and future perspectives. *Technovation*, *31*(1), 2–9. doi:10.1016/j.technovation.2010.10.002

GSMA Intelligence (2015). *Data*. GSMA Intelligence website.

Juncker, J.-C. (2014). *Speech: Time for Action – Statement in the European Parliament plenary session ahead of the vote on the College*. speech/14/1525, Strasbourg, 22 October.

Juncker, J-C. (2015). *Speech: State of the Union 2015 – Time for honesty, unity and solidarity*. speech/15/5614, Strasbourg, 9 September.

Katainen, J. (2015). *Speech: Circular Economy: boosting business, reducing waste*. Speech at the Circular economy stakeholder conference, Brussels, 25 June.

Landes, D. (1969). *The Unbound Prometheus*. Cambridge, UK: Cambridge University Press.

Mazzucato, M. (2013). *The Entrepreneurial State: debunking public vs. private sector myths.* London: Anthem Press.

Meyer, B. (2011). Macroeconomic modelling of sustainable development and the links between the economy and the environment. Study for the European Commission.

Moedas, C. (2015a). *Speech: Trust in Europe: The contribution of the social sciences and humanities.* Speech at the Trust Conference, Brussels, 29 October.

Moedas, C. (2015b). *Speech: Open Innovation, Open Science, Open to the World.* Speech/15/5243, Brussels, 22 June.

Moedas, C. (2015c). *Speech: The EU approach to science diplomacy.* Speech at the European Institute, Washington, 1 June.

Oreskes, N., & Conway, E. (2010). *The Merchants of Doubts.* London: Bloomsburry.

Rossetti di Valdalbero, D. (2013). *Europe: Towards a socio-ecological transition.* Brussels: Belgian royal Academy of Sciences.

Rossetti di Valdalbero, D. (2014). *The nexus between man and technology.* Brussels: Belgian royal Academy of Sciences.

Rossetti di Valdalbero, D. (2015). *Quelle Europe en 2050.* Brussels: Belgian Royal Academy of Sciences.

Schinas, M. (2012). *The EU in 2030.* Brussels: Centre for European Studies.

Schumpeter, J. (1942). *Capitalism, Socialism, and Democracy.* New York: Harper & Row.

Serres, M. (1992). *Le contrat naturel.* Paris: Flammarion.

# Chapter 3
# Crowdsourcing New Tools to Start Lean and Succeed in Entrepreneurship:
## Entrepreneurship in the Crowd Economy

**Priti Ambani**
*The Next Billion, New Zealand*

## ABSTRACT

*The very essence of the new entrepreneur is shattering tradition. On the heels of the new social internet, we are seeing the rise of the solo-entrepreneur or intrapreneur who embraces globalisation, failure and successes and collaboration. Powerful networks brought together by crowdsourcing are supplying the tools to start lean, innovate and solve complex problems. This chapter by Priti Ambani explores the changing ecosystem and the effect of networked crowds on starting lean and succeeding with entrepreneurship.*

## INTRODUCTION

### The Changing Paradigm of Work

Jobs are over and the future is income generation, says innovation strategist Heather McGowan (2014). The old lifecycle path of childhood dependence, followed by a learning phase directed to gaining employment, leading to "productive" years of jobs and career building and the ultimate capping of the cycle with retirement and fall back on social security -- is fading. Instead today's life path follows more of a wave pattern that oscillates between phases of actively learning new skills, being productive in society through a range of short term assignments and realistically even unemployed for periods of time. What is being witnessed today is a rise of a new life path that is peppered with life-long learning opportunities, short and medium term monetizing engagements that builds social capital. So new-age productive workers are building a diverse portfolio of skills, experiences, network tools, excess capacity and physical assets that aid in passive and active income generation, and last a lifetime.

DOI: 10.4018/978-1-5225-0568-6.ch003

## The Changing Definition of Entrepreneurship

A good place to start is the very essence of who is an entrepreneur. A person who sets up a business or businesses, taking on financial risks in the hope of profit. In this ever changing, hyper connected world, kids out of college aren't the only people trying their hand at entrepreneurship. Technological changes are happening at such a high rate that even established global enterprises are reinventing their game, strategy, product offering and business model, essentially facing the risk just like a newbie entrepreneur. According to the Global Entrepreneurship Monitor (2013), nearly 50% of the world's entrepreneurs are between the ages of 25 and 44, with 25 to 34 year-olds showing the highest rates of entrepreneurial activity.

## Rise of the Self and Hence the Solo-Entrepreneur

The Merriam-Webster dictionary defines an entrepreneur as "one who organizes, manages and assumes the risks of a business or enterprise." Though the definition frames the entrepreneur in terms of a business or entity, the changing paradigm of work as described above has added another nuance. An entrepreneur in today's age manages and assumes the risk of oneself. This is seen in the rise of the solo-entrepreneur who may not have an annual salary or a set job in the traditional sense but is monetizing their time through assets and talents that are rising through the new paradigms of work.

## The New Mindset of Permanent Beta

Authors Reid Hoffman and Ben Casnocha (2012) are calling for a new-age thinking of constant self-improvement in their book 'The Startup of You'. They say that, 'we are all works in progress.' Evolution is key to continual growth and innovation. This forms a key mindset of the new age entrepreneur. With an ever changing technological and social framework, staying in a state of permanent beta is no longer an option and is seen as the core DNA of successful entrepreneurs. What seemed like a great idea 5 years ago may not be relevant in today's framework of technology, social and cultural changes. A startup, called Airbnb that allows people to rent unique accommodations from ordinary citizens is on track to pull in $900 million in revenues in 2015 from $0 in 2008. But that is only half the story, services like Airbnb that have disrupted a sector are creating the need for support and tertiary businesses that support both renters and listers, like professional photographers, insurance and more.

Like great companies, people have to innovate to stay relevant. Great people, like great companies, are always evolving. They're never finished and never fully developed. Each day presents an opportunity to learn more, do more, grow more. Permanent beta is a lifelong commitment to continuous personal growth. It is the mindset of every entrepreneur of life. Per the publication, Re-imagining Higher Education, Deloitte University Press (2014), "by 2020, the knowledge college students acquire will have an expected shelf life of less than five years." The next generation of young professionals, irrespective of the career they choose must be agile, seek opportunity in problems and have an entrepreneurial mindset. They must also be able to articulate the value they provide as a business model

In this book chapter, the author attempts to present a view of entrepreneurship that will relate to changemakers in any stage of pursuing the entrepreneurial mindset, whether as a solo entrepreneur in their dorm room or an executive or intrapreneur reinventing her business in a c-suite office.

## The Factors Affecting Entrepreneurship

According to Richard Branson (2015), serial entrepreneur and founder of the Virgin group of companies, the biggest impact to entrepreneurship in the last 25 years has been technology. He also thinks that entrepreneurs have a big role to create societal impact. The Internet has helped aggregate knowledge, success stories, education, networks, mentors, solutions, ideas and even funding. Not only has the backbone technology improved, but even the user interface and consumer products resulting from the technology has disrupted and democratized the playing field. The Great Recession of 2008 spurred a rethink and helped innovators get out of their comfort zone. A poor job market lead to creative freedom that manifested in creating a career path that was customizable, with plenty of flexibility with a hyperlocal focus.

- **Globalization:** Business ideas are moving from products and services to experiences - with a global flavour. Access to information and technology has made the world flat and entrepreneurship is changing because of it. Global education, perspectives and knowledge is helping create a diverse startup culture that is better off with it.
- **Performance over Product:** The digital natives, the millennial generation and younger are after experiences while shunning product. Why pay for the product when what is needed is performance? This new value system that prefers access over ownership is forcing a re-think on what was previously the norm while seeking authenticity, transparency and sustainability.
- **Failure is No Longer Taboo:** The fear of failing has been downgraded. One reason for this that it is easier than ever for entrepreneurs to make a start - a small start and test ideas and products before scaling or becoming heavily vested. Smart and lean startup methodologies have buffered the impacts of failure. Facebook founder, Mark Zuckerberg (2009) says that, "if not breaking stuff you are not moving fast enough." The other side of the equation here is that ideas fail, assumptions crumble and startups fail. The millennial generation is also ready to fail fast and view failures as learning phases and move on to the next idea. Failure is also prominent in reality TV like Shark Tank that highlights the small number of really successful startups and all of the things that could really go wrong in the business.
- **Rapid Prototyping:** Technology has also enabled rapid prototyping, whether in the form of a 3D printed, first model of a product or launching a quick version of an app or platform. Analytics and data can give entrepreneurs a quick understanding of probabilities of success, help fine tune strategy and in many cases even scrap the idea.
- **Networked Crowds Are Tearing Down Barriers:** The barriers to entrepreneurship through creation, execution and capital have been removed due to technology and the networked crowds coming together. The crowd or collaborative economy that was seeded from the direct response to the great recession has democratized access to talent, ideas, funding, tools, physical resources. A true entrepreneurial movement in itself, the collaborative economy has both provided the means for entrepreneurs to test and grow their ideas rapidly.
- **Remote Work Is Spurring Creativity:** Rigid work is out and flexible, creativity is in. The lines between the home and work, vacation and deadlines, employees and owners, students and professionals are all blurring. Coffee shops and shared workspaces are replacing traditional brick and mortar floors with cubicles and how? Sitting in a coffee shop in Auckland, New Zealand can still give you global access. Virtual and flexible work is not creating lethargic workers. In fact, greater control over work parameters increases productivity and creativity.

- **Cloud Services:** Cloud services is an important cost-saving innovation that is boosting startups globally. Access versus ownership mentality is allowing business to operate effectively without maintaining their own servers and other expensive computational hardware. Cloud-based, online workspaces can provide the efficiency when working with remote teams while superior video conferencing and mobile data options provide easy connectivity.
- **The Promise of Impact through Exponential Solutions:** Creating a service that impacts a billion+ people globally is impactful. Within the next five years, the number of people online will double to 5 billion or more. Peter Diamandis (2015), chairman and CEO of the XPrize innovation competition and the founder of many high-tech organizations, says, "the best way to become a billionaire is to help a billion people." This challenge is enticing to the new-age entrepreneurs. Technology, mindsets and access to global knowledge is presenting opportunity for radical solutions to the world's most pressing problems.
- **Social Reputation Powered by Social Media:** Social reputation ratings and systems are helping to bypass the need for more traditional measures of achievement like educational degrees and formal work experience, helping to level the playing field for innovators with unconventional backgrounds and experiences. These reputation systems may become soon replace the FICO credit rating scores of tomorrow, as value systems other than the monetary system emerge.

## WHAT DOES IT MEAN TO START LEAN IN ENTREPRENEURSHIP?

Eric Ries, the author of probably one of the most widely read books on starting lean in entrepreneurship, 'The Lean Startup' (2011), presumably has a well-oiled publishing machine after the stellar success of his first book. Despite that Ries chose to crowdfund and crowdsource his second book, starting in March 2015, exclusively on the rewards crowdfunding platform, Kickstarter. The second book titled, The Leader's Guide (2015), will be "a four-color, 250-page, hardcover and digital book that will only be available to Kickstarter backers. It will look at how different companies are adopting Lean Startup methodology and help readers apply it in their own businesses." But Ries is not stopping at crowdfunding. The backers, exclusively will be encouraged to submit their own ideas and experiences for the book via an online community portal and will be involved in the draft and editing process even suggesting changes.

At the end of the campaign, 9,677 backers pledged $588,903 to help Ries bring his book to life. Ries (2015), writes on his blog, "Although it's fun to look at the vanity metrics of how much money we've raised and how many backers have joined the community, the real purpose of this campaign is to share ideas." Moreover, the campaign was also a publishing experiment as Ries intends for this campaign to throw light on how he can collaborate with the backer community as part of the research process for his next book, 'The Startup Way.'

### What Does Starting Lean in Entrepreneurship Mean in Context of the Lean Startup Methodology, Designed by Eric Ries?

The Lean Startup methodology (2011) backs the process of creating an MVP. An MVP or a minimum viable product is a barebones or minimalistic prototype of your business idea to test out your business plan and hypotheses. It includes the following

- Release early and iterate quickly.
- Fail often and fast.
- Communicate quickly and actively with a community of users during every phase of the business idea, development and launch.
- Validate models before a large scale up of teams or product features.
- Continue the circle of growing the idea, product iteration, feedback and launch.

The above method is not the end-all method to starting lean. The focus on minimalism should not take away quality, vision or passion. The methodology to adopt also rests on the business type. Startups that are focussed heavily on product can find it challenging to remain lean on product development. Other life goals may be harder to validate and there are other ways to startup lean.

The phenomenon of access over ownership has also taken on a new business connotation for starting lean. Entrepreneurs see the value of choosing performance over product. Can we leave behind the tradition of owning products when we only need perfomance? Seems like a simple, no-brainer of a question. But answering "yes" means that a complete rethink of the norm. By bringing the unified power of the Internet and connectivity the startup playing field is leveled and this provides entrepreneurs with resources to start lean while harnessing crowds and communities. This brings us to the new crowd economy.

## STARTING UP LEAN WITH THE CROWD ECONOMY

### What Is the Crowd Economy?

The crowd or collaborative economy is a dynamic ecosystem of productive people who participate through a digital platform with a purpose to achieve mutually beneficial goals. It is built on the principles of crowdsourcing that is playing a major role in the convergence of technology, business, ideas and resources to make the world a global stage of infinite collaborations.

Ross Dawson (2011), thought leader on crowdsourcing has defined crowdsourcing simply as, "tapping the minds of many." What crowds can share goes beyond minds or ideas. The concept has moved to spaces, funds, resources, skills and more. With crowdsourcing the community is the starting point. Epi Ludvik Nekaj, Founder and CEO of Crowdsourcing Week, on Virgin (2014) says, "crowdsourcing is not at all about mob behavior but very targeted cooperative solutions that be harnessed by anyone. People-powered platforms are forging interconnections between users that are breaking down the barriers between creators, producers and end users. By empowering people, organizations are finding new, previously unimagined pathways and solutions to complex problems." And this is especially true in the case of startups.

Most importantly, crowdsourcing enables solo entrepreneurship or the "startup of you" mindset. Once this mindset is achieved, entrepreneurs whether working solo or with a founding team can find it immensely beneficial to take an idea to a functioning business.

### What to Expect When You Marry a Lean Startup Strategy with Principles of Crowdsourcing

The million-dollar question when building a startup is "If we build it, will they come?" Reliance on crowdsourcing and crowd platforms will help entrepreneurs answer the question. Below are the outcomes that entrepreneurs and intrapreneurs can expect when the principles of crowdsourcing are applied to a lean startup mentality:

- The solo-entrepreneur is empowered with crowds - do more with less.
- Iterate fast and cheaply.
- Engaged community at every stage.
- Access to new ideas and perspectives.
- Direct access to emerging trends.
- Exposure to new partners and collaborators.

## CROWDSOURCED TOOLS FOR ENTREPRENEURS

This section of the book chapter takes a look at the various ways crowdsourcing can directly help startups create value and realise their ideas with a collaborative mindset to start and stay lean.

### Crowdsourcing a Business Plan, Ideas, Knowledge, and Strategy through Online Communities

Crowdsourcing is a becoming a sought after option to generate ideas, solicit feedback and get access to different cultural and social perspectives, all while sitting at one's desk - through online communities. Online communities are groups of individuals with a common purpose or background who are interacting through a web-based platform on an on-going basis. The common purpose that binds these individuals could be common beliefs, hobbies, interests or evangelism for a brand, idea or cause. The unique characteristics of these communities make them powerful resources for an entrepreneur.

Online communities can take on number of different avatars and can be used for varied applications. Here is an opportunity for entrepreneurs to get inventive and use online communities as an inspiration for ideas and input.

- **Mass Collaboration:** A form of collective action when large numbers of people from different backgrounds work independently on a single project. These projects are often modular in nature, and are self-regulating. These interactions can produce shared understandings and outputs e.g. Wikipedia. Contributing to collective projects within their interest verticals, individuals can obtain a grasp of a holistic and varied viewpoint that can be valuable in their growth as an entrepreneur.
- **Open Innovation:** Another application of crowdsourcing where the use of inflows and outflows of knowledge is used to accelerate innovation within organisations through cooperation, collaboration with external stakeholders through a more open organisational boundary e. Nine Sigma, OpenIDEO or 100% Open. One inventive way for entrepreneurs to gain a foothold in an area of

interest is to participate in open innovation challenges while building a portfolio of work. This kind of contribution helps cement expertise in relevant networks and build credibility.

- **Crowd Intelligence:** Shared or group intelligence and insights that emerges from the collaboration, collective efforts, group activity and/or competition of many individuals e.g. Netflix, Quora, Kaggle.

An online community like Quora or even Wikipedia has a global community of contributors- which can be a great thing. But one important factor to consider when using crowd intelligence sites is the user demographic. Most of the top platforms have a global presence with a wide range in demographics so while it's great to get a varied perspective, it is wise to be aware of biases.

- **iStart:** A program of the Kauffman Foundation, iStart connects entrepreneurs, mentors and advisors to each other to help bring potential business ideas to market. Their Business plan competitions allow aspiring entrepreneurs to test their ideas in a safe space, while mentors and advisors get the chance to offer leadership and growth opportunities for scalable business ideas.
- **Ennovent:** From a global community of contributors and innovators, Ennovent seeks the most relevant solutions in sustainable development in energy, food, water, health and education that work for rural and remote India. Ennovent's Discover Challenges helps clients to design and run public competitions to globally source innovations.
- **IdeaConnection:** Open innovation challenge site for new inventions, innovations and products.
- **Innocentive:** Online platform for open innovation, crowdsourcing and the innovation contests.
- **NineSigma:** Open innovation service provider, connecting clients with a global innovation network of experts.
- **Crowdicity:** Crowdicity is the easier way to harness the ideas and insights of your employees, partners, customers or anyone whose experience can help your organisation do things better.
- **Ideascale:** IdeaScale is an innovation management platform that uses crowdsourcing to help you find and develop the next big thing.
- **Kindling App:** Kindling is idea management and innovation software used by teams to discuss ideas, solve problems, and pursue opportunities.
- **Spigit:** Unleash your employees, partners, and customers for idea and innovation management. Only Spigit can identify the best ideas, predict the business impact, and enable both sustained as well as breakthrough innovation with crowds of any size.
- **Quora:** Quora is a question-and-answer website where questions are asked, answered, edited and organized by its community of users.
- **Stack Overflow:** Stack Overflow is a question and answer site for professional and enthusiast programmers. It's 100% free.
- **Yahoo Answers:** Yahoo! Answers (formerly known as Yahoo! Q & A) is a community-driven question-and-answer (Q&A) site or a knowledge market launched by Yahoo! In 2005 that allows users to both submit questions to be answered and answer questions asked by other users.

## Crowdsourced Insights and Data

Traditional market research is being replaced by platform companies offering unique insights from target demographics through crowdsourcing.

- **Listen and Learn Research:** We listen to what people say in social media around the world, to help companies learn about their customers, products or services.
- **Wufoo:** An online form builder that allows individuals or companies to build online forms easily, collect data, manage and report responses and even collect payments.
- **Uservoice:** All-in-one product management platform makes it easy to collect feedback, prioritize features and nail product releases while utilizing crowdsourced customer feedback.
- **Get Satisfaction:** Get Satisfaction is an online customer engagement community platform that connects companies with their customers in a digital setting to glean insights and feedback that make it a win-win for all parties involved.

## Open Data

Utilizing and capitalising on big data presents an exciting opportunity to entrepreneurs and startups. Whether one is looking for data sets to validate an idea or intelligently build a product, there is a high chance that this data is already freely available. Open data can fuel entrepreneurial thinking in a number of ways:

1. Fine tune product for market segments.
2. Come up with big problem solving ideas.
3. Come up with new product offerings.

In a similar vein open source software tools provide an exciting starting point for innovators, as these initiatives come with a built in community for feedback, answers and collaboration. Entrepreneurs looking for open source data or software to validate their ideas or look for business ideas have many programmes going for them.

- **Open Data Institute:** The Open Data Institute is playing the role of a catalyst in the open data movement. It functions as an industry evangelist that helps unlock supply, generates demand, and creates and disseminates knowledge to address local and global issues. The Open Data Institute's initiatives include convening world-class experts to collaborate, incubate, nurture and mentor new ideas, professional coaching and mentoring.
- **Data.gov:** This is the home of the U.S. government's open data site, that offers hundreds of thousands of data sets that is easy to digest, credible and search enabled.
- **Worldbank Open Data:** This important database offers free and open access to social and economic data on development in countries around the globe.
- **IBM DeveloperWorks Open:** IBM's open source community is helping IBM's sharpest minds collaborate with developers outside of IBM. Here is an opportunity to do a deep dive on IBM's work, re-use their techniques, and iterate on disruptive open technology innovations.
- **SourceFourge:** SourceForge is an Open Source community resource dedicated to helping open source projects become successful. The platform uses crowdsourcing to create a useful resource for open source software development and distribution. SourceFourge is a platform where IT professionals come to develop, download, review, and publish open source software.

- **GitHib:** GitHub is an important example of how people build software in the collaborative age. With a community of more than 11 million people, developers can discover, use, and contribute to over 28 million projects listed on its platform. Github uses a powerful collaborative development workflow that includes code review, and code management for open source and private projects.

## Crowd Finance

Crowdfunding has emerged as an important alternative finance source for entrepreneurs to validate their ideas, gain a social following and receive funds. Crowd financing is one of the most interesting social experiments that we have seen in the last decade. Its growth has been astounding. The 2015CF - Crowdfunding Industry Report by Massolution cites that early adopters of the current model of crowdsourcing raised $880 million in 2010, while $16 billion was crowdfunded in 2014. That number is set to be beaten in 2015 when crowdfunding is estimated to grow to over $34 billion. In comparison, the VC industry invests an average of $30 billion each year.

The reasons for the success and adoption that crowdfunding is seeing are many but these are also what makes crowdfunding attractive to entrepreneurs.

- *It's democratic.* With crowdfunding, entrepreneurs could pitch their ideas to a relevant community without having to wait for VC or angel funding. So it's empowering to entrepreneurs who can put their idea in the best form in front of an audience. Crowdfunding also helps level the playing field, reduces biases in funding decisions as the pitch and campaign is online. Global platforms help thrust successful campaigns into an international spotlight.
- *It's more than finance.* Crowdfunding relies heavily on social networks. A rich social presence will be helpful but a lack of it isn't a deterrent. With crowdfunding the possibilities of a campaign, with a idea that resonates with the public, going viral is very much a possibility. A crowdfunding campaign typically runs for a few weeks and this allows for entrepreneurs to connect with followers and fine tune their pitch even when live - something that is impossible with a 3-minute pitch in front of a VC! Crowdfunding is "slow" finance where the story, personal connection and innovation can shine through. It is hence much more than just finance. A crowdfunding campaign is a real social motivator, helping founders to nurture a community even before a product is launched.
- *It's validation for the business and the idea.* Startup founders who view crowdfunding only as an instrument to raise capital are shortchanging the opportunity. Not only is a crowdfunding an excellent way to validate market, and business idea, it is also an excellent way to finetune messaging and product. Keen and sharp entrepreneurs have walked with excellent input from their crowdfunding community, their first customers. Though the barriers to entry are low in crowdfunding, this is not the place to throw out half-baked ideas. Campaign founders should remember that they are entering the social domain with crowdfunding where reputations are built or broken easily.
- *It's customisable.* The beauty about crowdfunding is that it is highly customisable. It can be the first form of funding for a startup or come in at any stage of business growth. As a point of validation, a crowdfunding round can lead to VC or angel funding and much more. In fact many angels and VCs are integrating equity crowdfunding into their investment strategy. Crowdfundings is not just for startup entrepreneurs. Even multinationals are embracing crowdfunding as a core business strategy to test out ideas and accelerate innovation. This chapter will take about multinational in crowdfunding in a later section.

- *Crowdfunding case study.* One campaign to note that presents many lessons for entrepreneurs crowdfunding is the Coolest Cooler Campaign on Kickstarter in 2014. The Coolest Cooler, a wonderfully innovative cooler broke all the rules for the orange icebox that people are used to seeing with upgrades like USB charger, built-in blender and Bluetooth speakers. Ryan Grepper changed the notion of the cooler and re-imagined it into being the star of any outdoor party.

The project broke all Kickstarter records and raised $13 million, the second highest on Kickstarter on after Pebble Time. Few things are striking about this campaign.

- *Failed the first time.* One thing to note about this campaign that the successful raise only followed an earlier failed attempt in November of 2013. Grepper made a tactical mistake. A launch for a cooler in November, the middle of winter didn't work out well.
- *Re-launch with renewed focus.* The second time Grepper launched, he had a better idea to run a dynamic campaign, learning about timing, rewards and constant communication. During his campaign he brought on hourly updates to the community, asking his backers to chime in with ideas and features while taking them along on the journey. It is also worth noting that this time around Grepper and his team were more prepared to answer questions from backers and leverage media coverage while running the campaign on the same time.
- *After the raise.* Will 62,000 backers post raise, Grepper had a mighty task ahead of him to fulfill orders in a timely fashion. Hence crowdfunding is only the start to one's entrepreneurial journey. Another point to note is that Grepper himself backed 27 other projects on Kickstarter. A great way to see the real value of the crowd economy while utilising it to its fullest potential is by immersing oneself in it.

## DIFFERENT MODELS OF FUNDING THROUGH CROWDS

Crowdfunding is the practice of funding a project, business, venture or cause by raising varied amounts of money from a large number of people, typically via the Internet. By its very definition crowdfunding can be broadly classified into 4 main categories.

### Rewards-Based Crowdfunding

One of the more popular models of crowdfunding, rewards-based campaigns, offer prizes or "rewards" in lieu of contributions. In many cased entrepreneurs use this model to pre-sell their product as a reward for pre-determined contribution amount. Rewards-based crowdfunding has grown sophisticated over the years with platforms like Kickstarter requiring clear and critical information before they accept a campaign on their platform. For example, requiring actual photos of a prototype and disclosing risks and challenges. These types of checks in place will prevent backers from being disappointed with a failed campaign and also push entrepreneurs to do their due diligence before launching a crowdfunding initiative. There are scores of rewards-based crowdfunding platforms globally. Rewards-based crowdfunding typically last between 1-3 months and the key for entrepreneurs is to choose the right platform for their project based on campaign type, vertical, community and offerings. For example, Indiegogo prides itself on showcasing creative projects or a platform like The Next Billion caters to focusing on women entrepreneurs globally.

## Equity Based Crowdfunding

In equity based crowdfunding, funders receive actual shares in the company in return for capital. Equity crowdfunding is legal only in a few markets and countries as policy makers prepare regulation to protect investors, reduce fraud while pushing innovation. Equity crowdfunding has a bigger barrier to entry and is suitable for mature startups who are ready for a more sophisticated review of business plans, prototypes and deliverables. All campaigns on equity platform undergo stringent screening before they are launched. Platforms like Crowdcube, Seedrs, CircleUp and OurCrowd are some of the leading players in this space. Equity crowdfunding platforms are also offering hybrid models that combine crowds and angel investors, which is an interesting evolution of the model. OurCrowd is offering a hybrid model that brings in experienced and curated investors into play who bring due diligence and confidence into the round while increasing successful raises.

## Peer-to-Peer Lending/Commerce

The peer-to-peer lending model is based on direct social lending between people without a financial or retail intermediary. Through lending, entrepreneurs can raise funds though in the form of a loan that has a payback period and interest rate associated with the loan. This model is heavily suited for entrepreneurs who want to raise capital but do not want part with equity in their startup. At the same time investors see lending as a lower risk alternative to equity crowdfunding, as they are guaranteed a payback on their contribution. A peer-to-peer lending campaign is suited for entrepreneurs who are already able to demonstrate a revenue stream and seek a loan against it. Investors find it easier to understand the need for debt in comparison to speculating success of an early stage startup, they achieve returns quicker as loans get repaid faster. According to a report by PwC (2015), "Peer Pressure - How peer-to-peer Lending Platforms are Transforming the Consumer Lending Industry," nearly $5.5 billion worth of loans were issued in the United States on peer-to-peer platforms in 2014. And the vertical is set to grow to more than $125 billion by 2025 with top names that include Lending Club, Funders Circle and Prosper.

With all the above models of crowdfunding, the social benefit is profound. Even more than funds, a crowdfunding campaign can attract a cult of followers, future collaborators and partners, rich feedback and opportunities. The key thing for entrepreneurs is to develop a mindset for collaboration and keep their ear to the ground.

# CROWDFUNDING AND MULTINATIONAL COMPANIES

Why would multinational companies participate in crowdfunding? For the money? Sometimes, but more importantly for innovation, validation, ideas and social collaboration. Crowdfunding is no longer only for the scrappy startup. Large multinationals, tech giants and consumer product companies are putting crowdfunding into their corporate strategy. The Veronica Mars movie and Spike Lee were all big names well before they launched their campaigns on Kickstarter. Sony launched the FES e-paper smartwatch on crowdfunding site Makuake in Japan. Multinationals are also funding campaigns through platforms like CircleUp and OurCrowd. Consumer giant P&G has partnered with U.S. crowdfunding website CircleUp to help get small-business products in front of the corporation's executives. According to

Bloomberg BusinessWeek (2013), CircleUp has a twofold aim: "to help brands raise investment funds through its website, and get them noticed by multinationals that might be interested in aligning with or acquiring them."

GE Ventures, the venture capital wing of General Electric, today entered a co-investment partnership with Israeli crowdfunding platform OurCrowd that will enable GE Ventures to co-invest with OurCrowd in select early stage companies.

## CROWDSOURCING PEOPLE AND SKILLS

Since crowdsourcing is built on top of social networks it lends itself well to "sourcing" people and skills. Using social media and self-forming communities to find relevant resources is a very efficient way for entrepreneurs to hire talent, seek collaborators, mentors or even a cofounder from an increasingly global talent pool.

### Networking and Mentoring

- **MentorCloud:** A cloud-based platform that helps organisations form a structured mentor network, to support another achieve their personal, academic, and professional goals. MEMBERS connect as subject-matter experts, mentors, mentees, and peers, around specific goals and areas of mutual interest.
- **MentorPitch:** Offers a structured way to manage mentoring and measure all mentoring activities.
- **LinkedIn:** The top "business -oriented" social networking platform with global penetration with users in more than 200 countries.

### Cofounders and Startup Team

As more people associate and align themselves on online communities, the Internet is a great resource to put together a team. Social reputation is becoming so ingrained that the society in general is used to the idea of "looking up someone on the Internet." So why not begin the search for the team members and co-founders on the Internet.

- **CollabFounder:** This is a platform to build an idea from scratch, where one can post an idea and find collaborators get the idea executed.
- **Cofounders Lab:** Can help entrepreneurs look for team members who can provide complementary skills, resources, or capital needed to jump start a business idea. The platform also holds Meetups across the US for in-person meetings to help teams form.
- **Founder Dating:** This platform offers an entrepreneur network, a cofounder network and a mentor advisor network all in one place, making it versatile to the different needs of a startup founder.

### Experts and Talent

Crowdsourcing can make effortless, the mammoth task of finding labour, services, solutions, ideas, or content by soliciting contributions from groups of people, typically online and geographically dispersed, rather than via traditional employees or suppliers through platforms like Amazon Mechanical Turk, Crowdspring, Freelancer and many more. The advantages include sourcing from a global pool of talent, comparing options on one platform, a wide range of choice of talent that range from the newbie to the experienced. Another important feature of using crowdsourced talent is the option to work on a piecemeal approach without committing to a full-time hire. Not only do these platforms offer talent to startups and businesses, they are also an ideal platform for solo entrepreneurs to launch their online business without any capital intensive investment.

- **Upwork:** The platform that was previously oDesk/Elance is the one of the biggest platform for hiring freelancers with cumulative earnings of $3.2 billion since 2009. On Upwork, startups can hire talent with over 2700 listed skills for both short- and long-term projects.
- **Toptal:** Hiring experienced talent is possible through the Toptal network that screens candidates for a smooth and professional experience. Toptal community also hosts frequent meetups and tech events.
- **Freelancer:** In addition to posting projects for freelancers to bid on, Freelance.com also hosts contests on its platform that can be setup to invite scores and even hundreds of entries. The poster picks the winning entry who wins the cash prize. Startup founders could hire for a job posting on the platform or run a competition to find the best talent for the project.
- **99designs:** A niche platform for freelance designers, 99designs hosts design contests to invite submissions. Clients can give feedback on submitted designs before they choose the best one. Platform like 99designs offer creative services that are valuable to startup entrepreneurs.

## Education and Skills

The Internet is a powerful enabler. Through technology, networked crowds, an education from one of the world's elite Universities is not unthinkable with many offering their best courses for people to learn online for free. These MOOCs (Massive Open Online Courses) are usually broken down into modules and enable the student to learn at their own pace.

- **Coursera:** One of the largest platforms for free MOOCs with 15 million+ learners and 1400+ courses, many free. The site also partners with 120+ universities to offer courses that range from full MBA coursework to astronomy.
- **Udacity:** Founded by Sebastian Thrun and Peter Norvig and owes its roots to a Stanford University experiment in which the founders offered their 'Introduction to Artificial Intelligence' course online to anyone, for free. Over 160,000 students in more than 190 countries enrolled. Now Udacity's goal is to bridging the gap between real-world skills, relevant education, and employment.
- **EdX:** A free MOOC platform developed through a partnership between Harvard and MIT and involving other top universities that offers courses online for free.

## Crowdsourcing Hardware and Design

Crowdsourcing is redesigning product design and development by making the tools used to create more accessible and democratic. Expensive design software like Adobe offer subscription and cloud based tools that are affordable and cutting edge. The maker movement has also taken off due to commercial availability of 3D printers. modular blocks for building hardware amongst other technological advances. Prototyping is a critical step when a product is involved as it can entrepreneurs an opportunity to rectify design that can be costly in the long run.

- **Ponoko:** Covers technical hardware services like laser cutting, 3D printing & metal machining services to turn designs or prototypes into custom physical products. With Ponoko, entrepreneurs can outsource building a product to a community of designers who would give them a quote for the making the physical model.
- **The Fab Cloud:** An "Industrial Grade" additive manufacturing or 3D printing marketplace. Through the Fab Cloud, inventors can select the optimal manufacturer, seek engineering and consulting services to improve designs and collaborate with partners to take a project to completion. No prior design experience is necessary to work with the Gab Cloud.
- **Plethora:** Calls itself an engineering powerhouse at your fingertips. The platform is trying to democratise access to powers of creating by building a cloud assembly line. Through Plethora, an inventor can get fast feedback on manufacturability and costing, and realise the physical product.

## Software Development, Testing, and Security

The next security expert for enterprise system or government data will not be on the IT team. A college freshman halfway across the globe could be pointing out software vulnerability. It's becoming easier to identify threats through new technologies as companies take advantage of crowdsourcing, machine intelligence, and cognitive/advanced analytics to detect and stay ahead of security breaches. Companies like Facebook and Google have been offering bug bounty programs for years but wide acceptance of crowdsourcing is helping smaller companies can take advantage of a wide talent pool to evaluate IT systems by offering incentives and bounties. Though risks in a crowdsourced approach to cyber security remain, many of the platforms offering this service boast stringent background requirements for their hacker communities.

- **Cobalt:** Enables technology entrepreneurs to run bug-bounty programs on the platform and invite their community of hackers to seek out vulnerabilities
- **uTest.com:** uTest is an open community dedicated to professional testers and software testing to promote and advance the testing profession.

## Legal

Hiring an attorney to weed out challenges in an early stage company is a real need. But expensive legal services can be a burden. Platform services like PrioriLegal, Hire an Esquire, UpCounsel provide matchmaking for entrepreneurs seeking a full-service law firm. A legal tech startup, Casetext aims to become a community site for attorneys to share knowledge and a platform that gives users free access to legal texts -imagine a Quora for legal back and forth.

## Translation and Transcribing

- **MyGenGo:** Offers crowdsourced human translators that anyone can connect with through an API. That means one can upload articles, blog posts, or any other content and have it translated by cloud and crowd powered amateur translators.
- **CastingWords:** Works on Amazon Mechanical Turk, where it hires people to microtask and transcribe short segments of audio.
- **Duolingo:** A very interesting tool that combines crowd power to both teach a language for free while offering translation services.

## Cloud-Based Operations and Digital Resources

Cloud-based workspaces take collaboration between internal and external teams to the next level and break down communication and innovation barriers due to geographical location. Applications like Dropbox, Asana, Basecamp, Zoho Projects, Google Docs and Evernote help entrepreneurs lead a flexible schedule, be nimble and collaborate on the go.

Open democratic layers of the community have created a number of open sources, shared resources that are magnified due to social media. Creative Commons is a non-profit organization devoted to expanding the range of creative works available for others to build upon legally and to share, an immense benefit for entrepreneurs. Appirio is a tech platform that connects business owners to app builders to help make the seemingly daunting task to building apps doable.

## Workspace and Physical Resources

The Sharing economy, that is defined as a collaborative economic model built around the sharing of human and physical resources that include shared creation, production, distribution, trade and consumption of goods, space, travel and services by different people and organizations. Working in the sharing economy strives to be resource efficient with shared working spaces offering all the amenities of a full blown office, like a physical address, office hardware, wifi, voicemail and reception services at a reasonable rate. Liquidspace an online workspace reservation platform seeks to list unused work spaces like its partnership with Marriott to offer unoccupied conference rooms as co-working spaces. Rocketspace, another player in the space wants to offer "office-as-a-service" and has created a thriving shared ecosystem that can be called a "technology campus" in Silicon Valley for startups.

## Crowdsourced Publishing

Crowdsourcing set to disrupt how authors publish turning small-time amateur entrepreneurs into media entrepreneurs. Amazon is disrupting publishing through products like Kindle Scout publishing program, which lets readers discover the next bestseller. Readers vote on their favourite excerpts from unreleased books to determine what does (and what doesn't) get published. Another program called "Write On"—is a community where readers and writers come together around the creative process to make good stories great and great stories better. How? Amazon will allow authors to publish a part of their draft and let other users comment and drop suggestions.

## RISE OF SOCIAL ENTREPRENEURSHIP THROUGH CROWDSOURCING

The Internet, globalisation, networked crowds and collaboration has helped entrepreneurs realise the world is flat and the opportunities are endless. But with this information age, empathy has also flourished and with that a mindset has emerged that anyone, anywhere can make a difference, do well by doing good. The social entrepreneurship movement that pursues a sustainable economic goal along with solid social manifesto has accelerated in this hyperconnected age. Social entrepreneurship involves efforts to develop ideas or pursue a business model that aims to solve social problems while crowdsourcing is productivity on social platforms enabled by the Internet. Hence there is a natural alignment here. Crowdsourcing as a way to bring about collective action for a cause is opening up new avenues for participation. The powerful tools of social productivity can be used on a global scale not only for economic benefit but for also solving exponential problems like climate change, resource depletion and poverty.

## WOMEN ENTREPRENEURS: THE NEXT GLOBAL OPPORTUNITY

The new rules of entrepreneurship re-written by a more collaborative society are putting women entrepreneurs in the driver's seat, slowly removing the roadblocks of bias and lack of opportunity. Moreover, women across the world are launching and operating new enterprises at a faster pace than ever, positioning entrepreneurial ventures spearheaded by females as an increasingly important source for new jobs. Today, roughly 126 million women are establishing or leading brand new businesses in 67 economies around the world, and at least 48 million female entrepreneurs and 64 million female business owners currently employ one or more people per Entrepreneur Magazine (Barkawi, 2015).

The Next Billion, a crowdfunding platform exclusively for women-owned businesses was founded on the belief that next billion-dollar opportunity is not China or India but women entrepreneurs globally. Not only is crowdsourcing leveling the playing field for women, it is actually giving women a leg up. According to research on crowdfunding, the success rate for women was 69.5 percent, compared to 61.4 percent for men on Kickstarter (Inc, 2014).

## CONCLUSION

Entrepreneurs can leverage crowdsourcing to push boundaries and execute an idea without it being a capital intensive exercise. This change in rules creates a positive feedback loop for innovation and creativity that in turns lowers the barriers for entry. But entrepreneurs and intrapreneurs making the cut with the help of crowds must realise that with great power to disrupt comes great responsibility to give back to the community, honour their trust and always create shared value.

## REFERENCES

Ambani, P. (2015). *Four ways crowdsourcing will change our lives in 2015*. Retrieved on October 10, 2015 from http://www.virgin.com/entrepreneur/four-ways-crowdsourcing-will-change-our-lives-in-2015

Barkawi, W. (2015). *The Dynamic New Trends Of Entrepreneurship*. Retrieved on October 10, 2015 from http://www.entrepreneur.com/article/247319

Chen, P., Strickler, Y., & Adler, C. (2012). *Kickstarter Is Not a Store*. Retrieved on October 10, 2015 from https://www.kickstarter.com/blog/kickstarter-is-not-a-store

Marom, D., Robb, A., & Sade, O. (2015), *Gender Dynamics in Crowdfunding (Kickstarter): Evidence on Entrepreneurs, Investors, Deals and Taste Based Discrimination*. Retrieved October 10, 2015 from http://ssrn.com/abstract=2442954

McGowan, H. (2014). *Jobs are Over: The future is income generation*. Retrieved on October 10 from https://www.linkedin.com/pulse/20140720200547-22726740-jobs-are-over-the-future-is-income-generation-part-4?trk=mp-details-rr-rmpost

Nekaj, E. L. (2014). *Forget the sharing economy. It's time for the crowd economy*. Retrieved on October 10, 2015 from http://www.virgin.com/entrepreneur/forget-the-sharing-economy-its-time-for-the-crowd-economy

PwC. (2015). *Peer Pressure - How peer-to-peer Lending Platforms are Transforming the Consumer Lending Industry*. Retrieved on October 10, 2015 from https://www.pwc.com/us/en/consumer-finance/publications/assets/peer-to-peer-lending.pdf

Ries, E. (2015). *Only on Kickstarter: THE LEADER'S GUIDE by ERIC RIES*. Retrieved on October 10, 2015 from https://www.kickstarter.com/projects/881308232/only-on-kickstarter-the-leaders-guide-by-eric-ries/description

Strickler, Y. (2014). *The History of #1 UPDATED*. Retrieved on October 10, 2015 from https://www.kickstarter.com/blog/the-history-of-1-updated

Tozzi, J. (2013). *P&G Taps Crowdfunding Site to Scout for Startup Brands*. Retrieved on October 10, 2015 http://www.bloomberg.com/bw/articles/2013-01-29/p-and-g-taps-crowdfunding-site-to-scout-for-startup-brands

# Chapter 4
# Crowdsourcing Social Innovation:
## Towards a Collaborative Social Capitalism

**Emanuele Musa**
*Babele, Romania*

## ABSTRACT

*The growing awareness of humanity's finite resources and recognition of the limitations of one-off projects are prompting step changes in development planning. Sustainable development addresses the limitations of current practices; its aim is to achieve the triple bottom line of economic prosperity, environmental quality and social equity, meeting the needs of present society without compromising resources for future generations. Collective intelligence is considered by both Charles Leadbeater (former advisor to Tony Blair) and MIT as one of the most powerful ways to tackle complex problems, like climate change. This chapter explores the principles of crowdsourcing, its applications and the main trends. It presents theories, practices and examples of the use of crowdsourcing to innovate in the area of sustainable development for the common good. It announces the rise of collective brain-power to the challenge of creating better and more effective forms of civic and social engagement to solve problems on a world scale.*

## INTRODUCTION

The growing awareness of humanity's finite resources and recognition of the limitations of one-off projects are prompting step changes in development planning. Sustainable development addresses the limitations of current practices and features at the centre of almost every political, environmental, scientific and economic discussion held today. Its aim is to achieve the triple bottom line of economic prosperity, environmental quality and social equity, meeting the needs of present society without compromising resources for future generations.

The 1992 Rio Earth Summit was the moment when sustainable development captured worldwide attention. It established why sustainable development is necessary; the project now is how to execute it. The 21st century is the era of mass innovation - more ideas, knowledge and information being shared by

DOI: 10.4018/978-1-5225-0568-6.ch004

more people than ever before. Can we harness collective brainpower to innovate in the area of sustainable development for the common good? Google, Wikipedia and Linux are already using collective intelligence to develop new solutions through online communities. Could a collaborative online community do the same for sustainable development?

## THE LIMITS OF CAPITALISM

At the beginning of the twenty-first century, the world seems more divided than ever.

The world GDP keeps growing year after year, so in theory, we are all getting wealthier. However, the gap between the richest and the poorest is increasing and the richest 1% of the population will soon possess half of the global wealth. We keep extracting, exploiting and consuming at a rate that was never so high before. We are increasingly polluting, despite the evidence of global warming, the alarming loss of biodiversity, and the uncountable number of lives that are wasted throughout the process to consumerism.

A "revolution" toward a large scale improvement of well-being and sustainable development is approaching.

The growing awareness of humanity's finite resources and the impossibility to achieve an infinite economic growth in a finite planet, are demonstrating the capabilities of the capitalistic model to reduce inequalities and address the most urgent challenges of our time.

While market forces demonstrated to be a non-sufficient mean to lead to the most efficient allocation of resources, governments and institutions have failed to meet the needs of the Common Good and bring society together to solve global issues:

- Every year nearly 12 million children die of mainly preventable causes, including diseases for which vaccines are routinely administered in many countries;
- Although access to safe drinking water increased from 61 per cent in 1990 to 71 per cent by mid-decade, 1.4 billion people in developing countries still lack such access. Furthermore, 2.7 billion people still do not have adequate access to sanitation;
- In the developing world, about 130 million children still remain out of primary school, nearly 60 per cent of them girls. Adult illiteracy remains high, affecting roughly 855 million people, nearly two thirds of them women;
- According to the World Bank, in 2011, 17 percent of people in the developing world lived at or below $1.25 a day.

Through intense pressure from commercial marketing, consumption became the desire for status and social distinction at the personal and group level. Considering the finite natural resources available, it would be impossible for the estimated 2.2 billion people currently living on less than 2$ a day, to ever match the consumption level of the richest group. If every person on earth would consume at the same rate of a US citizen, we would need 5 planets in order to accommodate the need for resources.

We are still using a linear system applied to a finite planet.

Increases in poverty and inequality and the decline in opportunities have had a serious adverse effect on the well-being of individuals, communities and even countries. Several critics have argued that the development orchestrated by the industrialized countries tended to replicate the forces of colonialism continuing the pattern of resource expropriation and economic control by the industrialized countries.

The paradox of a global economy increasingly unified, and a global society increasingly divided is the most dangerous threat that weighs on the planet, because it makes difficult, if not impossible, the cooperation necessary to solve the most urgent problems of our time.

## WHAT IS THE ROLE OF BUSINESS IN SOCIETY?

To understand the role of business in society, we should go back a few centuries.

Until the 17th century, people owned businesses or worked for businesses, but they were just people, as the businesses did not exist independently of the people who owned them – thus their individual ethics were directly reflected on these businesses.

When the first corporations came into existence, they were granted charters for specific short-term projects, like building a bridge or a railroad. Once they fulfilled their purpose, they were disbanded.

But over time, the law changed and corporations no longer had to be turned off once their project was complete. They began to live on indefinitely, with a much more general purpose, profit (Korten, 2001).

Unlike people, who are driven by all kinds of motivations – doing the right thing, love for family, peers, the planet – publicly traded corporations are now required, by law and the markets, to maximize value for their shareholders – making as much profit as possible (Leonard, 2007).

The idea that a corporation's purpose is to maximize financial gain for its shareholders was first articulated in Dodge v. Ford Motor Company in 1919: Dodge v. Ford Motor Co. – 170 N.W. 668 (Mich. 1919). Over time, through both law and custom, the concept of "shareholder primacy" has come to be widely accepted. With such definition, there will always be a trade-off to be made between profits and the general interest of society.

Thus, there are many reasons why the traditional business sector cannot become the drive behind substantial change. If we have a look at the current situation throughout the world, we see a business environment which is shaped by a history of unhealthy competition, manifold cases of adulteration for the sake of profit or the ongoing issues of child labor for the sake of reducing labor costs (Bakan, 2004).

Reports about the horrible working conditions in the factories producing our iPads (Duhigg & Baraboza, 2012) or the toxic chemicals contained in our everyday products are easily available. These examples show quite well how the drive for higher dividends is heavily corrupting the social sense of entrepreneurs.

*We, the consumers, we didn't choose to put toxic products on the shelves, or to allow slave labour in factories around the world. We didn't choose to fill stores with electronics that can't be repaired and have to be thrown away. We did not choose a world in which some people can afford to live green, leaving the rest of us to be irresponsible planet wreckers. (Leonard, 2007)*

Another false myth of liberal thinkers is the capability of the market to automatically adjust itself according to a continuously changing environment. E.g. *when the issue of Global warming will become an absolute priority, then the market will invest in greener technology that will take our CO2 emissions down.* The counter argument is that the planet does not respond to such linear thinking. Transgressing one or more planetary boundaries may be deleterious or even catastrophic due to the risk of crossing thresholds that will trigger non-linear, abrupt environmental change within continental- to planetary-scale systems.

Thus, even if we would drastically reduce our environmental impact, this would not bring us back to the same environmental conditions of the 17th century.

The short term logic of the capital market is undermining our capability as a society to foster common good initiatives. The urgent environmental problems, and the incapability of this type of democracy and of the market alone to handle them, request a totally fresh approach.

## THE NEW WAY

Common good, is broadly understood as the overall social condition that enables individuals or groups to attain their fulfillment more easily. Unlike the utilitarian approach, which focuses on the greatest good for the greatest number, the principle of the common good is geared toward the benefit of all (John Paul II, 1991). As a reaction of the public institutions failures to answer to the global need of common good, citizens are getting "smarter", more capable of interpreting issues and discerning between options on their own than simply accepting the views of media and political elites. In the global environment, several companies have come under pressure from civil society and non-governmental organizations (NGOs) to be more responsive to the range of social needs in developing countries, including addressing concerns about the working conditions in factories or service centers, and attending to the environmental impacts of their activities.

Companies are an essential element of our society as they are deeply involved in the creation (or destruction) of common good, by providing workplaces adherent to law, ethical standards, and international norms. Their importance has increased over the last century to a point in which 52 of the 100 biggest economies on earth are now corporations (Anderson & Cavanagh, 2005).

Thus, for society to thrive, it is fundamental that companies embrace responsibility for the impact of their activities on the environment, consumers, employees, communities, stakeholders and all other members of the public sphere.

Today's corporate literature envisage a new economy that puts safe products, happy people and a healthy planet first; where businesses proactively promote the public interest by encouraging community growth and development, and voluntarily eliminating practices that harm the public sphere, regardless of legality.

Can we review the way we create companies, and measure success beyond profit? The key enabler for this new economy is sustainable development, which addresses the limitations of current business practices and promotes the deliberate inclusion of public interest into corporate decision-making. Its aim is to achieve the triple bottom line of economic prosperity, environmental quality and social equity, meeting the needs of present society without compromising resources for future generations.

The 1992 Rio Earth Summit was the moment when sustainable development captured worldwide attention. It established why sustainable development is necessary; the project now is how to embed its principles in the market.

## A NEW BUSINESS SECTOR

Typically, when we think about "sustainable business," we concentrate on corporate social responsibility (CSR): energy efficiency, reduced carbon footprint, recycling and reuse, fair treatment of employees, and charitable giving, among other considerations.

However, there is a growing market of mission-driven companies that are dedicated to being socially responsible from their inception, unlike most (though not all) corporations that pursue CSR for marketing purposes or to cut costs and increase profits.

The realm in which these mission-driven enterprises operate has come to be known as "social entrepreneurship," a term widely credited to Bill Drayton, founder of the social venture philanthropy, Ashoka.

While the term 'social entrepreneurship' is relatively new, initiatives to promote positive social change are not new. – earliest writings extend back to the late 1990s (Emerson & Twersky, 1996). For example, the Greystone Bakery was founded in 1982 by Roshi Bernie Glassman, with the explicit mission to train and hire unemployed local residents, who otherwise would have struggled to obtain employment elsewhere. The Grameen Bank was founded in 1983 to provide rural illiterate women with access to micro-credit facilities to launch new businesses.

"Every time I see a problem, I create a business to solve it", says Professor Mohammed Yunus, the founder of the Grameen Bank, in his books. He describes Social enterprises as businesses designed to meet a social goal not to create profit for its owners.

The strength of being a social enterprise lies partly on its predisposition to bring lessons from business and apply them to answering social need. They bring the self-sufficiency of for-profit businesses and the incentives of market forces to bear on global social problems in a way that neither pure capitalism nor pure charity has been able to match.

Gregory Dees, one of the "fathers" of the field of social entrepreneurship (Dees, 1998) that social entrepreneurs play the role of change agents in the social sector by:

1. Adopting a mission to create and sustain social value (not just private value);
2. Recognizing and relentlessly pursuing new opportunities to serve that mission;
3. Engaging in the process of continuous innovation, adaptation, and learning;
4. Acting boldly without being limited by resources currently in hand;
5. Exhibiting a heightened sense of accountability to the constituencies served and for the outcomes created.

Entrepreneur and social entrepreneur have in common the strong motivation to pursue their vision relentlessly and realize their ideas. However, what makes social entrepreneurship distinct from traditional business entrepreneurship is its focus on the social/environmental mission. This is true, no matter what legal structure the social entrepreneur chooses.

For-profit social enterprises put mission before profits, typically using their excess revenues as a means of scaling the reach of their mission. Nonprofits are increasingly finding that they cannot rely on philanthropy to sustain themselves, much less grow. Thus, they are pursuing earned income strategies that leverage the organization's excess capacity and capability. By law, the earned income they generate must be reinvested in the enterprise and its mission. Hybrid social enterprises, which combine features of both for-profits and nonprofits, use this legal structure to expand potential revenue streams, all aimed at increasing and sustaining the organization's ability to pursue its mission (Lyons, 2012).

Measuring the performance of a social enterprise is much more complex than it is for a commercial business, which can simply measure financial success. Measuring and monetizing lives saved, quality of life increased, and environmental damage mitigated (among other impacts), while possible, and are exceedingly difficult to do. However, there are great examples that showcase the potential of this sector to benefit large numbers of people.

## A FEW GREAT EXAMPLES

WaterHealth International (http://www.waterhealth.com) is solving the problem of clean drinking water: 3.6 million people die each year from water-related disease. The social enterprise uses a franchising system providing village entrepreneurs with UV filtration technology that allows processing and selling clean drinking water to remote villages at a low cost. Through its network of franchises, WaterHealth provides access to pure and safe drinking water to half a million people in 4 different countries.

There are many charitable water projects, but they have difficulties in pursuing their mission, as they do not have money for maintenance, for example for fixing the pumps when they break down. On the other hand, WaterHealth through the franchising system, have a network of entrepreneurs that pay for maintenance to troubleshoot equipment. Thus the franchising system allows these initiatives to be sustainable, the model is not applied only by WaterHealth International, but also by other social enterprises that handle the problem of clean water delivery, for example 1001 fontaines (http://www.1001fontaines.com/en).

Another great example of social enterprise is the one of Sulabh International (http://www.sulabhinternational.org/): an India-based organization that works to promote human rights, environmental sanitation, non-conventional sources of energy, waste management and social reforms through education. Sulabh International counts 50.000 volunteers and is the largest nonprofit organization in India.

Sulabh was founded by Bindeshwar Pathak in 1970, with the mission to rescue the untouchables (or scavengers: the poorest cast in India) from the sub-human occupation to remove raw (fresh and untreated) human excreta from buckets or other containers that are used as toilets or from the pits of pit latrines. The employment of manual scavengers is officially prohibited in India since 1993 but is still taking place to this day.

Bindeshwar first developed a scavenging-free two-pit pour flush toilet (Sulabh Shauchalaya) as well as a safe and hygienic on-site human waste disposal technology: which enabled the users to throw the excreta themselves in a clean way, without having to make use of the Untouchables. Sulabh created several solutions that were positioned to serve both the poor and the rich families, so that the high-end products could compensate the very low price proposed to the poorest families.

The social business was further developed through a new concept of pay-&-use public toilets, popularly known as Sulabh Complexes with bath, laundry and urinal facilities being used by about ten million people every day.

The innovation is an open-sourced technology which treats excreta and generates:

- Bio-gas, which is converted into electricity to power the building);
- Bio-fertilizer, which is then sold at market price;
- Treated Water, which is used for agricultural use.

The exceptional environmental impact of Sulabh is further complemented by the set-up of English-medium public school in New Delhi and also a network of centres all over the country to train boys and girls from poor families, especially untouchables, so that they can compete in open job market – and finally become *touchables*.

Sulabh is one of the best example of social enterprise that is economically sustainable, it maximizes environmental quality and social equity, using a business approach to solve a complex problem that governments and institutions have failed to address.

But the greatest innovation of all is that Sulabh's toilet block blueprints are open source, thus other organizations and people that are affected by the same problem, can now use the solution and improve it. Innovation is not anymore limited to the organization and its activities, but it can be further developed by other stakeholders, who can participate in the development process.

## THE IMPORTANCE OF STAKEHOLDERS PARTICIPATION

There is another feature that distinguishes social entrepreneurship from business entrepreneurship: social entrepreneurs are held to a higher standard of accountability. Business entrepreneurs are accountable only to their customers and shareholders, while social entrepreneurs are accountable to a much larger group – their stakeholders, "any group or individual who can affect or is affected by the achievement of the firm's objectives." according to Freeman's (Freeman et al, 2010) widely accepted definition. This group includes a part investors, employees (including volunteers in nonprofits), direct beneficiaries, also the local community and the society.

Stakeholders have the chance to influence the decision-making process, providing opportunities to further align business practices with societal needs and expectations, helping to drive long-term sustainability. They also must be taken into account by the organization when facing complex conditions in the operating environment.

Companies like Vodafone (2014) engage their stakeholders in dialogue to find out what social and environmental issues matter most to them about their performance in order to improve decision-making and accountability. Such practices are key to mitigate risks, helping the practitioners to compete in an increasingly complex and ever-changing business environment, while at the same time bringing about systemic change towards sustainable development (Jeffrey, 2009)

There are different stakeholder engagement approaches:

- From pull communication (*one-way engagement*: Information is made available, stakeholder choose whether to engage with it).
- To consultation (involved, but not responsible and not necessarily able to influence outside of consultation boundaries. *Limited two-way engagement*: organization asks questions, stakeholders answer).
- To real partnerships (shared accountability and responsibility. *Two-way engagement* joint learning, decision making and actions).

These engagement approaches are very well illustrated in the diagram realized by Stakeholdermap (see Figure 1).

*Figure 1. The relationship between stakeholder influence/power and stakeholder engagement approaches*
*Stakeholdermap, 2015. Source: http://stakeholdermap.com/stakeholder-engagement.html.*

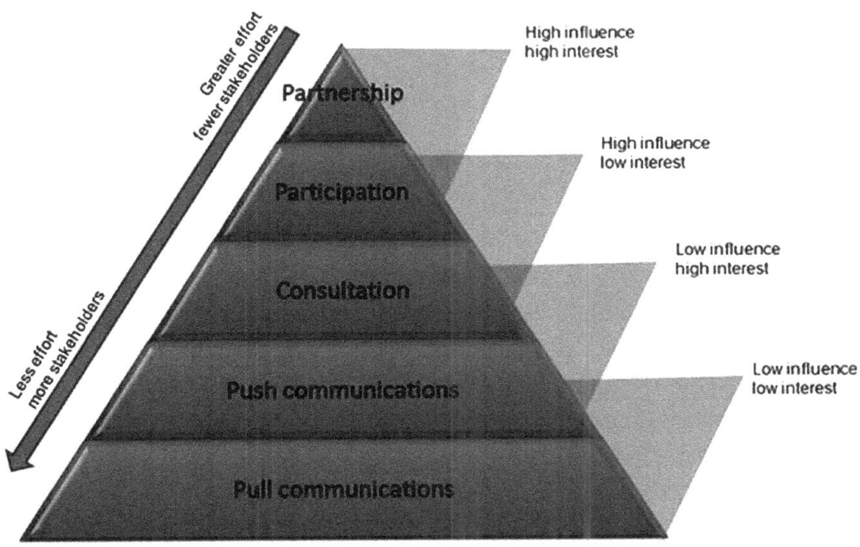


Through stakeholder engagement, processes organizations have positive economic results and creating win-win situations. Through the better understanding of the stakeholder needs and desires in consultation processes and through participation and partnerships, organizations can identify and create win-win situations with those stakeholders. The underlying idea is that partnerships between businesses and other sectors can drive sustainable development.

Anticipating the potential in stakeholder dialogues at an early stage and following a step-by-step guide can lead to successful shared value creation. A great example of shared value creation is Gram Vikas (http://www.gramvikas.org), a social enterprise and rural development organization headquartered in Orissa, India which assures access to basic education and adequate health services. The entire value creation process of Gram Vikas' involves "participatory decision making, shared responsibility taking and equal opportunities" with various stakeholders, and particularly disadvantaged groups. The results are that each of these groups are assuming ownership of the solution and its delivery, and thus the success rate is very high.

The trend towards cross-sector collaboration and stakeholder engagement is clear. The question is whether we are moving towards a convergence of values and whether shared value creation (expanding the total pool of social and economic value) will become the norm? Could the gap between social entrepreneurship and entrepreneurship be closed by actually making so that every enterprise is responsible, transparent and engages its stakeholders in the decision making process?

## THE AGE OF OPEN INNOVATION

Individually we are a drop, together we are an Ocean.

The 21st century is the era of mass innovation - more ideas, knowledge and information being shared by more people than ever before.

Knowledge has become the key resource in the post-industrial society (Bell) (Bell, 1974), thus reliance solely on internal innovation processes has become insufficient. Open innovation represents the logical result of dramatic, social, technological and environmental change and it describes how organizations work with large groups to achieve greater results than any team of experts working alone.

For Henry Chesbrough, one of its pioneers, open innovation is:

*... the use of purposive inflows and outflows of knowledge to accelerate innovation. With knowledge now widely distributed, companies cannot rely entirely on their own research, but should acquire inventions or intellectual property from other companies when it advances the business model. (Chesbrough, 2006)*

The process has become increasingly essential, if not inevitable.

"Wisdom of the Crowd" processes have taken place in various forms and in many areas for decades. But it is now, in the information or digital age, that they are being exploited at such a fast pace.

Innovation challenges, hackathons, external product development are all new phenomena that can best be summarized under the concept of open innovation.

The phenomenon that paved the road for a completely new approach to working together was the Open-source culture. The term open-source refers to a software development strategy in which the source-code is made available to a community. In this way everyone can make changes and improvement to the software. An example of the success of this strategy is the globally established computer operating system Linux, which benefitted from the ongoing analysis, review, testing and contributions of a large and diverse group of people from all over the world.

Wikipedia is another famous example of open source culture. It has over 80.000 contributors who have written more than 15 million articles in over 250 languages.

Traditional encyclopedias such as Microsoft Encarta or Britannica, did not see the crowdsourcing phenomenon coming, and continued to source their articles from few experts.

After some resistance, Microsoft decided to abandon its encyclopedia, acknowledging what everyone else realized long ago: it just couldn't compete with a free, collaborative project, where volunteer editors quickly update popular entries. Encarta could not simply keep-up to the pace and was embarrassingly outdated.

*The open source approach to collaboration enables people from all-over the planet to engage and create together* (Rayner, 2001). Open Innovation is also changing the way organizations work. A few corporations immediately realized the potential. LEGO, in its darkest days, tapped into abundant external talent to help them succeed in the face of some of their greater challenges. Back in 2003, the company was on the verge of Bankruptcy. Under the lead of a new CEO, Jørgen Vig Knudstorp, the company successfully reached a passionate group of smart and enthusiastic people, who were very familiar with their products and were just waiting to contribute: their customers.

They had hoped to get support from 100 of these fans; they were inundated with responses from 10,000.

The community shared creative ideas, and helped the company prioritizing which one were the best and had the greatest potential to become profitable. LEGO overcome its crisis and today it still develops this process of creative collaboration, by involving passionate people outside of their internal structure.

We are shifting from a world in which everyone seems a competitor, towards a world in which the upside of opening-up and engaging with people and ideas from elsewhere outweighs the risk of sharing confidential information. Innovation can come from any source, and the organizations that are succeeding in inspiring external talent to engage in their crowd-storming process, are gaining a strong competitive advantage compared to those who still uniquely rely on internal resources.

Companies like Apple and Google also depend on the effort of outside talent in the form of apps and open source software.

After the launch of Iphone in 2007, Apple switched the focus from the phone functionalities towards applications (or apps) that can be run on the iphone to give it new capabilities. Apps demonstrated in a countless number of ways that the iphone can become much more than a phone: through apps you can read a Restaurant customer review, find a cab, buy a book, split a restaurant bill, find a hotel or translate speech into a foreign language. Only in 2012, Apple paid $4 billion to Apps developers and created 210.000 jobs related to the app economy in the United States.

In the same way, Google and Apple depend on open source software that has been developed, used, tested and improved on an open-cooperative environment. While Android derived from the Linux operating system; Apple used a package of open-source products called Darwin.

According to former P&G CEO A.G. Lafley: "Innovation is a social process. And [it] can only happen when people do that simple, profound thing – connect to share problems, opportunities and learning. To put it another way, anyone can innovate, but practically no one can innovate alone".

The paradox of open innovation lies in the conflict between the potential benefits of collaboration and the prospects of knowledge leakage and misappropriation of the results of the process. The process is based on 2-way interaction that may cause the organization to lose control of specific information. This knowledge, in the hands of a competitor, could compromise the competitive advantage potentially gained from the open innovation process. However, recent research focused on the significance of value creation through stakeholder engagement (Svendsen, 1998) offers a model that addresses this inherent conflict in open innovation processes.

Stakeholder participation removes the structural tension of open innovation according to Gould, R.W, (2012). Consideration of the social, organizational, and ethical benefits of engagement with relevant stakeholders enhances the concept of open innovation, helping to move beyond its solely practice-based origins.

## THE LOVE STORY BETWEEN OPEN AND SOCIAL INNOVATION

The concept of open-innovation is very much associated with the business sector only; and in such a way Chesbrough defines open innovation as a process taking place mainly in commercial research and development, reducing it to product development.

However, seeking ideas and solving problems is just one of the many facets of open innovation and there are several reasons for expanding this paradigm shift to social innovation.

In fact, open innovation and stakeholder engagement towards collective social responsibility describe similar organizational processes: organizations reach outside their boundaries to access and share essential information with their stakeholders and with the crowd. Yet the two concepts, and their associated languages and discussions, have remained isolated from each other.

There is a vast potential to co-create social value through open and transparent networks that involve a diverse range of stakeholders, thanks to the ease of online collaboration tools and social media.

A few relevant initiatives started to emerge in the social sphere, creating a completely new field called open social innovation.

As previously mentioned, stakeholders' participation brings even more value when it goes beyond just listening to the people who can affect the organization, and a process of co-creation is put in place. Modern information technology is a key enabler for empowering stakeholders and facilitates their participation in the development process: people can share their knowledge, exchange ideas; each one comes with incremental improvements, and together can benefit from the collective wisdom of a crowd.

A fantastic example of the power of open-social innovation is the R&D-I-Y, *research and develops it yourself*: a process that has been applied to the Windowfarms network (Pearl, 2011).

Britta Riley is the cofounder of Windowfarms.org, a New York based company that makes hydroponic platforms for growing food in city windows, designed with the help of more than 1,800 enthusiastic collaborators from all over the world.

Britta took her inspiration from NASA, which uses hydroponics to explore how to grow food in space. She reasoned that many apartment windows have less than stellar conditions for growing plants, especially in a Northern winter. Conditions in any particular window would limit what could grow there—but perhaps hydroponics could contribute to food security on earth.

Britta decided to open source the project. She published the design on the web and invited anyone from anywhere in the world to improve the system. With no intellectual property issues, it was open to co-developers. Collaboratively, they have developed a system that grows a salad a week in an apartment window and allows an individual to cut their carbon footprint nearly in half.

On a global scale, the project has taken on a life of its own. Enthusiasts in Finland are working to customize the system with LED grow lights--also developed collaboratively--so that they can continue their gardens during the long, dark winter. Other contributions include air pumps to replace water pumps and optimum nutrients for strawberries that result in fruit throughout a New York winter. Britta says that the real reward from working with this company is the joy of collaboration.

Many other examples of stakeholder's participation through open-innovation deserve the right attention.

People co-create concepts through open content systems such as Wikis – these web applications have led to fascinating examples of collective intelligence. Wikipedia has inspired many organizations to create their own versions or copycats. This is the case of Energypedia, a project of German development agency GIZ that combines local and global knowledge to collect experiences and best practices in water and sanitation issues.

In the wake of the Fukushima nuclear disaster, Japanese citizens' set-up a network of volunteers publishing radiation levels they measured themselves after that trust in official data had collapsed.

Volunteers and political activists also take advantage of the Internet and tech devices such as mobile phones, tablets, etc. to monitor events and issues taking place around the world. Ipaidabribe.com, an anti-corruption project, allows ordinary citizens to send messages denouncing cases of corruption.

However, the best example of open-social innovation is the one of Fold it, in which an effective online tool which combines advanced gamification techniques with crowd collaboration, to contribute to Cancer and HIV/AIDS research.

## THE FOLDIT EXAMPLE

Foldit is an online puzzle video game about protein folding that was developed by Adrian Treuille. It is part of an experimental research project developed with the University of Washington's Center for Game Science in collaboration with the UW Department of Biochemistry.

The public beta version was released in May 2008 and has 240,000 registered players.

The objective of Fold it is to fold the structures of selected proteins as well as possible: one of the hardest and most expensive problems in biology today.

The protein biosynthesis is reasonably well understood, as is the means by which proteins are encoded as DNA. Determining how the primary structure of a protein turns into a functioning three-dimensional structure – how the molecule "folds" – is more difficult; the general process is known, but predicting protein structures is computationally demanding.

Foldit is an attempt to apply the human brain's natural three-dimensional pattern matching abilities to this problem. By analyzing the ways in which humans intuitively approach these puzzles in the game, researchers improve the algorithms employed by existing protein-folding software. As more players complete more puzzles, the researchers can create a better understanding of these protein structures and craft new medicines to promote better health and cure disease.

While at the beginning, gamers were playing on folding proteins to which the solution was already known, now Fold it is being used for real-world scientific problems.

A report by an international team of researchers from the U.S., Poland, and Czech Republic in Nature Structural & Molecular Biology unraveled that Fold-it players have solved the crystal structure of Mason-Pfizer monkey virus (M-PMV) retroviral protease (Khatib et al, 2011).

Retroviral proteases have critical roles in viral maturation and proliferation and they are very important for antiretroviral drug development for diseases such as AIDS. Researchers have for over a decade been unable to solve the structure despite using many different methods. Even recently, the protein-folding distributed computer program Rosetta that uses thousands of home computers' idle time to compute protein structures, was not able to give an answer. The Foldit players using human intuition and three-dimensional pattern-matching skills, however, were able to solve the problem within days.

## COLLECTIVE SOCIAL ENTREPRENEURSHIP

Social entrepreneurship has been considerably growing in the past 30 years, but it still remains a small part, only 6.5% of the entire economy in Europe (Seforis Consortium, 2013).

On the other hand, the explosion of global challenges in areas such as "climate change and environmental degradation; inequality and poverty; lack of access to basic healthcare, clean water and energy; mass-migration and international terrorism" requires modern sustainable enterprises to apply a more radical involvement of all their stakeholders.

Thus, to achieve true impact, social enterprises need to find a way to go beyond making progress on their own; they need to lead collaborations with others (Milway, 2014). Creating a sustainable ecosystem would require individual social enterprises to collaborate not just within their sector but also across sectors. Montgomery, Dacin, and Dacin (2012) term such collaboration "collective social entrepreneurship."

According to Robert Wayne Gould (2012) this can be realized through a combination of open innovation and stakeholder participation.

In the previous paragraphs, we have seen how Collective intelligence can be extremely effective at tackling complex problems. On the other hand, we have analyzed the importance of stakeholders' engagement to support social entrepreneurial organizations to produce long-lasting impact for the common good.

Thus, to initiate a new societal paradigm based on open-source collaboration, all stakeholders need to be empowered to voice their concerns and participate more actively in the decision making process.

The participation, investment and inclusion of all members of society in the development process is the key to help people fulfilling their human potential, but also to ensure that full advantage is taken from a country's human resources, as well as to promote peace and stability.

One of the main organizations that are democratizing the participation of stakeholders in the development process is IDEO.

## ABOUT OPEN IDEO

OpenIDEO is an open innovation platform on which global communities can take part in solving any of the various challenges presented on the platform. It is an online international community enabling people to collaborate in developing innovative solutions to pressing social and environmental challenges. Everyone can participate: veteran designers, critics, academics, and curious lurker.

Similar to a formal innovation process, ideas are presented as "challenges". One such challenge initiated by Jamie Oliver to raise children's awareness of the benefits of fresh food so they can make better choices. Challenges make progress through various creative phases: inspiration, concepting, evaluation, refinement and implementation.

The first step is called inspiration, the process starts with research to develop empathy and understanding on the topic: capturing people's needs and experiences before diving into solutions. Contributors are encouraged to submit inspirations in the form of images, stories, and visuals. During the next phase of the project, concepting, contributors are asked to post a solution - in other words, how they would make the concept a reality. People are invited to comment, ideas owners to incorporate the feedback received, and then refining their ideas again. This iterative approach is based on the philosophy of learning through building – trying out ideas with real people as quickly as possible.

Next, members are asked to rate and comment on concepts in the evaluation phase because each concept will gain merit depending on the community response. Only the best concepts carry forward.

In the next step of the design thinking process, called refinement: IDEO facilitators tweak the evaluated ideas, adding a highly focused design approach to it. This phase is to be understood through Tim Brown, writing on his blog:

*The idea of crowdsourcing innovation is, in my view, still a big experiment. Conventionally the question has been whether the crowd can outperform the internal team. Our view is that small teams are good for some things and the broader community is good for others. The goal of OpenIDEO is to find out whether it is possible to orchestrate collaboration between the two to achieve better results. (Brown, 2010)*

The results of Open Ideo are impressive: within 5 years they managed to post 29 challenges, and over 80.000 people participated submitting over 6000 ideas. Out of them, 300 projects are in the development stage.

## OPEN LEAN INNOVATION

Another interesting trend that is converging with the crowdsourcing universe is the Lean startup methodology, which was coined by Eric Ries from his study of Lean Manufacturing techniques created by Taiichi Ohno and Shigeo Shingo of Toyota.

The lean thinking became particularly relevant after the Dot-Com bubble of the late 90's, in which companies like Pets.com and Webvan could raise huge amounts of capital, spending it as fast as they could raise it, only to collapse when common sense re-asserted itself and the bubble burst.

After the bubble, venture investors spent the next three years doing triage, sorting through the rubble to find companies that weren't bleeding cash and could actually be turned into real businesses. Startups began to recognize that they weren't merely a smaller version of a large company. Rather they understood that a startup is a temporary organization designed to search for a repeatable and scalable business model. This meant that startups needed their own tools, techniques and methodologies distinct from those used in large companies.

The "lean startup," favors experimentation overelaborate planning, customer feedback over intuition, and iterative design over traditional "big design up front" development. Although the methodology is just a few years old, its concepts—such as "minimum viable product" and "pivoting"—have quickly taken root in the startup world, and business schools have already begun adapting their curricula to teach them.

The unit of progress for Lean Startups is validated learning - a rigorous method for demonstrating progress when one is embedded in the soil of extreme uncertainty. Validation comes in the form of data that demonstrates that the key risks in the business have been addressed by the current product. The lean startup asks people to figure out the right thing to build – the thing customers want and will pay for – as quickly as possible.

Modern entrepreneurs need feedback on their solution. They want to know when it's going to fail. They have to know who will use it, and who won't. They need to know how much they're willing to "pay" for it. That information won't be found "pitching" a solution.

Thus, the startups that ultimately succeed go quickly from failure to failure, all the while adapting, iterating on, and improving their initial ideas as they continually learn from customers.

Each stage of a startup development is highly iterative, customer feedback is essential to learn everything: from the problem that the product should solve, to the best communication or distribution channel, as well as the revenue model and the pricing. This continuous collaboration with the customers is key to reduce the inner uncertainty within the innovation process.

As in the case of open innovation: there is a paradox between the benefit of opening a project to the feedback of the customers, and the fear that several entrepreneurs have that their idea will be stolen, thus they often prefer to develop their project in secrecy.

However, several economists and successful entrepreneurs support the concept of sharing business ideas. For example, Nilofer Merchant in her article Harvard Business review "Let your ideas go" (Merchant, 2012), or Patrick Hull in Forbes Magazine "Talk about your idea, it won't get stolen" (Hull, 2013). They strongly believe that ideas cannot be stolen, because the hardest task in creating a startup is not having the idea, but implementing it. Timing and Execution are the essential elements to succeed.

If we apply the concept of Lean startup to social innovation initiatives, then it becomes fundamental for social entrepreneurs to involve their stakeholders in the process of validating the organization strategy - promoting full transparency from the financial performance to the resulting social impact. At the end of the day, young social enterprises face the same challenges of traditional startups: they often lack the resources to succeed (knowledge, network, and skills) and they operate in conditions of extreme uncertainty - which according to Forbes causes 8 startups out of 10 to fail in the first 2 years of activity. Thus, by opening up and enabling the mentors, peer entrepreneurs and stakeholders to contribute with their knowledge in the business modeling process of social businesses, we can considerably increase the chances of success of these organizations and scale their impact (Potter, 2014).

Furthermore, this open approach to social innovation will be the key to enable NGOs, to shift their revenue model (based on traditional philanthropic donations) towards the one of sustainable social enterprises. A recent OECD report showcases the decline in official development assistance, reinforcing the need for NGOs to decrease their dependence from external financial support. The opportunity for social projects to crowd-source consulting from their stakeholders in order to evolve their strategy is probably the greatest hope that we have to revitalize the third sector.

A great example of stakeholders' participation to validate the strategy of social entrepreneurs through crowdsourcing is the open innovation platform Babele.

## BABELE: CROWDSOURCING BUSINESS MODELS FOR SOCIAL ENTREPRENEURS

Babele (https://www.babele.co) is a social enterprise launched in the fall of 2013, with the objective to harness collective intelligence to help social enterprise validate their key assumption and validate a sustainable strategy.

Babele has a unique open source approach for supporting social entrepreneurship and creating social impact. It has been inspired by the emerging paradigm of more open, collaborative and adaptive organizations. Instead of creating social business strategies in secrecy and thus reinventing the wheel without learning from others, it offers an advanced system to share ideas, get inspiration from others, co-create business strategies for social good and replicate them in other parts of the world. The platform provides a business modeling framework, enabling entrepreneurs to create, debate, learn and experiment with projects in a collaborative way. Peer-entrepreneurs, citizens and mentors can participate by submitting ideas and giving feedback on each precise element of the business strategy of the social entrepreneur.

The ecosystem hosts a community over 850 social business model and collaborators from 103 countries. The platform combines the principles of social entrepreneurship, crowdsourcing and lean startup, engaging crowds of stakeholders in problem-solving - supporting the business development process of social enterprises from all-over the world.

Also the business model of the social enterprise is highly influenced by the open-source culture, making the platform available for free to all the social enterprises that fully open their business model to everyone, and offering a premium service for organizations that wish to customize the privacy access to the strategy to different groups of constituencies.

Furthermore, Babele embraces the philosophy of creating the network of networks for social innovation.

The key risk for the social innovation sector is to suffer from scarce effectiveness in addressing local and global problems due to the high market fragmentation.

While individual initiatives and organizations (Universities with social entrepreneurship classes, social incubators, foundations and support networks) are flourishing throughout the world at an incredible rate (e.g. Ashoka, Echoing Green, Schwab foundation, Skoll foundation, NeSst, etc.) on the other hand, they are failing at joining forces and sharing best practices.

Babele is actively working to connect the dots between these networks and organizations, offering crowd-mentoring communities to manage their support programs for local social enterprises.

The organizations can stimulate peer-learning among their own projects, engage with their mentors, but also exchange with other organizations. The final result is a network of communities that can interact with each other: social innovation actors, entrepreneurs, incubators, universities, municipalities, citizens, or CSR driven companies.

This approach is fundamental to consolidate the social innovation market, incentivize openness and transparency and support the replication of the best initiatives that have the potential to address the most urgent challenges of our time (see Figure 2).

## BEYOND R&D THROUGH THE POWER OF NETWORKS

Robert Fabricant discusses if Social enterprises can be seen as R& D labs in an insightful article in the Harvard Business Review *Meet Your New R&D Team: Social Entrepreneurs (*Fabricant, 2013). According to him, social entrepreneurs can have a "R&D function" for learning how to serve underdeveloped markets; they are innovative and find solutions to dedicated social issues. Once their solution has a proven impact, they should scale-up. In order to do this, they need more resources. Therefore, partnering with corporate CSR initiatives may offer them the necessary means to grow. There are many examples; Robert Fabricant describes four of these partnerships in his article.

This point of view is not new. What Robert Fabricant mentions are in fact "hybrid strategies", which means creating partnerships between social entrepreneurs and multinational companies. Oliver Kayser, after 18 years with McKinsey and 5 with Ashoka, has created Hystra consulting, a company specialized in these hybrid strategies.

*Figure 2. Interactions within the Babele platform between independent projects, projects within public and private communities, mentors, citizens and peer-entrepreneurs*
*Babele, 2015.*

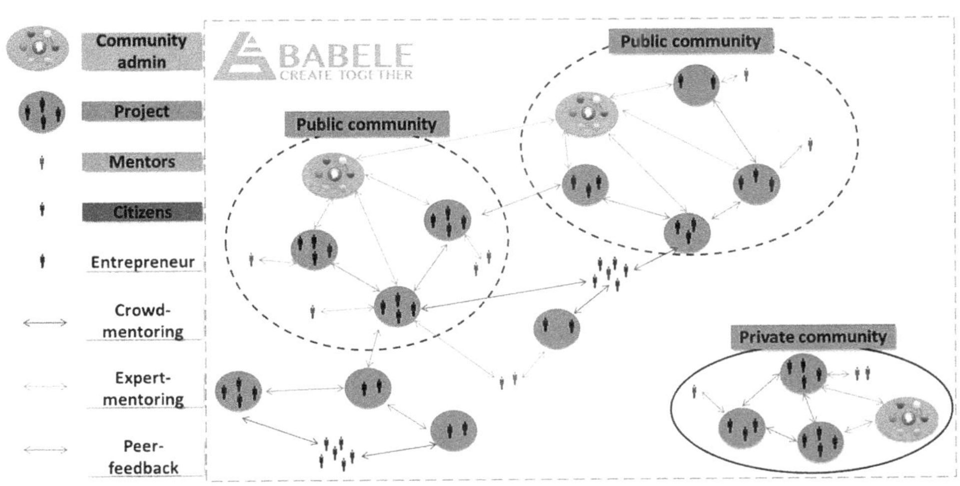

*Across the world, social entrepreneurs have been experimenting with a mix of social and business strategies, gaining new insights into market needs and coming up with innovative solutions to intractable global problems. It is now time for concerned and conscientious corporate leaders to do what they do best in order to help social solutions achieve the scale needed to change the world. (Kayser, 2009)*

Both Robert and Oliver recognize that social entrepreneurs bring innovative solutions to social issues, and therefore are doing "R&D". However, they also acknowledge that social enterprises need to partner with multinational companies in order to grow their scale and scope.

If we continue with the metaphor where social enterprises are the "R&D Lab" for Europe social challenges, then corporations are the "production" department, which deploys the prototype conceived upstream.

Their point of view remains in the current paradigm, where only the corporations have the force to scale-up from the prototype model to consistent production. They are the only ones capable of doing this because social enterprises are too small and do not have enough resources.

Now imagine a new paradigm, where social enterprises would not only be the "R&D lab" but also the "production" department. They would be interconnected within an efficient network, where they collaborate, experiment and exchange in a transparent way.

This statement might sound absurd but it will become less hard to believe through the following comparison. In the 80's we had super-computers ("corporations") that were extremely powerful in comparison with any standard computer in the market ("social enterprise"). Now with the power of the internet ("our efficient network"), many interconnected PCs have become more powerful than a supercomputer.

A new ecosystem can be created to gather the competencies and resources to make it happen where social enterprises will not only be the "R&D lab" of Europe's social issues, but they will be the engine solving these social issues and promoting a new kind of growth.

## OPEN GOVERNANCE

Open social innovation is contaminating several other sectors beyond social entrepreneurship.

Civic engagement is another key trend that combines the collective wisdom of the crowd (citizens) with the collaborative creation of sustainable initiatives that have the potential to improve the quality of life of local communities.

Governments are increasingly realizing that they can (and need) to communicate with citizens in a different way. Citizens can help provide better solutions for cities, but for that, they need to be included in decision-making processes. For example, in the German city of Nuremberg, citizens were asked to locate the noisiest areas of their neighborhood. By jointly identifying these spots, the city administration did not only get a different picture of the problems but could also work on much better solutions.

Multi-stakeholder governance appears to be the future way of developing public policy, bringing together governments, the private sector and civil society in partnership. The movement towards this new governance paradigm has been most marked in areas involving global networks of stakeholders: too intricate to be represented by governments alone. Nowhere is this better illustrated than on the Internet, where it is an inherent characteristic of the network that, laws, and the conduct to which those laws are directed, will cross national borders. This momentum has developed to bring multi-stakeholder governance to the Internet.

The United Nations e-Government Survey 2008 indicates that governments are moving forward in e-government; the UN e-participation index indicated a constant upward movement with 189 countries online in 2008.

More than 200 municipalities and public institutions are estimated to have initiated participatory budgeting in the world.

## MEU RIO

A magnificent example of this phenomenon is the online platform MEU RIO that allows the citizens of Rio de Janeiro to have a say in what's happening in the city.

Alessandra Orofino, founder of the initiative, explains in an interview that their main work is the one to translate public policy issues into a language that is understandable to broader society and young people. The site gives people an opportunity to act on things they think are important, and allows both organizers and users to identify areas for change and action. Meu Rio has a team dedicated to researching public policy so they can mobilize people effectively.

As an example, in 2013, an eight-year-old student wrote to Meu Rio about her school, the Escola Municipal Friedenreich. It has around 300 students, and is one of the best public schools in the country along with specialized staff and facilities for the disabled.

The city had decided to demolish a school to build a parking lot for the nearby Maracanã Stadium. There was no plan to rebuild the school or transfer the students, and the parents only found out through the local news.

So the organization set up a campaign to save the school and got 20,000 signatures on a petition. The campaign started attracting media attention, but even the secretary of education didn't know what to do. So Meu Rio decided to try a new tactic. They set up a webcam at an apartment across the street from the school, and monitored the school 24/7 through a website with a live feed from the camera. People could sign up to be a "guardian" of the school and watch the feed, and if bulldozers showed up, those watching could press a red button to contact Meu Rio, which would send out text messages to followers in order to physically protect the school. Around 3,000 people signed up to watch the school, and using analytics, Meu Rio discovered that for the two months of the campaign, not a minute passed that someone wasn't watching the school. Public officials realized it would be a PR disaster to demolish the school, and gave up.

## TOWARDS PURE PARTICIPATORY DEMOCRACY

People participation can go beyond idea brainstorming and problem solving. The next step of stakeholders' engagement is the co-ownership of the common good solutions that are collaboratively conceived and implemented in the local communities.

A very small village in the middle of Lapland (Sweden), is living on the next edge of sustainable development and crowdsourcing - practicing as no-one else co-creation for the common good.

The majority of Vuollerims inhabitants have understood that, while in a period of abundance, competition might be the best strategy; in a period of scarcity, collaboration is certainly the approach that offers the best possible outcome.

Being in such a peripheral part of the world, can present several challenges, for example the migration of young minds towards the bigger and more attractive cities in the south of Sweden.

To address these challenges, the people of Vuollerim have come together and worked on a new welfare system that involves citizens in first person and embraces entrepreneurship as well as taking initiative, this initiative is based on collaborating for the common good, rather than focusing on individual benefits.

The district of Vuollerim has about 800 inhabitants, 60 companies and 40 non-profit associations. All of these businesses and organizations have jointly agreed (by writing it in their statutes) to aim for local development and the best for the village.

All started in 1999: when the first village-owned company was born in order to save the hardware store that had closed down and was about to be demolished. The Vuollerim's economic association is composed by more than 100 members, who recognised how essential the hardware store was for Vuollerim and its surroundings. Together they bought the property, which is now running as a multifunction store called 'The Greenhouse', in the middle of the village.

Another great example is VIVA, which manages the Inn: a year-round open hotel. VIVA is co-owned by 150 citizens: that have built and decorated the Hotel in collaboration with several other villagers.

This democratic collaboration among citizens has been extended also to the Educational field.

Vuollerims Charter School, "A good small school", is a village-owned company, which was founded in 2009 and has over 100 co-owners. The reason to start a charter school in the district of Vuollerim was to ensure there was a complete primary school education available in the village, now and in the future. This had become a concern after the municipality's decision to close down the elementary school in Vuollerim for the ages 6 to 9.

Another incredible collectively owned company is The Village Team, which provides innovators in all ages with the skills, the experience and the expertise necessary to take their idea off the ground and develop unique, locally anchored, sustainable enterprises. These days The Village Team business is in the start of implementing a crowdsourcing experiment in the open innovation platform Babele: a tool for crowdsourcing ideas and co-developing smarter solutions in a collaborative way.

The Village Team contributes as a "mentor" and sounding board for the development of new business ideas. They choose the projects depending on the individual competences and interests of the members, as well as on the local needs that are anticipated for the village. Currently, the company engages in producing local music, renting out conference halls and vehicle storage, as well as offering copying and tire services.

Another village company is Lapland Vuollerim Welcomes You AB. It is a community based tourism company, providing genuine 'everyday life' experiences. Lapland Vuollerim has been nominated for Best Outdoor Product of the Year, World Responsible Tourism Award and Tourism for Tomorrow Innovation Award and it is a socially responsible business owned by villagers, reinvesting 100% of their profits into further growth of the company and into the local economy.

What is absolutely fascinating of this model is that all the above mentioned companies strive to enhance both economic as well as social well-being and growth for the district. All profits for each village owned company are reinvested in the district in one way or another

This collaborative welfare model enabled Vuollerim citizens to work together towards a common goal: be better off as a community rather than having a village with Winners and Losers.

The village is on its way to a brighter future by retaining jobs, youth and investing in education. This model is a true HOPE for all the peripheral areas in the world that are suffering from similar challenges. It can be done everywhere! The secret is to work together.

## HOW CROWDSOURCING AND SOCIAL INNOVATION CAN FAVOR PARTICIPATORY DEMOCRACY

Crowdsourcing and design thinking can be smartly combined to funnel the participation of societal stakeholders, who can take an active role in the development of civic projects for the common good. The ideal open innovation platform to achieve for participative local democracy should incorporate elements of design thinking - as a way to funnel the creativity of the crowd toward the development of the most effective and feasible solution - and a sustainable planning framework in order to use collective intelligence to define all the key elements of the project (from financial variables until its expected impact).

Below are the key phases that the ideal participatory democracy platform should contain:

### Phase One: Research Phase

In this stage, stakeholders work together to create a clear picture of the status quo:

- Identify the key local problems that negatively affect the livelihood of the municipality and gather all the relevant data (quantitative and qualitative) to describe the problem and how it occurs;
- Collectively assess the root-cause of the problems and analyse the resulting negative impact through an 8-dimentional framework, which key variables are: Political, Economical, Social, Technological, Environmental, Demographical, Cultural and Legal;
- Gather success stories, solutions and best practices coming from other municipalities and contexts.

### Phase Two: Ideation Phase

In this stage, stakeholders co-propose the solutions with the potential to address the problem.

The ideation framework has to be basic, in order to enable creativity in the crowd-storming process.

This will be composed by a short description of the solution, a presentation of why and how the solution has the potential to address the problem, necessary resources and implementation complexity.

### Phase Three: Feedback and Screening Phase

In this stage, stakeholders dig deeper into the ideas with insights that will encourage idea development.

Stakeholders will also be able to vote their favorite ideas while justifying their choice;

By the end of this phase, the 10 most voted ideas will proceed to the following phase.

### Phase Four: Planning Phase

In this stage, the selected projects will be developed extensively through a collaborative project planning methodology.

Stakeholders will develop each idea into a complete implementation plan, which will include key activities, needed resources, key partners involved in the implementation, realization timeline, as well as a complete section with financials budgeting and expected social/environmental impact.

## Phase Five: Refinement Phase

In this stage, stakeholders will examine each idea and give feedback to each aspect, in order to further refine the proposed plans.

- Mentors and advisors are allocated to each project-topic based on their key interests and competences;
- Project proposers can incentivise proactivity and collaboration by rating the work done by their peers;
- The companies, organizations and institutions that could be involved in the implementation will help validate timeline and budgets for the realization of the projects;
- The municipality's will participate by sharing valuable data to assess each project feasibility;
- By the end of the phase, the community votes the best projects that will be implemented.

## Phase Six: Implementation Phase

In this stage, the winning ideas are announced and an agreement is established among the organization that will participate in the project implementation.

- A transparent and easy-to-digest reporting will display the project progress and performance against established Key Performance Indicators;
- A blog with problems and challenges will be shared with the community, in order to benefit from the collective intelligence process also in this final stage.

If it is true that Democracy is a process that must constantly be reinforced at all stages by the internal participation and action by the institutions of the state as well as the international community, then it is fundamental to create new platforms to enhance the discussion amongst principal societal stakeholders, in order to spread awareness of the things which are known, discuss a strategy and get societies together to solve the global issues. Through this process, the crowd can engage and create meaningful solutions for the common good of the municipality, bringing a new fresh perspective to the current model of democracy.

# CONCLUSION

The paradox of a global economy increasingly unified, and a global society increasingly divided is the most dangerous threat that weighs on the planet, because it makes difficult if not impossible, the cooperation necessary to solve the other problems.

While market forces demonstrated to be a non-sufficient mean to lead to the most efficient allocation of resources, these governments and institutions have failed to meet the needs of the Common Good and bring society together to solve global issues:

Our society has promoted homo economicus, selfish person, maximizing its interests, focusing solely on profits. This has led to a critical unsustainable situation. More and more people realize that we need to create a new type of system. People are not the rationally self-interested actor lurking at the core of mainstream economic theory (the homo economicus: a mercenary ready to crush their neighbors in order to maximize its own interest).

We can also be homo reciprocans, if society promotes a development model based on sustainability, then people will act accordingly (Berlingen et al, 2015). We can see these values reflected in the new organizational models that pioneers are creating through social entrepreneurship and impact enterprises, Sulabh, Windowfarms, FoldIt are just some examples.

The demand for common good is influenced by the increasing level of awareness of the citizens that understand the importance of common good initiatives aimed at leveling inequalities and fight poverty. The world's short-sightedness about policy-making, together with the ageing of the population have led to a crisis of the welfare state, which has questioned the current pension, healthcare and education systems within the most developed countries. As a reaction of the public institutions failures to answer to the global need of common good, citizens are getting "smarter", more capable of interpreting issues and discerning between options on their own than simply accepting the views of media and political elites.

To solve the biggest challenges of our time we need a bottom up, collaborative and open sourced approach: where good ideas working in one place are easily replicated, the impact of a solution is easily reported and transparency is highly encouraged.

Stakeholder engagement combined with open innovation for social good projects allow achieving greater results than any organization working alone.

## REFERENCES

Anderson, S., & Cavanagh, J. (2005). *Field Guide to the Global Economy* (2nd ed.). New Press.

Bakan, J. (2004). *The Corporation: The Pathological Pursuit of Profit and Power*. New York: Free Press.

Bell, D. (1974). *The Coming of Post-Industrial Society*.New York: Harper Colophon Books.

Berlingen, F., Gauthey, M. A., De Grave, A., Filippova, D., Guedira, A., Léonard, A., & Tordo, M. (2015). *Société collaborative, la fin des hiérarchies*. Paris: Rue de l'Echiquier.

Brown, T. (2010). *Blog post*. Retrieved from http://designthinking.ideo.com/?p=482

Chesbrough, H. W. (2006). *Open innovation: The new imperative for creating and profiting from technology*. Boston: Harvard Business Press.

Dees, G. (1998). *The Meaning of Social Entrepreneurship*. Kansas City: Kauffman Foundation and Stanford University.

Duhigg, C., & Baraboza, D. (2012, January 5). In China, Human Costs Are Built Into an iPad. *The New York Times*. Retrieved from http://www.nytimes.com

Emerson, J., & Twersky, F. (1996). *New Social Entrepreneurs: The Success, Challenges, and Lessons of Non-Profit Enterprise Creation*. Roberts Foundation.

Fabricant, R. (2013). Meet Your New R&D Team: Social Entrepreneurs. *Harvard Business Review.* Retrieved from: https://hbr.org/2013/03/meet-your-new-rd-team-social-e

Freeman, R. E., Harrison, J. S., Wicks, A., Parmar, B. L., & De Colle, S. (2010). Stakeholder theory – The state of the Art. Cambridge.

Gould, R. W. (2012). Open Innovation and Stakeholder involvement. *Journal of Technology Management & Innovation*, *7*(3).

Hull, P. (2013). Talk about your idea, it won't get stolen. *Forbes Magazine.* Retrieved from: http://www.forbes.com/sites/patrickhull/2013/06/13/talk-about-your-idea-it-wont-get-stolen/#2688ac5972d1

Jeffery, N. (2009). *Stakeholder Engagement: A Road Map to Meaningful Engagement.* Doughty Centre, Cranfield School of Management.

John Paul, I. I. (1991). *Pontifical Council for Justice and Peace.* Academic Press.

Kayser, O. (2009). *Ashoka to Partner with Hystra to Increase Corporate Engagement in Scaling Up Social Innovations.* Retrieved from https://www.ashoka.org/press/5809

Khatib, F., DiMaio, F., Cooper, S., & Kazmierczyk, M. (2011). Crystal structure of a monomeric retroviral protease solved by protein folding game players in *Nature Structural &. Molecular Biology*, *18*, 1175–1177.

Korten, D. (2001). *When Corporations Rule the World.* Kumarian Press, Inc and Berrett-Koehler Publishers, *Inc.*

Leonard, A. (2007). *The Story of stuff.* Retrieved from http://storyofstuff.org/wp-content/uploads/movies/scripts/Story%20of%20Stuff.pdf

Leonard, A. (2007). *The Story of change.* Retrieved from http://storyofstuff.org/wp-content/uploads/movies/scripts/SoChange_Annotated_Script.pdf

Lyons, T. S. (2012). *Understanding Social Entrepreneurship: The Relentless Pursuit of Mission in an ever changing world.* New York: Routledge.

Merchant, N. (2012). Let your ideas go. *Harvard Business Review.* Retrieved from: https://hbr.org/2012/06/let-your-ideas-go

Milway, K. (2014). How social entrepreneurs can have the most impact. *Harvard Business Review.* Retrieved from: https://hbr.org/2014/05/how-social-entrepreneurs-can-have-the-most-impact/

Pearl, B. (2011). *Britta Riley: A garden in my apartment.* Retrieved from http://opensource.com/life/11/12/britta-riley-garden-my-apartment

Potter, D. (2014). Why the Future of Social Innovation Is Open. *Stanford Social Innovation Review.* Retrieved from: http://ssir.org/articles/entry/why_the_future_of_social_innovation_is_open

Rayner, T. (2011). *The Coalition of the willing*. Retrieved from http://coalitionofthewilling.org.uk/

Seforis Consortium. (2013). *The State of Social Entrepreneurship, Key facts and figures.* Retrieved from http://www.seforis.eu/upload/reports/1._Key_Facts_and_Figures_of_Social_Entrepreneurship.pdf

Svendsen, A. (1998). *The Stakeholder Strategy: Profiting from Collaborative Business Relationships.* San Francisco: Berrett-Koehler Publishers, Inc.

Vodafone. (2014). *Corporate Social Responsibility report 2014.* Retrieved from https://www.vodafone.com/content/sustainability/our_vision_and_approach/managing_sustainability/stakeholder_engagement.html

*World Development Report*. (n.d.). Washington, DC: World Bank.

# Chapter 5
# The World's Oldest Profession – Now and Then:
## Disruption of the Commercial Banking Model

**Steven A. Cinelli**
*REMARQ, USA*

## ABSTRACT

*Modern banking found its roots during the Renaissance period casted by the European merchant banks. Their success was due in large part to their aggregation, absorption and deployment of information about borrowers, structures and markets. In the 21st century, banking again is being advanced due to insights developed by vast amounts of information and data, this time gathered and managed through new technologies and models, in quest of efficiency, improved risk management and improved portfolio performance. New entrants into the business of banking operate outside of existing regulatory structures, and may enjoy a level of competitive flexibility compared to existing commercial banks. Might this portend the end of the so-called modern commercial banking model, or might it serve as a strategic imperative for the banks to adapt to innovation?*

## THE WORLD'S OLDEST PROFESSION: IN TIMES OF CHANGE

While the title of this chapter may have caused a heart to skip a beat, we are not referencing the happenings in a brothel where the practice of carnal frivolity may prevail. A quip by Rudyard Kipling in the 19th century laid claim to this descript, but in paging through history, we find that one industry may have even longer lineage, with such industry currently in the midst of one of its most significant transitions.

We are speaking of banking and the art of the money lending. Ascribed to Babylonia over four millennia ago, the practice of deposit taking and granting credit further evolved within ancient Greece and the Roman Empire. As some have placed the Code of Hammurabi (1800 BC) as the basis for use of women as chattel, commercializing the "other" old profession, the Code also laid out the basis for economic

DOI: 10.4018/978-1-5225-0568-6.ch005

transactions, seemingly a predicate for the practice of lending funds (Graeber, 2011). Even Asia developed a system of credit extension during select dynasties around 1000 AD. The modern age of banking is generally viewed as being borne from the European merchant banking families who saw opportunities to broaden their trading of goods and commodities to the extension of credit to other merchants desirous of trading themselves. Dating back to the 14th century and throughout the 19th century, families such as the Medicis, Fuggers, Warburgs, Barings, Hambros, Abs, and Rothschilds evolved from funding their own activities, to becoming financiers for others, including the largest of industrial companies and sovereigns. From the Renaissance period through the emergence of the modern commercial bank, such merchant banking activity drove economies. As Jared Diamond phrased it in his iconic tome, *Guns, Germs and Steel: The Fates of Human Societies*, "As society develops, so does the economic structure. Banking and finance are fundamental parts of the 'societal genome'" (Diamond, 2007).

The success of the merchant banks during yesteryear is characterized by a prevailing attitude, one of which knowledge was everything, and those with it, carried the day and had their way. Whether that might encompass comprehending trade routes, financing structures, commodity prices, margins and risks, capital strength of a customer, even anticipating outcomes, the merchant bankers applied their skills as "merchants" first, and then how to profitability finance an independent project or party. Essentially, their knowledge of "conditions" proved to be both a value-add, but further a competitive edge over other potential financing parties, and sometimes even the businesses they financed. Walter Bagehot, a Victorian London banker, stated, "We must not let in daylight upon magic", emphasizing the laconic nature of bankers to share information (Wechsberg, 1966). They used this to their advantage, underscoring that information was the elixir for optimal financial decision making.

In its essence, finance is truly an information business. Potential users of capital convey information to providers of capital and through an iterative process, an outcome of risk-adjusted finance emanates. Investing, trading, and lending results are determined by "good, better and best" information. Roll this observation forward to the 21st century, we witness the extensive power of information technology in banking and finance to generate, decipher, improve, expedite, analyze and regurgitate reams of data points to render broader yet tailored, robust yet quicker decisions, which are intended in the end to improve financing outcomes. Simply, it's all about the information.

## A Challenge

More so than other industries, banking and finance have historically sought applied technology to enable improvement in existing processes and reducing costs, including that of risk assessment. The challenge is that much of the earlier applied technology has been developed and adapted through many years, creating a spiders' web of legacy systems, needing to be continually serviced.

Think about the analogy of telecommunications. Throughout much of the Americas, Europe, and Asia, there has been a constant evolution of telecommunications, but these developed countries, even with current wireless means, still have considerable investments in wiring, cabling, pipes, and poles, all supporting part of the telecom infrastructure, which remains costly and inefficient, and experiencing the law of diminishing returns. Consider now the developing regions, like Africa, parts of India and China, void of such legacy structures, which are launching immediately into a modern telecom infrastructure, and in the end, might enjoy considerable advantage.

Reverting to banking, nimble financial technology startups, not laden with the past baggage, are forging new territories and methods to support the banking customer in ways and measures that excite the future, while concerning those invested in the past. Is the industry characterized as a zero-sum game, in watching new players advance at the expense of the old guard? In certain ways, as will be discussed, this may be true, but overall, the monolithic institutions recognize the need to embrace and collaborate with the nascent players, for their own competitiveness and profitability, and even their basic viability.

## In the Shadows

A term now used regularly is that of "shadow banking", which former Federal Reserve Chairman Ben Bernanke described as a diverse set of institutions and markets that, collectively, carry out traditional banking functions – but do so outside, or in ways only loosely linked to, the traditional system of regulated depository institutions (Wikipedia, n.d.). Examples of components of the shadow banking system included securitization vehicles, asset-backed commercial paper conduits, money market mutual funds, markets for repurchase agreements, investment banks and mortgage companies.

The provision of credit by regulated depositories has been challenged with new entrants into the market. These new players are benefiting from the use of technology, to gather requisite facts and data, and in ways use algorithmic approaches to decision making. The ability to create a transactional process with a customer and interact efficiently through online means has created a value proposition for these newcomers.

Besides technology, the newcomers benefit by the lack for regulatory oversight and legal and financial overhang which amongst other things require banks to maintain certain reserves and increasing levels of capital to support losses. The new marketplace platforms generally don't take credit risk, rather laying off the investment to institutional investors, garnering transaction fees and possibly ongoing monitoring fees. But, there is growing interest, possibly trepidation, by the regulators about the marketplace lending platforms, and whether there should be a level of equivalent oversight. Do they subscribe to the same or lesser underwriting strictures, and is the level of sought out transparency sufficient?

But the concern is not just one of oversight, rather the new players are impinging on viability of the basic banking model. Commercial banking has used a fractional reserve model, whereby it accepts deposits and holds a small percentage of such deposits as reserve (for deposit withdrawal needs), and then to lends the rest to borrowers. Banks thereby intermediate between savers and borrowers earning a spread between loan and deposit rates, utilizing a highly leveraged balance sheet. To mitigate the risk of excessive deposit withdrawal beyond the reserve level, banks are heavily regulated by their respective governments, which typically provide forms of deposit insurance and access to a central bank serving as a lender of last resort. Over the last eighty years, particularly in the US, additional regulations have been laid forth in response to excess and imprudent practices, which in many ways attempt to support the functioning of banks while reducing systemic risk to the financial markets.

According to a recent Goldman Sachs report on Shadow Banking (Nash & Beardsley, 2015), there is considerable concern within the conventional commercial banking trade. Certainly there is the competition in many of the product and service areas that banks have typically dominated, but noteworthy is that such areas have been some of the biggest profit centers of the banks. And as profits are the key contributor to enhancing such banks' equity capital levels in the wake of mandatory deleveraging, one may conclude that the regulated commercial banking model may be further challenged suggesting that the historic model (spread lending) be short-lived.

In many of its core business lines, competition from the new players is quite threatening. Among areas of insurgence are:

- Consumer lending,
- Small Business lending,
- Highly leveraged finance,
- Mortgage banking,
- Commercial real estate finance,
- Student loans.

Consumer lending has transitioned over the last fifty years particularly with the advent of general purpose credit cards, dating back to the late 1950s, when Bank of America introduced BankAmericard and a supportive merchant processing system. During the early 20$^{th}$ century, many large scale retailers, such as Sears and Montgomery Wards provided in-store credit cards, enabling their clients to essentially "buy now, pay later", which not only increased business but enhanced client loyalty. BankAmericard created opportunities for both the individual consumer by availing a broadly used credit product, and for merchants which might benefit with the extension of credit to their clients, while not having nor couldn't support the necessary infrastructure. For the convenience, banks charged considerable interest rates on the outstanding balances on such credit cards, while additionally monetizing the effort by supporting the merchants which paid merchant processing fees. Consider that in the 2000's, consumer have paid well over $1 trillion in credit card interest. Overall, the bank card system has become one of the great profit centers for numerous institutions. In time, Bank of America began licensing the card to other banks, eventually evolving the entire program into what is now known as Visa International. As an alternative to Visa, other institutions collaborated on developing MasterCard, another general purpose card. Now nearly 75% of consumer lending in the US is conducted by revolving credit bank cards. This sizable market has become a target for a new breed of lenders.

Presently, exclusive of secured consumer lending programs (e.g., auto and real estate), as well as student loans, the size of the US consumer debt market amounts to just shy of $900 billion (Federal Reserve Bank, n.d.), with the top 10% representing the focus of the peer-to-peer (P2P) lending firms, such as Lending Club and Prosper. While over 80% of credit cards that are held by strong credit quality borrowers ("prime"), the P2P platforms have seen an opportunity to provide better credit pricing for such borrowers, and without the same costs and capital requirements of the banks. Apparently, the cost of origination of new borrowing relationships is less for such platforms as compared to the banks, but this observation is suspect, as many of the consumer-lending banks have deployed online and technology enabled methods also, and in many ways, given their respective size, have a comparative cost advantage to the non-banks. And given the direct-to-consumer nature of marketing by the P2P segment, sales and marketing costs may represent nearly 60% of revenues, recognizing the expense of web-based outreach and snail mail efforts. Both deploy credit scoring systems for such small borrowings, typically without any direct intercourse with the borrower, reducing the costs and improving the margin of such borrowing activity. It is also worth stating that "P2P" has become much the misnomer, as the funding sources for such platforms are largely institutional now, with hedge funds and other yield-seeking investors provide the preponderance of the funding for these and other non-bank credit products.

With the non-bank entrants into the consumer lending market, Goldman envisions that a substantive level of profitability may be bled from the traditional banks, as the platforms underprice the banks, with consumers refinancing high interest credit cards with lower rate amortizing term loans. By financing such consumer borrowings with investor funding rather than interest bearing deposits, and the requisite reserve provisions, the platforms purportedly have an advantageous business model that benefits both debtor and creditor (capital provider) from a pricing standpoint.

While we witness the growth of the non-bank financing initiatives, might we posit that the appetite for volume comes with an inclusion of adverse credit risk, ultimately leading to a reassessment by the P2P funding sources? In assessing credit risk, we are already beginning to watch higher level of delinquencies in the P2P sector, with overall losses growing.

A second area of growth impinging on the traditional banks is the SME (small to medium enterprise, or small business) lending market. Extending loans to businesses is a large and profitable market for the banks, but tends to have relatively high origination and servicing costs, and is subject to scale. With bank net interest margins at relatively low levels (~70bp ROA) coupled with more restrictive underwriting, small SME loans (under $250,000) are of questionable profitability, laying a pathway to growing platforms like OnDeck Capital and Kabbage. Large non-bank financial institutions, such as GE Capital, have been active in the SME market for many years, and have provided an alternative to banks, but their size too has limited their ability to move down market. Interestingly, current SME lending seems to be price inelastic, in that the alternative lenders are charging significantly higher rates than banks, yet are growing their business rapidly. As stated in OnDeckCapital's S-1 filing in 2014, its average interest rate exceeded 50% per annum. For a small business, credit availability outweighs the cost of credit. As expected, on a portfolio basis, an exceedingly high gross-yield can absorb increased levels of losses, which are beginning to appear.

The highly leveraged transaction (HLT) and commercial real estate (CRE) lending markets have similarly been broadened with non-bank participants. Attracted by high fees and the ability to syndicate loans to other banks, HLTs and CRE activity have been exceedingly attractive for the banks. But with risk concerns, as well as mandated increases to reserve requirements, bank lending in these areas has been subjected to more scrutiny, at the same time alternative asset managers have begun taking advantage of the increasing void. Many private equity fund managers have created loan funds to address their needs for financial leverage. And the overall institutional investment community has gravitated towards such investments as they convey their disillusionment with traditional asset classes. Rather than relying on returns from over-priced public equities and anemically yielding fixed income offerings, investment managers have sought out ways to improve overall returns, which has become not just a desire for incremental "alpha", but as a mandate given, say, persistent asset-liability shortfalls within pension plans. Surely, private equity and venture capital play into the mix, but high yielding debt instruments have been somewhat satisfying the appetite for those current income seekers. Given that alternative managers have more flexibility in terms of structure, pricing, terms and conditions at the expense of the handcuffed-banks, we should see more participation in these traditional credits outside the banking system.

We do note that similar metrics apply to the CRE space as to the SME market, in that smaller sized loans have become the focus of the non-bank players, particularly for real estate crowdfinancing (RECF) activities. In many ways, the 21$^{st}$ century version of the 1980s real estate syndication model, RECF platforms have carved a niche for short to medium term secured loans, which the banking industry has gradually become lukewarm towards.

One of the largest lending spaces, for which the modern banking industry was in part built to address, is that of home financing or mortgage lending. From the establishment of the Federal Home Loan Bank in the 1930s, and the liquidity outlet of the Federal National Mortgage Association (*aka* Fannie Mae), banks played a much exclusive role in availing funds to enable home ownership. Through the years, this became a multi-trillion-dollar annual financing market, and with Fannie Mae and later Freddie Mac (Federal Home Loan Mortgage Corporation) serving to purchase loans from the originating lender, banks could replenish their balance sheets for further lending. In the 1980's, the private capital markets saw an opportunity to participate in the ultimate investing in such mortgage loans through pools, creating the infamous mortgage backed securities (MBS) market, relieving and/or competing with the government sponsored enterprises (GSEs). The banks were still the primary originator of such loans, but this has changed at rousing speed.

Over the last ten years, the mortgage ecosystem has shifted dramatically, and most of the activity has drastically dampened bank involvement. Presently, nearly 40% of new US mortgages are originated and processed through non-banks, with that number continuing to escalate. Q1 2014 mortgage origination volume for the banks was the lowest since 1997 (Freddie Mac, n.d.), with that figure lowering with refinance activity coming to a crawl and new bank underwriting guidelines becoming stricter, while non-banks have stepped up their volume. These non-bank originators also sell loans to the GSEs or possibly some private market buyers, and with their expansion, the banks' participation and profit thereon is waning. While the MBS market has not recovered from the financial crisis of 2008-2010, the reliance on the GSEs has increased, and even for banks that portfolio their loans, we see that banks hold less than 15% of the near $10 trillion of mortgage loans outstanding. We also observe that other forms of revenues in the home loan ecosystem, such as servicing, has been moved to non-bank entities. In both the US and EU, new players are jumping in, from LendInvest in the UK to Sindeo as crowdfunding platforms to the major originators such as Quicken, Freedom, Stearns, Penny Mac and Prospect, all significantly building their volumes.

Another growing industry that historically was facilitated by the banking industry has been the student loan market. Now largely conducted by the US government under its Direct Loan program, the market is approximately $1.2 trillion in size. This was at one time a lucrative lending segment for the banks, where banks would work with student and their parents, ultimately funding full loans or the amounts over stated government loan limits. With a combination of increased loan sizes by the government programs and increased scrutiny in private loan activity, many of the banks have pulled back from this market. The singular approach to underwriting and pricing within the current student loan market has created an opportunity to refinance existing loans off the government's and bank's balance sheets, with companies like Social Finance (SoFi) and CommonBond demonstrating a command in the space. Though the banks have largely reduced their role, the refinance activity has further eroded bank market share and earnings opportunity. To date, SoFi has originated and securitized over $2 billion of student loans refinances, funded by private sector investors.

What is fascinating is that the products provided by the marketplace platforms are the same as their banking brethren, delivered in a new wrapper and manner. Whether amortizing consumer and mortgage loans, short- and long-term commercial real estate loans, or fixed and variable rate C&I (commercial and industrial) loans, customer engagement has predominantly been moved online, and from a pure product standpoint, differentiation seems to lie solely in price differential, either on the high side or low side. With few exceptions, new forms of finance have not appeared, just delivery and price, which the latter can always be trumped. We see consumer loan customers reduce the cost of high APR bank cards by moving to the P2P market, while SME borrowers access credit anywhere they can find it independent of cost. Will the performance of loans originated by non-banks mirror that of banks, or given the nuance of the market segmentation prove out differently?

Beyond delivery and pricing, we also must consider how the originator generates a return. Certainly, there are advantages held by the non-banks in terms of overall funding costs. They rely not on their own balance sheet to hold the loans, i.e., funding with deposits (liabilities) that carry their own funding cost. And, as the banks bear the credit risk, reserves need be set aside, and with the desire of regulators to strengthen bank balance sheets, this "cost" effects overall profitability. This is not to say such lending is not profitable, as normally it should be. But size and scale determine the ultimate return on invested capital by the credit grantor. And that is the challenge within the changing market – profitable bank business is being assumed by the shadow players which enjoy a different funding model and business structure, and are willing to perform in markets left somewhat idle by the banks.

### Regulatory Overhang

Over time, banking regulation, legislatively adopted, has habitually been applied to correct or restrict certain behaviors and activities of banks, during times of unfettered risk taking and apparent calamitous outlooks. Whether the Glass-Steagall Act, the Gramm Leach Bliley Bill or the Dodd Frank Wall Street Reform Act, actions have been taken by legislators to both scold bad actors for egregious practices (*ala* subprime lending) and to shore up the health of the financial system. The concept of "too big to fail" has resonated on Capitol Hill and Wall Street, suggesting a fear that overzealous banking practices, precipitating failure of a large institution, can bring down an entire economy. Rarely, if ever, has there been financial legislation within the US (and abroad) that has been scripted to positively expand the market and create opportunities for banks and borrowers alike. One exception to the rule may be the Jumpstart Our Business Startups (JOBS) Act of 2010, which was an admirable yet somewhat innocuous attempt at directing funding into small businesses, while availing new investment opportunities for the unaccredited investor community. Clearly, there are staunch advocates and evangelists for the intent of the JOBS legislation, but as of this writing, demonstrable advances to small business capital formation, due specifically to the Act's provisions, are limited at best.

Given the financial structure of a commercial bank, the exposure of the federal government in terms of guaranteeing risks of failure through deposit insurance, banking has increasingly become a risk averse proposition. Leveraging equity capital at 12-18x in order to provide attractive equity returns to shareholders may need be reconsidered. Banking is a business of risk taking, with a resultant appropriate return for the risk entertained. As governments desire to maintain a strong financial system to support banking needs and overall economic growth, they generally will place (regulate) controls on practices to minimize risk of the industry in general. These restrictions in a market of growing credit demands gives way to opportunities for the new players.

## Technology Applied Then and Now

Technology has evolved banking and its approach to consumer and business customers in numerous ways. We have seen the branches giving way to ATMs and mobile banking, consumer loans being subsumed by bank credit cards, and commercial lending being tailored and repriced with interest rate swaps and securitization. But these are internal initiatives of the commercial banks, where they have adjusted and advanced their offering to still serve the client. We have been witnessing non-banks stepping into the shoes of the banks, providing digital offerings and methodologies as not just to compete but deliver new ways of engagement, with different models and even different clients. Many institutions have now become compelled to pursue next generation efforts as a strategic imperative. We have discussed the how technology has enabled new providers of credit to engage more efficiently with borrowers, which the banks are witnessing, but technology can be applied even further.

According to a research paper by Santander Bank's InnoVentures (2015), data capture through the "internet of things" methods can be applied to improving risk management while reducing operational costs in asset based lending and trade finance. Imagine being able to assess the quality, quantity and diminution of the value of assets being financed, such as trucks or manufacturing equipment by gauging ongoing usage on a remote basis, measuring distance, or weight or process flows. Or regarding trade finance, which has notoriously been a labor and paper intensive effort, consider GPS data informing on the movement of shipments, as well as the quality at the time of delivery. Applied technology could serve to expedite the process, with documentary letters of credit and acceptances, with multiple confirmations and verifications, moving into digital form.

Taking pages out of the tomes of the merchant bankers, accessing and managing data and information is critical, particularly when it can be deployed for improved insights, decision making and risk management. Digital technology has precipitated an abundance of data, particularly for an information business like banking, but generally banks have made little use of it, particularly on gathering insights. Consider the sophistication of online retail and social media companies that constantly collect data not merely to interact currently, say, to sell a product or service today, but rather to understand behavior patterns and affording a continuous feedback loop, which if done well, should enhance the lifetime value of its customer. Similarly, larger corporates, like Nike, Proctor & Gamble, Heineken, and Universal Studios have even launched crowdfunding initiatives, not because they need to raise funds for a new product offering, but to interact and gather feedback early on with the end customer so that when a product is released, the receptivity is drastically improved. The same philosophy should be engaged by the banks, one where enlightenment not mere cost reduction from technology is the core tenet, as this has yielded opportunity for the younger breed of financial institutions. Some of the new fintech players are using the philosophies of non-financial models, such as how to derive greater lifetime value of a customer, by assessing developing financial service needs scenarios, even as most have been secular in single product strategies.

## Transaction Processing: What about Bitcoin?

While the term cryptocurrency has become part of the 'fintech' nomenclature, a more descriptive term is a distributed ledger network, which records ownership of a "value" through a shared registry. If you consider transaction processing, whether through commercial banks, card processors, central banks, equity and other exchanges, there generally is a clearing house function, where certain parties certify ownership and clear the transaction. Even with technology applied, this can still result in delays, inaccuracies, and increased risk. With distributed ledgers, transactions can be cleared instantaneously, with better accuracy and lessened settlement risk, while significantly reducing infrastructure expenses. And for checks and balances, transactions are openly verified by participants, so having a central clearing authority is rendered obsolete. When applied to voluminous levels of transactions, this would enhance efficiency, verifiability, and post-review support. According to the World Bank, this could eliminate tens of billions of dollars of infrastructure costs in payments, trading and compliance over the next ten years.

Might this portend for banks to engage in next generation ledger networks, or might new institutions outside the banking industry garner such transaction activity, further challenging the controls of a central banking system?

## Global FinTech Investing

Investment in financial technology has been a growth market. With under $1 billion deployed in 2008, the market has witnessed nearly a twelvefold increase in investments to over $12 billion in 2014, according to Accenture. Beyond venture capital firms, many of the largest financial institutions whether banks or insurance companies are allocating increased capital in hopes of "digitally" advancing their future. In certain ways, banks are having a difficult time grappling with fintech, as adaptation to innovation has not been their long suit. Surely there are institutions, such as Wells Fargo, that have had long standing innovation efforts, but many applications are internally developed, with limited deployment. Still, we are impressed with notable institutions such as Citigroup, UBS, Santander, Capital One and BBVA which have staunch commitments to not being left behind, reaching out externally to partner with next generation developers and providers.

As an information business, banking and credit provision can be aided by technology in terms of efficiency, risk assessment and management, and, with new non-bank business models, there is the opportunity to create viable and valuable business franchises. And as we have also seen, the interest in fintech has buoyed valuations of many of these incipient endeavors only to question is there too much value being ascribed too early. Do the young entrants, such as Lending Club or SOFI or OnDeck Capital deserve the valuations they are receiving in their capital raising endeavors, certainly when compared to their analog competitors? Should conventional multiples on revenues, or asset levels or even earnings be the commonality, given that the base business in the lending of money, which generate revenues and earnings in similar manner? Or might there be too much capital chasing good, but not great, models, which are not really game-changers but mere distant cousins of the current banking participants? But looked at differently, there has always been a premium placed on innovation and break out opportunities. Might the score keeping now be altered, and not critique the lofty values of new players, but rather, ask ourselves when an innovation premium be placed on the banks playing the forefront? Time will tell.

## An Impersonal Relationship?

Reared in banking, one is taught the "Five C's of Credit": Capacity, Capital, Collateral, Conditions and Character. To be specific, they are defined as:

- *Capacity* is the ability to repay the loan in terms of income or cash flow compared to the level of obligations the borrower has to service;
- *Capital* refers to the overall financial strength of the borrower, i.e., net worth or other assets, intimating a fallback position to cushion against a default;
- *Collateral* is used as security in support of a loan, that a lender can ultimately foreclose on to recoup its loan

- *Conditions* refers to both the purpose and terms of the loan (interest rate and payments) but more specifically the general business and economic climate that the borrower operates within which could affect the ability of the borrower to repay; and
- *Character* refers to a more subjective assessment of the borrower, i.e., has he conducted himself in past affairs appropriately and in a trustworthy way, will he honor the obligation, how is he viewed by his colleagues and customers. In other words, it becomes "reputational capital".

An acute observation over the last two decades, with the deployment of technology and the scaling of lenders, particularly large banks, there has been a move towards more algorithmic credit assessment, with credit scorecards assigning values for particularly elements of a borrower's so-called credit worthiness. Certainly the use of credit scores, such as provided by Fair Issac & Company, commonly referred to as FICO, which assigns values to past credit payment history. Further, these scoring methods give points for time of employment, whether someone owns his home, one's educational achievements and other "objective" variables. With the centralization of credit decision making far away from where the banker and borrower conduct discourse, the credit scoring model is highly "objective", with decisions being made from on high, and rendered within minutes. True, this has expedited the credit decision process, but one of the five "C's" has been removed from the equation it appears - the assessment of "character". Is this important any longer?

Years past, when seeking a loan, an aspiring borrower would approach his "human" banker and engage in discussion about the funding needs, his ability to pay the loan back, his overall financial picture, and both an objective and subjective review would be made by the banker in determining whether to extend a loan. The human banker seems no longer in the mix, though there may be client intercession with customer support personnel, whom doesn't have decision making authority. The process has become centralized and formulaic, and with technology, the capture of information and even the decision making is now automated. Fine for the objective analysis, in a review of numeric values, but the personalization or subjective aspects of the decision has been lost. The assessment of "character" seemingly no longer applies. Maybe for large volumes of business, objectivity and insuring portfolio diversification should be the agenda, not getting bogged down with subjective assessments, which could challenge the model and create inefficiency. But then again, are we failing to capture a determinant of end result?

Think of it more broadly. Having a decision-making banker sitting across the desk from a would-be borrower, who knows that the banker does have the ability to approve or decline the credit application. When a positive decision is made, do you think that the borrower has developed some level of personal rapport with the banker, different than have a decision rendered online or via mail? When a client is asked "who's your banker?", is the reply John Smith of the Main Street branch of Bank of Kalamajoo; or rather is it merely the name of an institution, such as Bank of America or Wells Fargo, or now Prosper, OnDeckCapital, or Lending Club. Has the art of credit provision, which advancing with technology, become so impersonal and purely data laden and algorithmic, that it may affect the outcome of the loan? Does the motto of the London Stock Exchange, "*Dictum meum pactum*", or translated "My word (is) my bond", resonate any longer? But in borrowing from amorphous, enormous and impersonal entities, do borrowers have the same level of moral commitment to make good on any borrowing? Or because it's Citibank with trillions of dollars in footings, how much of a problem for it is non-collection on a $10,000 loan? In the US, we take it even further when coming to the advocacy of non-recourse lending, where borrowers can "walk" from a shortfall in a secured loan collection? Does the impersonalization change the credit risk?

Consider the nature of general communication in the digital age. There are over 200 million emails sent and received daily and over 10 billion texts sent daily. Has the new paradigm of digital messaging caused changes in personal rapport, relationships, even writing skills? Just maybe.

So the digitization of credit provision, while efficient and maybe in some ways insightful with much more data to be measured, have we lost something in credit quality in terms on the softer side – from both the lender and borrower perspective. Is the framework of lending still one of the historic moral imperative or as we have seen, an increased level of moral hazard, with the shifting of risk away from the credit grantor? And in so doing, save the concern of damaged credit scores, are we raising the level of moral turpitude?

We began this tale with the relating of the oldest professions. The characteristics have become increasingly striking as time passes. In days past, "relationships" were aspirational, rather than the ephemeral nature of a mere tryst. An individual would have a single paramour, or suitor, or partner, with whom things and experienced were shared and over time, the depth of such relationship would lay the foundation of a sturdy bond, and needs and desires are satisfied consistently by one's partner. But, whether by disappointment or distraction, or even curiosity, a relationship can become threatened by something new, delightful in appearance with expected satisfaction, and generally engage for a focal purpose. Yes, what was once acclaimed to be a "relationship", has now become merely a series of transactions, with most clients having many conscripts for particular needs. Relationship banking has given way to transactional activity. The ability to move from one to the other, paying as you go, without having solely "one bed to sleep in". Maybe the mantra of the technology generation serves to incite - it's not about product, but about customer experience. And with that maybe a "relationship" is crafted.

The discussion of the "relationship" is really an opposite way of assessing the future of commercial banking as we have come to know it. As commented, this is an information business in transition. From the accepting of deposits to the lending of funds, information technology has created many alternate means of conducting the business of banking but without the accoutrements of the ivory tower. From sending and receiving money through one's PayPal account to managing account balances via mobile devices to seeking a loan for one's business or to purchase a home through an online platform: all these activities were formally the role of one's bank (and banker), where a customer indeed had a relationship. No longer.

And even further, given the increasing levels of legislative application and regulatory scrutiny, the industry itself is becoming challenged in its competitiveness with nascent fintech participants. Surely, there are significant resources being deployed by the banking industry to advance itself, but we wonder whether the highly leveraged, fractional reserve business model, in an age of accelerated digitization reducing costs and means of intermediation, has continued longevity. And with desires to avoid another crisis as we experienced only years ago, the industry and its overseers have hunkered down. Given the risks inherent with insured deposits being used for deploying risk capital generating ever diminishing margins, there have been proponents of new banking models, including "narrow banking", where banks would accept deposits but invest them only in liquid and gilt securities, rather than loans. Other financial intermediaries would take up the lending business, as we have already been witnessing. All told, so many functions, services and offerings of banks are now being tendered from outside the basic structure, which begs the question, has the commercial bank model exhausted its purpose?

So the environment is one of potential tectonic shifts. New players are advancing, not bogged down with regulatory shackles, nipping away at core business lines of the world's oldest profession. Certainly we don't expect to see the demise of the banking model as we know it (at least in the near term), but it is becoming clear that like the telecom industry which evolved throughout the 20th century, affording micro- and macro-economic opportunities globally, a new age of banking should yield similar effects. In markets of the banked, underbanked and unbanked, new technologies applied to the banking process should contribute to broader and more advanced access to financial services and yes, evolving the societal genome.

## REFERENCES

Diamond, J. (1997). *Guns, Germs and Steel: The Fates of Human Societies*. New York: W.W. Horton & Co.

Federal Reserve Bank. (n.d.). *Consumer Credit*. Retrieved 1 August 2015 from http://www.federalreserve.gov/economicdata&research/statisticalreleaseandhistoricaldata/consumercredit

*Freddie Mac*. (n.d.). Retrieved 15 September 2015 from http://www.freddiemac.com/economic&housingresearch/mortgageoriginations

Graeber, D. (2011). *Debt: The First 5,000 Years*. New York: Melville House.

Nash, R., & Beardsley, E. (2015). *The Future of Finance Part One, The Rise of New Shadow Bank*. Goldman Sachs.

Innoventures, S., & Wyman, O., & the Anthemis Group. (2015). *The Fintech 2.0 Paper: Rebooting financial services*. Retrieved from: http://santanderinnoventures.com/wp-content/uploads/2015/06/The-Fintech-2-0-Paper.pdf

Wechsberg, J. (1966). *The Merchant Bankers*. New York: Pocket Books.

Wikipedia. (n.d.). *Shadow Banking*. Retrieved 1 August 2015 from: http://en.wikipedia.org/wiki/ShadowBanking

# Chapter 6
# The Crowdfunding Market, Models, Platforms, and Projects

**Melek Demiray**
*Istanbul Technical University, Turkey*

**Sebnem Burnaz**
*Istanbul Technical University, Turkey*

**Yonca Aslanbay**
*Istanbul Bilgi University, Turkey*

## ABSTRACT

*The changing structure and intensity of competition on one hand, the ongoing pressure on creating differentiated offerings lead to seek out new ways to improve technology as well as new ways to perform business. Technological developments enable market participants or other interest parties to engage in the process of both innovation and production, resulting with an online social network market model of co-creation. This chapter focuses on describing different aspects of recently developing crowdfunding market models where new projects are offered and funded by large number of individuals from general public. The current extent and dynamics of the growing crowdfunding market, various models of crowdfunding, crowdfunding platforms and the projects that are crowdfunded are analyzed through the chapter by giving examples and comparative comments. The analysis of the market is concluded by the foreseen future trends.*

## INTRODUCTION

Crowdfunding is a rapidly emerging system for funding a wide variety of projects from art-based campaign to technology related initiatives. Although, artistic and social projects have become successful in the earlier period of crowdfunding, recently campaigns concerning small and medium sized start-up companies, software programs, real estate and other kind of business have become popular (Review of Crowdfunding Regulation, 2014). Crowdfunding enables individuals to actualize their original ideas and social-based projects that they cannot find capital easily from other kind of financial sources. That

DOI: 10.4018/978-1-5225-0568-6.ch006

is why it is defined as "funding escalator" for start-up companies and support to form democratic and collective financial system (Unleashing the potential of Crowdfunding in the European Union, 2014). Today, crowdfunding market includes profit and non-profit projects in the area of business, technology, art, film, music, publishing, journalism, real estate and social cause (Milosevic, 2015; Martínez-Canas et al., 2012). Business and entrepreneurship remained as the most popular crowdfunding category, collecting \$6.7 billion in 2014, which represents 41.3% of total crowdfunding volume. Social causes at 18.9% (\$3.06 billion), films and performing arts at 12.13% (\$1.97), real estate at 6.25% (\$1.01 billion), and music and recording arts at 4.54% (\$736 million) rounded out the top five categories. (Massolution's 2015 CF Industry Report, 2015).

This chapter focuses on the recently developing crowdfunding online communities, as free basis, open communities for innovative co-creation, that the new projects are offered and funded by large number of individuals from general public. Beugre and Das (2013) define the term "crowd" to include a group of people or a community, a social entity organizing resources to exploit an opportunity that adds economic or social value. Crowdfunding mainly consists of collecting funds through small investors in return for tangible or intangible benefits (Ordanini, Meceli, Pizeli, & Parasuraman, 2011). Currently, 1250 crowdfunding platforms operate in the world. The crowdfunding market size rises from \$6.1 billion in 2013 to \$16.2 billion in 2014, increasing by 167%. It is expected that it would reach \$34.4 billion in 2015 which is almost two times more than the existing volume (Massolution's 2015 CF Industry Report, 2015). The chapter sheds a light to the current extent and dynamics of the growing crowdfunding market, introducing different models of crowdfunding, crowdfunding platforms and the projects that are crowdfunded. The analysis of the market is concluded by the foreseen future trends.

## THE CROWDFUNDING MODELS

In crowdfunding, participants of the crowd benefit through different motivations such as getting free products, experiencing the pleasure of the process, participating an action group, etc. which bring different models of crowdfunding into the scene. The landscape of crowdfunding models is mainly grouped into two major categories as the nonfinancial return models and the financial return models, each comprising two types (Crowdfunding Final Report, 2014). These crowdfunding models that the funders are engaged are described as donation-based, reward-based, lending-based, and equity-based, respectively (Buysere, Gajda, Kleverlaan, & Marom, 2012; Bretschneider, Knaub, & Wieck, 2014). Even though these are currently well accepted models based on the basic definition of Crowdfunding's Potential for the Developing World Report (2013), this chapter will also include a newly emerging crowdfunding model as civic crowdfunding. Table 1 refers to the general characteristics of the models analyzed.

### Typology of Crowdfunding Models

#### Donation-Based Crowdfunding

Donation-based crowdfunding is aimed at supporting projects where contributors do not expect any financial or non-financial return (Kuti & Madarász, 2014). The projects can range from pure charity-based projects to personal ones or from academic or research based projects to art related ones. This type of model ties the donors to the project without getting any direct return for their donations, nor any

*Table 1. Characteristics of the major crowdfunding models*

| Model | Features | Pros | Cons |
|---|---|---|---|
| Donation-based | Philanthropic: funders donate without anticipating financial return | No risk | Entrepreneurs have difficulty collecting large amount of capital. |
| Civic | Focused on common good | No risk | In some cases, only goverment agecies and non-profit organizations are allowed to launch campaign. |
| Reward-based | Funders get a reward or pre-selling of a product created. | Low risk (met the goal and fraud risk). | No real potential for financial return and no accountability mechanism |
| Lending-based | Funders get a debt instrument with a fixed rate of interest on a specified schedule. | Fix rate of return enables entrepreneurs to reach capital. | Start-ups' high-failure rate risk of loss. Existing/established, cash flow positive enterprises may use this method. |
| Equity-based | Funders get equity or profit sharing instruments | Unlimited potential for financial revenue. Preferred large numbers of funders. | Potential loss of investment. Securities laws may be complex. |

Crowdfunding's Potential for the Developing World, 2013; Davies, 2015.

physical or monetary reward (Cholakova & Clarysse, 2015; Mollick, 2014). Some of the reward based models also allow unrewarded requests for donations, but it is found that only 22% of all crowdfunding initiatives were requests for donations, with no rewards offered (Bradford, 2012).

The contributions made based on the donation model are mainly donations. Funders stand as philanthropists since they receive nothing in return for their support (Marom & Sade, 2014). However, although the contributor's motive is mainly altruistic in funding a project, the receiver's motive may not be alike; donations may fund for-profit enterprises. On the other hand, the success of a donation-based crowdfunding campaign is highly related with the perceived benefit to the common good or emotional appeal (Crowdfunding Final Report, 2014).

Underlying motivations for the donors or the funders to support this type of crowdfunding vary. While donors of smaller amounts may get just recognition, donors assuring higher amounts might be rewarded with a product, customization of a product or some form of personalized recognition (Crowdfunding's Potential for the Developing World, 2013). From a psychological point of view, when a donor is part of a project, he receives personal satisfaction, especially when he gets thanks/gratification and recognition from the project managers (Meric, Bouiass, & Maque, 2015).

The donation-based model is mainly used by non-governmental organizations (NGOs) to collect donations for specific projects. The major motive is social in this case and funders are eager to provide high amounts if they believe that they support a specific project. They also become more loyal when the NGO keeps them updated about how the project develops and evolves further. This is in most of the time a key to build long term relationships (Buysere et al., 2012). Recently, a new type of crowdfunding called as civic crowdfunding, has been originated from the donation-based model and received considerable attention and interest.

## Civic Crowdfunding

Civic crowdfunding has emerged in the broadest sense from the crowdsourcing that serves to bring together a "crowd" to make incremental contributions toward a final project, product, or service. Studies about crowdsourcing contribute to conceptualize civic crowdfunding through identifying elements common to crowd activity, and describing who constitute "the crowd" (Stiver, Barroca, Minocha, Richards, & Roberts, 2015).

Civic crowdfunding is mostly related with the projects and platforms that work toward to crowdfund parks, community centers and social services. In some cases the outputs of the civic crowdfunding could be standard public goods, while others could be private or quasi-public related projects such as roads, tunnels and bridges. These projects offer an opportunity for the crowd to co-produce civic goods (Davies, 2014). Therefore, civic crowdfunding can be defined as funding projects that "provide services to communities" (Stiver et al., 2015, p.260) or as using crowdfunding for projects that offer community or public related benefits (Davies, 2015). The use of the term dates back to 2012, especially by platforms such as Spacehive and Neighbor.ly (Davies, 2015). For example, the Mansfield Business Improvement District was able to collect £38,000 (€43,921) through Spacehive (Hollow, 2013).

Civic type is a growing crowdfunding trend for its ability to direct citizen funds toward specific projects, as well as for its capacity to link project initiators with municipalities, organizations, and individual citizens who may be interested to make contributions to the projects. It is an excellent alternative source of funding at a time of limited government budgets (Stiver et al., 2015). In addition, civic crowdfunding has great potential for non-financial benefits such as facilitating networking, and encouraging collaboration between citizens and government (Stiver et al., 2015). A project subject to civic crowdfunding makes use of a shared resource that is accessible to the community either as a public asset, a community-owned resource or a public-private partnership, and may or may not involve government (Davies, 2014; Van Veelen, 2015).

While a 'civic' project focuses on the good being produced, it is important to note that not all civic crowdfunding platforms allow individuals and community groups to initiate projects. For example, some projects can only be created by government-authorized entities. Among the platforms that allow initiation by any individual or group, civic crowdfunding provides a new tool for the community organizations wanting to intervene over their members' environment (Davies, 2015).

Civic crowdfunding can produce very different outcomes and it is argued that the consequences are positive rather than negative to the public. The goods produced as the outputs of the civic crowdfunding campaign are expected to be goods that can be consumed equally by members of a community, regardless of their contribution to the production of the good. Therefore, the expected output of a civic project becomes a public good or a common pool resource. On the other hand, the relationship between civic crowdfunding and government has been interpreted in differing ways. For some it represents a direct threat to public funding of services, while for others it is a new means of public–private engagement that encourages a more constructive relationship between communities and government (Davies, 2015)

Although the research shows that civic goods are typically non-excludable, such as traditional public goods or common pool resources, the share of projects containing private goods is 20%. This relatively high proportion shows that it is considered that the quality of a civic project relates to the goals of the organization or person involved, rather than the production of a specific type of good, since many of the private good projects involved fundraising for the maintenance or organizational costs of non-profit organizations (Davies, 2014).

## Reward-Based Crowdfunding

This crowdfunding model is the most recognized and common so far (Martinez-Canas, Ruiz-Palomino, & Pozo-Rubio, 2012). Funders support a project with the expectation of something material or immaterial in exchange. The offer could be in the form of non- monetary tangible (e.g., product) or intangible (e.g., reputation, identity) rewards to contributors (Cholakova & Clarysse, 2015). Kickstarter being the pioneer and Indiegogo as the follower, are the two most well-known platforms where reward-based crowdfunding model works most efficiently.

A variety of motivations can lead to reward-based crowdfunding such as providing support to startups, growing existing companies or financing creative projects. The reward-based crowdfunding develops through the well-defined stages: it starts by the project initiators when they send all relevant information, and continues by introducing sales offers. The goal of the project, the investment needs and the deadline appear on the project website. It is apparent what the funders will receive, either real products or services in exchange. If the project fails to receive the granted amount, supporters may be refunded.

The owners or initiators of specific projects use reward-based crowdfunding to collect donations for their project. The rewards are usually much lower than the donation amount, so the funder may be satisfied with a return with a lower economic value than the original investment. The perceived value of the reward can be much higher than the actual economic value, for example, entrance tickets for a concert or special VIP tickets or a personal meeting with the musicians as such rewards for a high donation (Buysere et al., 2012).

The different reward categories established by the initiators can be distinguished based on the amount of support or the level of contribution. Supporters offering more money are classified in a different category and can expect a proportionately higher reward. The four most common reward types are:

1. Copies of the object (e.g., the actual product, an assembled version of a DIY kit);
2. Creative collaborations of various kinds (e.g., the funder might appear as a hero in the comic, or she may be painted into the mural);
3. Creative experiences (e.g., a visit to the film set, a phone call from the author, dinner with the cast, a concert in the backer's backyard); and
4. Creative mementos (e.g., photos sent from filming location, explicit thanks in the closing credits of the movie, etc.) (Kuppuswamy & Bayus, 2014).

If the project initiators behave sensibly in selecting right combination of different reward-levels, it is likely to get a higher average donation than with a pure donation-based approach (Buysere et al., 2012). The reward-based model can take the special form of pre-sales crowdfunding when contributors get in return something symbolic, like the opportunity to participate in the cultural experience they finance (e.g. appearing as an extra in a film), or a product that was developed and produced with the funds raised (Unleashing the potential of Crowdfunding in the European Union, 2014). Examples can include being credited in a movie, having creative input into a product under development, or being given an opportunity to meet the creators of a project. Alternately, reward-based crowdfunding treats funders as early customers, allowing them access to the products produced by funded projects at an earlier date, with better price, or with some other special benefit. The "pre-selling" of products to early customers is a common feature of such projects that are usually similar to entrepreneurial ventures, such as projects producing new software, hardware, or consumer products (Mollick, 2014).

In some instances, a new product or service could be shown online and the funders may be requested if they are interested to order and pay in advance. This type of crowdfunding is mainly called as pre-sales or pre-purchasing as a form of reward for the contributor. This could be also a source of market research and demand estimation and also may finance the project in advance. Funders that participate in these crowdfunding campaigns do it because they want this product or service to be made. An important incentive to support this type of campaigns is to obtain the product at lower a price (Buysere et al., 2012).

The reward and the pre-sales crowdfunding models are similar to each other, and often appear together on the same websites, that is why they deserve to be analyzed jointly. In the reward model the project owner often offers a symbolic reward to the investor that is not comparable to the funding. The reward could be small or it could be something with presenting more prestige, such as acknowledging investor's name in a movie (Bradford, 2012). In the pre-sales model the supporters receive the product that the entrepreneur is making. For example, if the entrepreneur is producing a music album, contributors would receive a copy of the album or the right to buy the album at a reduced price after it is completed (Marom & Sade, 2014). Like with the reward model, contributors do not receive a financial return such as interest, dividends, or part of the earnings of the business. (Crowdfunding Final Report, 2014).

## Lending-Based Crowdfunding

This model of crowdfunding is built on offering funds in the form of a loan. Contributors provide the funds temporarily with the expectation of refund. In some cases, investors are given interest on the funds they loan. In other cases, they receive only their principal money (Marom & Sade, 2014). The main benefit of lending-based crowdfunding is that it permits a company to borrow money from a group of people without engaging with a bank. Some of the platforms here act like a middleman and make the repayments to the investors, where other platforms perform the intermediary role in order to relate the borrower (investee) and lenders (investors) (Buysere et al., 2012).

Lenders as investors can be promised interest on the funds they loan or they may be only given the return of their principal money. Kiva is the most prominent lending-based crowdfunding platform without offering interest. It deals with its so-called "field partners" that are microfinance lenders instead of lending money directly to entrepreneurs. Entrepreneurs send their loan demand to the Kiva website, where potential lenders can see these demands and find projects for funding. Kiva gives this money back to the field partners and credits lenders with the refunds the entrepreneurs make. Lenders only receive their principal back and the field partners use the interest to cover their operating costs. In contrast with Kiva, Prosper and Lending Club are the two leading lending sites that offer interest. Not all of the loans are for business purposes. The majority of the loans on these sites are for personal expenses, but the amount of the small business lending is growing (Bradford, 2012).

Some platforms have been removed from institution-to-peer type of lending especially with the investors preferring more to work with peer-to-peer lending (Kuti & Madarász, 2014). Peer-to-peer lending (P2P) can be defined as the use of an online platform that matches lenders/investors with borrowers/issuers to deliver loans. The borrower can either be an individual or a business. This model is found beneficial since it enables the lenders to provide money for very small parts of the overall loan. These loan parts are then combined by the online platform and when there is enough to cover the required loan, the loan is paid to the borrower. The borrower then pays back the loan with interest, which is arranged by the platform. The interest is paid to the lender until the loan matures; the borrower pays it back early or the borrower defaults. This interest rate depends on the borrower's evaluated risk, but it is usually higher than the savings rates available to the lender but lower than a traditional loan available to the borrower (Kirby & Worner, 2014).

In fact, the lending model of crowdfunding itself is often called peer-to-peer (P2P) lending. Although it carries some characteristics of crowdfunding lending, the main difference is that the lenders and borrowers usually do not know each other. P2P is mainly accepted as an attractive new financing model for loans; the main motivation for the funder is to get a higher financial return. The borrowers prefer this model because they look for a loan with a lower interest rate than the one they can get from a bank. The risk factor determines the interest rates and is calculated by P2P platforms or by independent institutions using mainly financial data. It can also be used by borrowers who can offer fewer securities. This model of crowdfunding is continuously growing; new lending is estimated at above €20 million per month for Europe. The main reason underlying this trend is low risk attributed with P2P since only the money provided by the funders is being lent out and there is no money creation within the platforms (Buysere et al., 2012). Many debt-based crowdfunding platforms fall within the P2P lending industry and identify with both (Crowdfunding Industry Report, 2013). Debt-based crowdfunding deals with the sale of debt tools in exchange for capital. However, it is different than the equity-based crowdfunding, debt mechanisms are more secure and draw the attention of investors who wish to see a more immediate return on their investment. This model enables businesses to reach loans easily and individuals to get higher returns. Some other lending-based platforms give the possibility to social borrowing, through providing loans to social projects with no interest being offered, such as businesses in emerging markets can receive micro-financing (Buysere et al., 2012). In that case of micro-financed loans, the lender may be more concerned with the social good promoted by the venture than any return generated by the loan (Mollick, 2014).

A variety of business models support P2P lending. For instance, in "client segregated account model", an individual lender is introduced to an individual borrower through an intermediary platform, and a contract is built between them. An intermediary platform is again existent in "notary model" between the lender and the borrower; the lender propose the loans, and then the loan is created once the amount of money required is reached. The key point here is that the loan is originated by a bank, the platform delivers a note which is accepted as security to the lender for the value of their contribution to the loan. "Guaranteed return model" allows lenders to make investment this time at a predetermined rate of return on the investment guaranteed by the intermediary platform. (Kirby & Worner, 2014).

Peer-to-peer lending is regulated in five different ways as it is presented in Table 2. These are as follows:

1. Exempt/unregulated through lack of definition;
2. Platforms regulated as an intermediary;
3. Platforms regulated as a bank;
4. The US model (federal regulation through the Securities and Exchange Commission (SEC) and state level, where platforms have to apply on a state-by-state basis);
5. Prohibited

Smaller peer-to-peer lending platforms may serve niche markets. These include for example platforms focusing on transactions in real estate financing, venture capital, business-to-business, graduate financing, funeral financing, art project financing, technological start-ups or consumer-to-consumer loans for transactions such as eBay purchases (Kirby & Worner, 2014).

*Table 2. Peer-to-peer lending business models*

| Regulatory Regime | Description | Countries Currently Using Regime |
|---|---|---|
| Exempt Market/ Unregulated through lack of definition | In these jurisdictions either the regulation has classified peer-to-peer lending as an exempt market or there is a lack of definition in legislation. | Brazil, China, Ecuador, Egypt, South Korea, Tunisia, United Kingdom |
| Intermediary Regulation | This regulates peer-to-peer lending platforms as an intermediary. This usually requires registration as an intermediary, and other regulatory requirements depending on the jurisdiction. | Australia, Argentina, Brazil, Canada (Ontario), New Zealand |
| Banking Regulation | This regulates peer-to-peer lending platforms as banks. | France, Germany, Italy |
| US Model | This is a two tier system. This requires the registration of peer-to-peer lending platforms with the SEC, as well as applying for a licence to conduct business on a state by state basis. | United States of America |
| Prohibited | Both peer-to-peer lending and equity crowd-funding are banned under legislation | Israel, Japan |

Kirby & Worner, 2014.

## Equity-Based Crowdfunding

In addition to other forms listed above, crowdfunding efforts may also consider funders as investors, giving them equity stakes in return for their (Massolution's 2013 CF Industry Report, 2013). So called equity-based crowdfunding offers a contract for shareholding or a revenue sharing plan to investors (Buysere et al., 2012; Cholakova & Clarysse, 2015) based on the share of the profits or return of the business they are supporting (Marom & Sade, 2014). Therefore, this type of crowdfunding model is the one that most clearly involves the sale of a security (Bradford, 2012). Thus, equity-based crowdfunding deals with the sale of equity for capital. An investor provides a business or entrepreneur with capital in exchange of a pre-determined percentage of equity or shareholding. This type of investment crowdfunding is most similar to angel or venture capital investing. It does not include debt offerings and return on investment is often higher risk than that of debt. This is mainly because of the fact that equity is pursued from the early periods of young start-ups that exhibit higher rate of failure compared to other more secure investment alternatives. For instance, investors while supporting an SME through equity must be ready to take on losses or wait for a longtime to get any return on their investment. (Crowdfunding Industry Report, 2013).

When an individual entrepreneur or a company aims to collect investment from a group of people instead of getting funds from a business angel or another private investor, this is typically called crowd-investing as a method of financing, or equity crowdfunding. Basically, the entrepreneur makes the sales of equity or equity-like shares to a group of investors through an open call on Internet-based platforms (Ahlers, Cumming, Guenther, & Schweizer, 2012). Therefore, equity-based crowdfunding involves the distribution of shares through the internet, where investors can acquire stock in corporations for a small amount of money. By this way, equity-based crowdfunding may allow even those companies to access funding, which have failed to get funds from angel investors, government programmes, or relatives (Kuti & Madarász, 2014).

Companies using equity-based crowdfunding platforms send offerings and information on websites and also use social networks to invite investors to evaluate the offering and indicate their desire to invest. The following topics are to be covered as the expected information: information about the business and the owners, the tentative business plan, type of equity security being offered, percent of the company being sold in the offering, amount of time remaining in the offering, progress made toward meeting funding target (Crowdfunding's Potential for the Developing World, 2013).

Motives to engage in equity-based crowdfunding for the funders may change. Some may be primarily interested in investing in projects that share their own values, that are locally appealing, or that improve employment in their public. Others may have actual information about what the project or company addresses, what the market wants, and aim to bring funds and expertise to the success of the project. Consequently, funders are considered as investors via those crowdfunding efforts, giving them equity stakes or similar consideration in return for their funding.

Equity-based crowdfunding was broadly legalized in the US by the Jumpstart Our Business Startups (JOBS) Act (2012), and previously legalized in other countries. However, the equity crowdfunding market is significantly influenced by the legislative environment of the home country (Heminway & Hoffman, 2010). Since it is subject to various regulatory issues, equity crowdfunding has been restricted in many countries for years; the U.K., Ireland, France, the Netherlands, Switzerland, and Australia are main OECD countries in which crowdfunding platforms are permitted to sell equity shares to small investors (Ahlers et al., 2012). In fact, until the passage of JOBS Act in United States, companies could not publicly solicit funds from unaccredited investors. Some relevant aspects of this Act are as follows:

1. Companies can raise up to $1 million from investors and up to $2 million if they provide audited financial statements;
2. Financial statements disclosure is not needed for enterprises in which the number of shareholders is below 1000;
3. Investors with annual income of less than $100,000 may invest up to the greater of $2,000 or 5% of their annual income in such enterprises.

For income above $100,000 per year, investment is permitted up to 10% of total salary but should not exceed $100,000 (CATA2012). The legislation is still in its developmental stages and the Securities and Exchange Commission in the United States is in the process of designing many of the rules (Mitra, 2012). Also, many of other countries still impose severe constraints; equity crowdfunding is generally limited to €100,000 in Europe (Buysere et al., 2012; Belleflamme, Lambert, & Schwienbacher, 2014).

The regulation of equity crowdfunding is performed by means of three approaches. First, the regulation may prohibit equity crowdfunding completely; second, equity crowdfunding is accepted as legal but the regulation of the industry creates high barriers to entry; and finally the regulation may allow the industry to exist, but with strict limits. Another approach to the regulation of equity crowdfunding by some authorities appears to reduce the regulations around the issuance of shares through equity crowdfunding for the main purpose of promoting SME growth (Kirby & Worner, 2014). Consequently, equity crowdfunding is relatively rare worldwide, constituting less than 10% of all crowdfunding investment (Massolution's 2015 CF Industry Report, 2015) making up a remarkably a small market. Problems could arise as the new retail investors enter this market since especially small investors may lack the experience to deal with the problems associated with start-up companies or illiquid investments, creating a high risk of instability and raising investor protection issues (Kirby & Worner, 2014).

The variety of equity-based crowdfunding instruments may pose a challenge for the retail investor, as it is a complex matter for less investment-savvy individuals and because the rights of the investor differ from platform to platform and country to country. The ultimate adoption of this type relative to other forms of crowdfunding is uncertain. Even in the absence of equity crowdfunding, investor model crowdfunding can take other forms, including, for example shares of future profits or royalties, a portion of returns for a future planned public offering, or a share of a real estate investment (Mollick, 2014).

### Comparative Analysis of the Models

Crowdfunding models are diverse in motivational aspects (Kuppuswamy & Bayus, 2014). For instance, donors of donation-based model desire to help the causes, supporters of civic project focus on public good, and funders' motivation of reward-based crowdfunding activities is to fund for actualization of creative ideas. In addition, although both lending-based and equity-based models are used to receive maximum monetary return, the risk of lending-based model is lesser than the equity-based one. On the other hand, in terms of emotional motivation, only donation-based model enables funders feel good mainly based on the altruism. Concerning the expected financial return, donation-based and reward-based crowdfunding models have no financial compensation, but some tangible rewards such as products or services produced (i.e. a DIY tool, a copy of CD or a poster of the supported singer) thanks to the campaigns are delivered to funders in reward-based model (Mollick, 2014). However, relating the financial return models, lending-based model has a fixed amount of interest revenue and investors of equity-based crowdfunding are benefitted from revenue of shares.

Capital acquiring processes by crowdfunding models have different levels of complexity based on their regulatory arrangements. (Hemer, 2011). Whereas donation-based model has the simplest procedure compared to the others, equity-based crowdfunding is the most sophisticated among all other kind of models. In addition, lending-based approach is more complicated than reward-based ones due to its stricter financial transaction structure.

Success factors of the campaigns vary in the crowdfunding models. For instance, failure probability and financial situation of the enterprises, uncertainty level of the business environment, and human resources play critical role to achieve successful campaigns in equity based crowdfunding (Ahlers et al., 2012). However, for reward-based model, good preparation, innovative ideas, extensive social network, update information (Mollick, 2014) and correct expression of the project description (Marom & Sade, 2013) increase success rate of the campaign. As for typology of the project initiators, while donation-based crowdfunding campaigns are initiated by individuals, charities and non-profit organizations, civic projects are launched by government agencies like municipalities and non-profit organizations.

## Characteristics of the Crowdfunding Models

### The Funding Mechanisms

Within all crowdfunding types, the payment is made through the platform supporting the model and various channels can be used based on the structure of the platform, such as paypal, credit card, virtual card, money transfer, etc. On the other hand, lending or equity-based crowdfunding platforms in the US and the UK have to get their licence from the Securities and Exchange Commission (SEC) and the

Financial Services Authority (FSA) respectively, to be able to perform their operations. Their services are approved by these institutions that are responsible for the regulation of the whole crowdfunding process (Isler, 2014).

Two main approaches of funding mechanisms have evolved to collect payments. The first and more accepted is the "all-or-nothing" model where entrepreneurs target a sum of money that must be collected within a pre-arranged period of time, and where if the target is not reached there is no financial transaction and promoters' credit card is not charged. The second, the "keep it all (keep-what-you-raise)" model stands on direct compensation where funds are paid immediately to the project and entrepreneurs keep all the money they raise even if the target is not met (Martínez-Canas et al., 2012). Although it is reported that the majority of crowdfunding platforms often operate an all-or-nothing model (Crowdfunding Final Report, 2014), some platforms also offer both fundraising mechanisms demanding to select one of them.

The main difference between two mechanisms is mainly based on whether the funding goal is reached or not in a predetermined time period. By all-or-nothing funding model the funds are returned to the funder if a funding goal is not reached, and the creator receives no funds (Gerber, Hui, & Kuo, 2013). Keep-it-all funding model allows creators to keep any amount of money they raise even if they do not meet their funding goals. If there are insufficient funds to meet the objectives, then it is up to the recipient to refund them to the contributors. Meanwhile, in both models, when the funding goal is reached, creators pay a percentage of what they raised to the platform (between 3-5%) and a payment processing fee (between 3-5%) to the web-based payment system (Hui, Greenberg, & Gerber, 2012).

All-or-nothing financing mechanism works in one of two ways: either funders are not charged until the entity reaches its fundraising target, or funds are held in a third party for a while and returned if the goal is not met. A successful example of all-or-nothing financing is Kickstarter platform. Kickstarter declares a success rate for its campaigns of 37% in 2014, substantially higher than typical fundraising efforts. They report that 81% of projects that raised more than 20% percent of their goal ultimately succeed in attaining funding, showing the significance of all-or-nothing campaigns in attracting supporters early in the process (Crowdfunding's Potential for the Developing World, 2013). Since no funds will be collected if the funding goal is not met, an over-ambitious funding goal may thus result in the fundraising effort failing. Once a project has reached its funding goal, it can continue to receive contributions until its deadline, enabling funded projects to exceed their original funding goal (Kuppuswamy & Bayus, 2014). If a project is successful, the money pledged by contributors is collected using Amazon Payments, and Kickstarter takes 5% of the funds raised, with Amazon charging an additional 3-5% for use of its payment system (Safner, 2015).

Like Kickstarter's all-or-nothing system which only transfers promised money to creators if the funding goal is met within a deadline, IndieGoGo uses also the "keep it all " funding model and transfers all pledged money upon the deadline, regardless of whether the funding goal is met (Safner, 2015). Keep it all financing allows an entity to retain whatever funds it raises by the time the campaign ends. Such funds typically are transferred immediately at the end of the fundraising campaign, typically which lasts between 30 and 90 days. IndieGoGo offers both fundraising mechanisms. They report that the probability of any campaign reaching its goal doubles once the first contribution is received, quadruples once the campaign reaches 10% of its goal, and more than quintuples once 25% of the goal is raised. For keep it all campaigns, the platform charges higher fees if the goal is not met (Crowdfunding's Potential for the Developing World, 2013).

Both Kickstarter and IndieGoGo take a part of the money collected. If a project reaches its funding goal, Kickstarter collects a 5% fee; if not, Kickstarter does not charge a fee. IndieGoGo allows project creators to draw on pledged funds immediately, whether or not the funding goal is reached, but the fee depends on whether the funding goal is met. IndieGoGo charges a 4% fee if the funding goal is reached and a 9% fee if it is not (Bradford, 2012).

The research reveals that more than 40% of all crowdfunding platforms take a commission on participation payment and charge an operation fee. Table 3 displays top ten crowdfunding platforms along with their model types, commission rates, and other related information as of August 21, 2014. Sixty-three of these platforms accept pay-pal as payment instrument and 45% require the investors to place money on third parties' escrow accounts before giving support to any project.

## The Funder Guarantees

The funders, as supporters of the crowdfunding projects are the essential actors of the crowdfunding system and need to make their investment decisions under clarity and credibility. In fact, crowdfunding has been proven to be inherently self-regulating by the characteristics and mechanisms of the model itself. The crowdfunding campaigns are evaluated by a large number of people and the wisdom of the crowd dramatically reduces the risk of unreliability. Besides, due to the short period of time between the funding and the getting the outcome of the campaign, it is possible to witness the use of the funds by the campaign owners or the promised reward. Finally, the flow of feedback via the social networks is easily spread, and by this way gives a kind of guarantee that the campaigners are under pressure to realize their responsibilities. (Crowdfunding Final Report, 2014).

The operations of all types of crowdfunding is potentially subject to different regulations in Europe, providing most of the necessary key points for member states to sufficiently regulate the sector. The Payment Services Directive may have an impact on all crowdfunding models, as most platforms manage funds of campaign backers in one way or another. The Financial Conduct Authority (FCA) regulation (2014) focuses upon the investor side of and guides the steps a platform must carry out in order to ensure

*Table 3. Top ten crowdfunding platforms, their model types, commission rates, and remarks*

| Rank | Platform | Model | Commission Rate | Remarks |
|---|---|---|---|---|
| 1 | Gofundme | Donation-based | %5 | Operation fee for participation 2.9% + $0.5 |
| 2 | KickStarter | Reward-based | %5 | Operation fee for participation %3-%5 |
| 3 | Indiegogo | Donation/Reward-based | %9 | If the funding goal met, 5% of the commission refund |
| 4 | YouCaring | Donation-based | %5 | Operation fee for participation %3-%5, outside the USA $25 |
| 5 | Causes | Donation-based | 4.75 | Operation fee for participation 2.9% + $0.5 |
| 6 | Giveforward | Donation-based | %5 | Operation fee for participation 2.9% + $0.5 |
| 7 | Patreon | Donation/Reward-based | %5 | Operation fee for participation 4% |
| 8 | Crowdrise | Donation-based | %5 | Operation fee for participation 2.9% + $0.3 |
| 9 | Fundly | Donation-based | 4.9% | Operation fee for participation 3% |
| 10 | FirstGiving | Donation/Reward-based | 7.5% | Operation fee for participation 2.5% |

Isler, 2014.

safety for those players (Crowdfunding Industry Report, 2013). Several member states such as Germany, the Netherlands, Belgium have sought to address concerns around crowdfunding with financial return through guidelines. Others like Italy, the UK, France and Spain are considering regulatory action to facilitate this new form of financing, while also aiming to adequately protect investors (Unleashing the potential of Crowdfunding in the European Union, 2014).

In most instances, the crowdfunding platforms typically do not accept any responsibility in case of failure and generally only verify whether a proposed campaign is suitable to the given platform. The complete failure of delivery in a crowdfunding campaign can be recovered by the current regulatory framework, for example the consumer protection rules can be considered to cover this risk and provide the campaign backers with sufficient tools to protect themselves from abuse or fraud. In case the company is based outside of the EU, taking action against the campaign owner may be legally and financially difficult.

Some platforms have started to assess the identity of the campaign owner to reduce the risk of fraud. Furthermore, in case of the pre-sales model, when the campaign backer orders products that are still in development progress, they should evaluate this risk properly before funding the project (Crowdfunding Final Report, 2014). Funders of pre-sales and reward crowdfunding sites are not offered any financial return; they are promised only a product or service that can be considered as a consumption item. A clear distinction is made between investment and consumption by The Supreme Court. Therefore, no investment contract is offered and because investors on reward or pre-sales sites do not present stock, notes, etc. that falls within the definition of security, federal securities law does not apply. Similarly, donation-based crowdfunding sites are not offering securities to investors. Contributors get nothing in return for their contributions, so they clearly have no expectation of profits, which is a requirement for something to be an investment contract (Bradford, 2012).

Equity-based crowdfunding usually involves securities since investors receive ordinary corporate stock in exchange for their contributions, they obviously purchase securities. Even if investors are given participation in the return of the business that does not involve corporate stock, these investments would still be securities. Contributors here have an expectation of profits or revenue sharing (Bradford, 2012). On the other hand, each type of equity-based crowdfunding comes with alternatives offering different rights to the investors. The wide array of instruments in equity-based crowdfunding is driven by differences in national regulations (Crowdfunding Final Report, 2014).

The approach to funder guarantee is more complicated for lending-based crowdfunding. The definition of security is not constrained to equity interests, it also involves some forms of debt, notes under another part of the definition. Some of the lending-based platforms, such as Kiva, do not give investors a formal note in return for their investments; others, such as Lending Club and Prosper, do. The term "note" appears in the definition of a security, but not all notes are securities (Bradford, 2012). In addition, deposit-taking for credit funding may be considered as a regulated banking activity for lending-based crowdfunding, under the Capital Requirements Directive. The complexity can be the result of different operative models. The platform is not always the party that collects the funds, nor the one that will provide the loan directly to the project owner; members may make transactions with each other directly, excluding the platform operator itself to act as a party. Furthermore, regulations regarding consumer loans may play a role again under the Consumer Credit Directive which has the main aim to guarantee that all consumers in the EU enjoy a high and equivalent level of protection of their interests and to create an open internal market (Crowdfunding Final Report, 2014).

## THE CROWDFUNDING PLATFORMS

### The Operationalization of the Platforms

The actors of the crowdfunding system are creators (entrepreneurs or initiators of the projects), funders (supporters or investors of the projects) and managers of the crowdfunding platforms (Gerber et al., 2013; Ordanini et al., 2011). The platforms which are intermediary online settings bring actors together to collect money via open call and give information about the initiatives. Creators find the opportunity to publicize and promote their projects to potential funders. With the help of the advanced technology, a myriad of people can use platforms economically and effectively acting as creators and/or funders (Kuti & Madarász, 2014). The design of the platforms' websites enables creators to launch the project easily and encourages funders to support campaign they find attractive as well as it provides communication among all actors of the crowdfunding system. Moreover, some of the platforms like Indiegogo, act as typically e-commerce website since they sell some products produced as outputs of the completed project on the platform to its members who are not necessarily to be funders in that particular campaign period.

All platforms have their own policies that are generally related with crowdfunding model(s) used and funding mechanisms including fee rates of the project. Although some of the platforms accept every initiative, others make pre-assessment of the projects depending on their own specifications. Platforms which assess ideas in advance evaluate the experiences of the creator and the availability of the business plan of the projects (Buysere et al., 2012).

### The Typology of the Platforms

The crowdfunding ecosystem comprises mainly two types of platforms: horizontal and vertical. Horizontal platforms like Kickstarter or Indiegogo offer a broad basis for all kind of initiatives. Indiegogo which was established in 2007 in the USA, serves at 24 different categories including animals, art, comic, community, dance, design, education, environment, fashion, film, food, gaming, health, music, photography, politics, religion, small business, sports, technology, theater, transmedia, video/web, and writing. Kickstarter, was founded in 2009 in the USA and has 15 categories. Horizontal platforms may serve only one kind of crowdfunding model like Kapipal. It is a donation-based crowdfunding platform to gather money for all kind of campaigns (Ordanini et al., 2011). On the other hand, there are some platforms which have enlarged their project categories over time because of their market demand. Verkami, for instance, was launched at the end of 2010 in Spain for funding art and cultural-based crowdfunding projects (Martínez-Canas et al., 2012). In time, besides artists and cultural entrepreneurs, performance organizers have started to attend to Verkami website to use it for their organizations. Therefore, crowdfunding platform of Verkami has had more extensive service categories from media development to community projects.

Vertical platforms are area, project or sector centric that attract communities with similar interest and motivations. ArtistShare, established in 2003 in the USA, can be regarded as one of the initial crowdfunding online settings (Crowdfunding Final Report, 2014). The structure of the platform is vertical since it gathers money for only music projects. In the music industry, vertical crowdfunding platforms are becoming common since they enable new and established artists to collect fund straight-forwardly from consumers for producing their albums and also fans find opportunity to be contributors in the production phase of the albums of their favorite artists (Ordanini et al., 2011). SellaBand, MyMajorCompany, Artistshare, Slicethepie, Akamusic, Megatotal, and Bandstocks can be cited as examples of crowdfunding platforms that provide connection between fans and music providers (Ordanini et al., 2011; Belleflamme et al., 2014; Galuszka & Bystrov, 2014). Artists publicize demos of their songs in the website of the platform and potential funders can freely listen to them. Fans may be supporter in the production of the album they want to fund. If the artist reaches the funding goal (e.g., $50.000 for SellaBand), he/she can get the money for producing the professional album (Belleflamme et al., 2014). Germany-based website SellaBand, one of the leading crowdfunding platform in European music industry was established in August 2006 (Ordanini et al., 2011). Up to 20 August 2015, more than 85.000 fans have become member of the platform and at least 80 artists and music bands have managed to record their albums thanks to collected money that was more than $4,000,000 through SellaBand website (www.sellaband.com). Depending on the type of the platform's model, funders may receive a particular amount of revenue of the album or some rewards like a poster of the artist or a copy of the CD (Ordanini et al., 2011; Macht & Weatherston, 2014). MegaTotal, which was established in 2007 in Poland, is an example of equity-based crowdfunding platform in the music sector (Galuszka & Bystrov, 2014). Almost 5,000 artists and 100,000 fans are the member of the platform. As of August 20, 2015, 112 projects have been supported successfully and approximately 60 albums released to the music market (http://www.megatotal.pl).

While vertical crowdfunding platforms have been commonly used in music industry, other industries like cinema, gaming, journalism, publishing, fashion, software, and sport have benefited from this kind of funding system. For instance, "Habemus Papem" and "Polisse" which were the candidates for Golden Palm prize at the Cannes Film Festival in 2011, were financed through PeopleForCinema.com., a French crowdfunding platform (Hollow, 2013). In addition to long feature films; short films, animation and even documentary films may find the possibility for actualization. Film makers not only request money for production of their film, but also for the distribution of it in the retail market (Braet, Spek, & Pauwelsa,2015). Moreover, funders without financial return expectations support journalists to search and publish their works they are interested in (Zheng, Li, Wua, & Xu, 2014; Macht & Weatherston, 2014). A crowdfunding platform which was launched in San Francisco, USA in 2008, has been used to finance projects of novice and experienced journalists. Journalists create a profile and post their stories including their plans and funding goals on the platform (Burtch, Ghose, & Wattal, 2012). After they reach the targeted amount, they publish their works on the platform's website. Fashion is another industry used by crowdfunding actors. In these kind of platforms, designers share images of their own design of products such as clothes, jewelry, shoes and if potential funders like the designed products, they invest them. The revenue from the sales of completed products are divided up among actors (Ordanini et al., 2011). In the area of software, Blender Foundation and Trampoline Systems are examples of vertical crowdfunding platforms (Belleflamme et al., 2014). Blender Foundation is a donation-based crowdfunding website that was launched in Nederland to finance public good software projects. Trampoline Systems is the first business which collect their capital via equity crowdfunding platform in the world (trampolinesystems.com). It was launched in 2009 to finance business applications and software solutions by aiming to gather £1 million from more than 100 funders (Ordanini et al., 2011). Crowdfunding system is also used in real estate industry where many investors who become the owner of a property, may receive rental income (Pyburn, 2014). Another unique vertical crowdfunding platform is MyFootballClub.co.uk which was launched in Britain in April 2007 to support professional football teams. The platform has gathered 35 pounds from almost 50,000 funders to buy Ebbsfleet United Football Club (Martínez-Canas et al., 2012). Currently, they collect 25 pounds from their members annually to support Slough Town football team of the Southern League Division 1 Central during the 2014-15 season (http://www.myfootballclub.co.uk/).

Independent of the scope, all kind of projects may be offered by both typologies, horizontal or vertical platforms. Project owners or funders pursue to enhance their experiences, community relations, returns through either of the two, considering the services offered to them.

## THE CROWDFUNDING PROJECTS

### Operational Dynamics of Projects

Crowdfunding projects are launched through campaign webpage on the crowdfunding platforms. The process in the crowdfunding campaign has several steps which are illustrated in Figure 1. The first step is preparation of crowdfunding project which includes also testing the campaign materials. In this phase the owner of the project describes clearly the aim of the project, its achievable funding goal and funding levels such as $1, $10, $50, $100, $500, and $1,000, the number of funders per level and the rewards/values promised to offer. Higher funding levels offer bigger rewards/values. Kuppuswamy and Bayus (2014) state that a lot of projects become unsuccessful due to their inappropriate funding goal. Also, it is found that attractive rewards play a critical role in the crowdfunding technology related project that becomes successful (Crowdfunding Final Report, 2014). In the beginning phase of the campaign, the creator prepares presentation material/video including explanation of how he/she will be successful in the project and spend the money, makes a business plan and specify deadlines of the relevant works.

The second step is publishing the project in the platform page aim of which is attracting the attention of the potential funders and spread the word about the work. The creator may benefit from the social media instruments to access wider audience in a shorter time. The third step is managing the project through evaluating comments and feedbacks of the potential funders, answering their questions, controlling the discussions on the platform page and giving information about the project improvements. Informing the (potential) funders is vital while conducting the campaign since it is revealed that success ratio of the project with updates is 58.7% whereas the success possibility of the project without updates is 32.6% (Creative Funding for Creative Media, 2014). In addition, potential funders desire to support the project that may become real, thus, in order to persuade them the creator need to be transparent. The last step is producing and delivering the promised values such as thank you message, rewards (i.e. a t-shirt, a poster, early edition of the product), and in some cases experiences and knowledge of the creators in the platform page. Hui et al. (2014) state that creators spend up to 7 hours a day like a full time employee on the crowdfunding project.

*Figure 1. Steps of the crowdfunding process*

## Typology of Projects by Models

Characteristics of the project and crowdfunding model compatibility is a key point to achieve success in the crowdfunding system. All types of models are not suitable for all kind of initiative categories such as SME, entrepreneurship, technology, art, film, music, publishing, media, real estate and social cause. Also, the amount of funding goal, the typology of the creator and potential funders of the venture play critical roles to determine the crowdfunding model used. Nonfinancial and financial crowdfunding models are frequently used for projects with different characteristics.

Donation-based crowdfunding model is typically used for social cause-related, medical, environmental, educational, public, journalism and religious campaigns by individuals, charities and other non-profit organizations (Buysere et al., 2012; Bradford, 2012; Creative Funding for Creative Media, 2014). Moreover, creative activities, personal beliefs and concerns like targeting to make scientific research about particular topic are subjects of donation-based model (Mitra, 2012; Buysere et al., 2012). There is a few pure donation platforms and donation-based projects are generally presented in the same platform with the reward-based ones like Indiegogo (Bradford, 2012). Similar to the donation-based campaigns, civic crowdfunding projects which do not include revenue expectation, aim to create common good assets both in higher level like bridges or buildings and in small range such as gardens or parks for underserved areas (Davies, 2014; Stiver et al., 2015). Moreover, some public services (i.e. training program or providing wireless internet network) might be outputs of the civic crowdfunding campaigns (Stiver et al., 2015). Governmental agencies like municipalities, private and quasi-public companies, non-profit organizations and individuals might initiate civic crowdfunding projects and such projects are generally launched in metropolitan cities (Davies, 2015; Davies, 2014). As another nonfinancial crowdfunding model, reward-based projects are typically used for creative campaigns especially in the music, film, gaming, publishing, fashion and other kind of artistic areas, technology-related entrepreneurship business activities like producing innovative consumer products and developing media campaigns to generate customized outcomes (Buysere et al., 2012; Safner, 2015; Creative Funding for Creative Media, 2014). Reward-based model is useful to test market demand of an innovative product (Buysere et al., 2012) and to spread out new ideas even concerned by niche audience. Generally, donation and reward-based models are suitable for smaller funding needed campaigns (Braet et al., 2015). Thirty-five percent of the campaigns in these kinds of models financed total amount of less than \$2500 and 28% of them funded with the amount of between \$2500-\$5000 in 2012 (Isler, 2014). Kickstarter and Indiegogo, which are the most popular crowdfunding platforms around the world, operate based on these models.

As for financial crowdfunding models, lending-based projects focus on SMEs, business extension, production, engineering and technology-related activities (Crowdfunding Industry Report, 2013). The level of financial credibility of the business is critical for this type of model so it is not suitable for start-ups (Ahlers et al., 2012). Lending-based crowdfunding projects are used as alternatively to conventional lending system for the companies with inconsistent expense and income period which desire to pursue sustainability in their activities (Buysere et al., 2012). Similar to the lending-based approach, equity based model is suitable for SMEs and initiatives in the industries of software programs, gaming, film, music, and publishing (Mitra, 2012). Unlike lending-based model, equity-based crowdfunding model is appropriate for innovative start-ups (Buysere et al., 2012). Moreover, food, beverage, consumer products, telecommunication and technology companies and e-commerce firms are benefited from this kind of funding mechanism (Crowdfunding Industry Report, 2013). It is reported that 31% of the projects on Crowdcube, equity-based crowdfunding platform in the UK, are technology related (Crowdfunding Final Report, 2014).

SMEs which are very critical for economic growth and employment creation for countries, collect money through lending and equity-based crowdfunding model in all stage of their business cycle. Figure 2 presents the data about the tendency of using the type of financial crowdfunding models by stage of the SMEs (Crowdfunding Industry Report, 2013).

Whereas the majority of the lending-based crowdfunding model users support established SMEs (with 4-8 years trading history) at 55%, users of equity-based model are typically young SME's (57%). Surprisingly, none of the established SMEs with more than 8 years trading background is benefited from equity-based crowdfunding model. Moreover, it is found that 90% of average funding amount for SMEs related financial crowdfunding models is more than £1000 per individual (Crowdfunding Industry Report, 2013).

Media, as another industry example using all kind of crowdsourcing models, has different amounts of funding by models. According to Massolution Industry Report (2013), market size of projects in media category are €143,085 for equity-based model and €3,540 for lending-based one. As for nonfinancial models, the distribution of the funding amount in media sector is similar and less in reward-based and donation based crowdfunding models as €1,732 and €1,054, respectively. When all categories are taken into account based on the same report, nearly 50% of the reward based projects are reported successful, whereas the highest success rate of the campaign belongs to lending-based model with 90%. Donation and equity-based crowdfunding campaigns have similar success rates which are about 45% and 43%

*Figure 2. Financial crowdfunding models usage rate by stage of the SMEs*
*Crowdfunding Industry Report, 2013.*

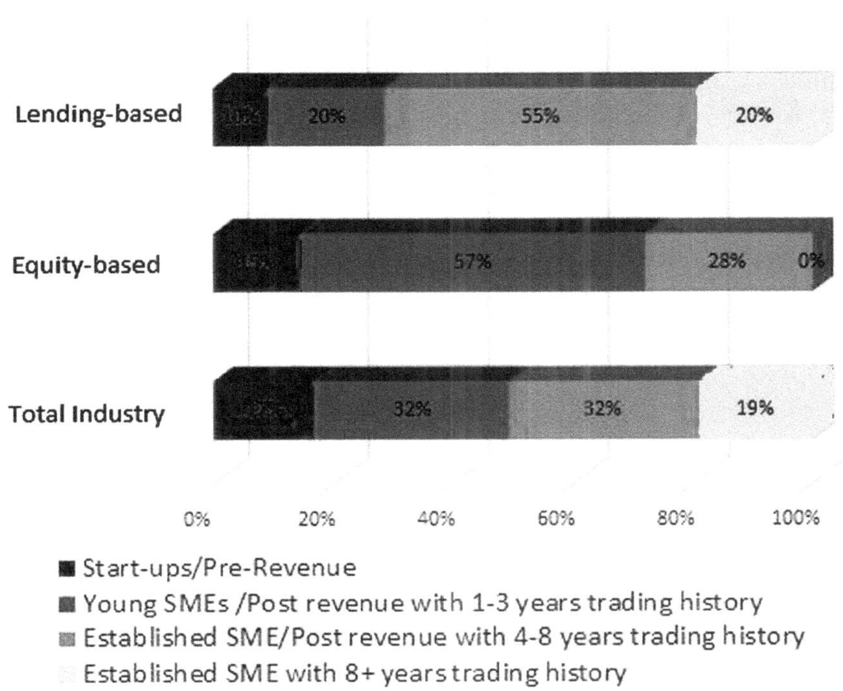

respectively (Massolution Industry Report, 2013). Finally, success rate of the civic crowdfunding project is around 63% (Davies, 2014).

The type of crowdfunding model could be determined by the amount of funding goal and the life stage of the enterprise (shown in Figure 3). Donation-based crowdfunding is generally used in idea development phase with a small amount of funding requirement. Lending (debt) and equity-based crowdfunding models are effective at idea actualization and relatively large amount of capital is needed at this phase.

## Categories of Projects by Platforms

The decision of the crowdfunding platform is critical for success of the project (Crowdfunding Final Report, 2014). The proper platform should be selected by the project creator based on the category and features of the project, type and characteristics of the platform and location (Creative Funding for Creative Media, 2014). Creators should take their potential market and funders into account and evaluate carefully the feasibility of the platforms for their projects. In this section, horizontal and vertical crowdfunding platforms are evaluated separately based on the project nature.

Horizontal crowdfunding platforms like Kickstarter, Indiegogo, and RocketHub have generally been used to launch "creative projects" to finance with return as a form of reward or without return (Gerber et al., 2013). Through Kickstarter, the world's largest crowdfunding platform, 22,252 creative projects have been actualized in 2014 (www.kickstarter.com). Five projects on Kickstarter are described in 25 best inventions of 2014 by TIME. Charity campaign and general business cost projects which are not defined as 'creative' cannot be launched on Kickstarter (Kuppuswamy & Bayus, 2014). In 2014, 3.3 million members from almost every country in the world have pledged approximately $529 million meaning the generation of $1,000 in a minute (www.kickstarter.com). There are 15 different categories

*Figure 3. Crowdfunding type based on the enterprise life stages*
*Crowdfunding's Potential for the Developing World, 2013.*

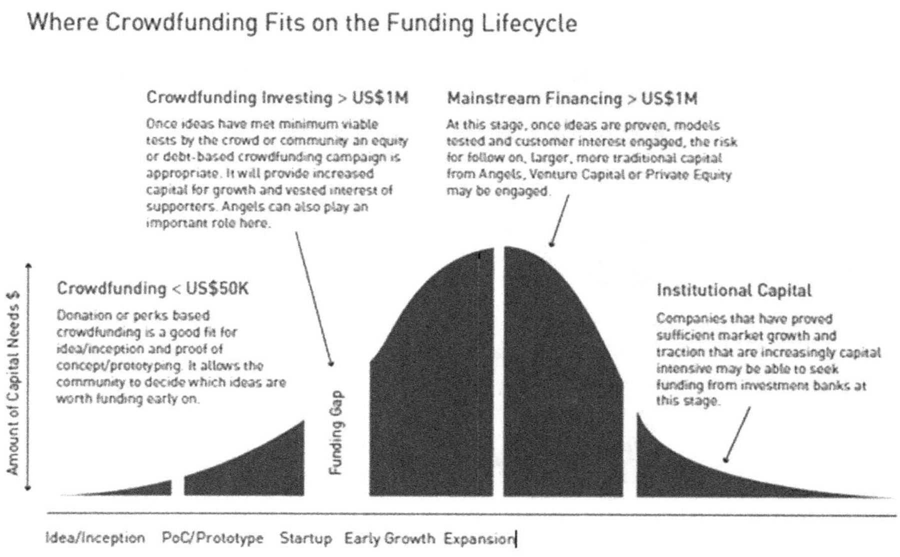

on Kickstarter and projects successfully funded and total pledged amount in 2014 by category are illustrated in Figure 4 (www.kickstarter.com).

Music, film, and publishing constitute top three categories. It is important to note that crafts and journalism were included only in the mid-2014. Technology, design and games are the most popular categories based on the total amount of fund pledged in 2014 on the Kickstarter platform. In addition, it is revealed that artistic categories on the Kickstarter collected more money than the National Endowment for the Arts in 2013 (Safner, 2015). Category of the projects also have an impact on success rate of the projects (Crowdfunding Final Report, 2014). Since its foundation, Kickstarter have almost 9.3 million total funders who have pledged more than $1.9 billion to support 251,609 projects. Only 91,256 projects successfully have been funded. Average success rate for all projects is 37% on Kickstarter in 2014 (www.kickstarter.com). In addition, it is found that more than 90% of successfully funded projects on Kickstarter are managed to sustain their activities at least a year after the campaign period and almost 30% of them have income annually about $100,000 (Kuti & Madarász, 2014). Figure 5 shows the number of launched projects and the success rate by category as of August 25, 2015 (www.kickstarter.com). Music, publishing, art and games are the most popular launched project categories.

Figure 5 shows that success rates vary across the categories. Dance, theatre, music, and comics have the highest success rates with 64%, 61%, 51%, and 50%, respectively. However, technology, journalism, crafts and fashion is relatively underperforming, but, vertical or other horizontal crowdfunding platforms may have different success rate by categories (Crowdfunding Final Report, 2014). Although success rate is low in the category of technology compared to the other categories, the amount of the gathered money of successfully completed projects, $282 million, is comparatively high. Similarly, the success rate of technology-related campaign is lower than the success rate of average campaign on Indiegogo since the users of world's leading crowdfunding platforms have focused on artistic campaigns like music, art, and theater instead of technological projects (Crowdfunding Final Report, 2014).

Successful campaigns in all categories can be conducted through both horizontal and vertical platforms. Concerning the horizontal platforms, one of the well-known example is the film project called Veronica Mars which has been launched in March 2013 on Kickstarter. Indeed, Veronica Mars was a famous TV

*Figure 4. Kickstarter projects successfully funded in 2014 by category*

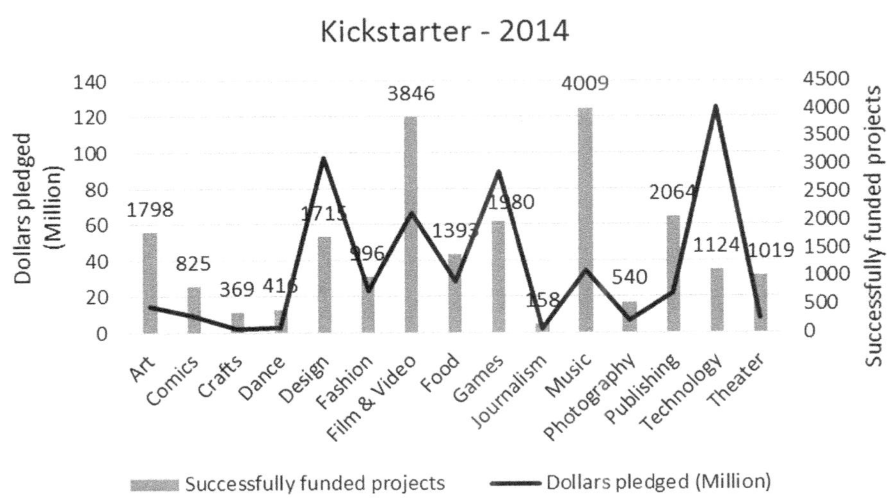

*Figure 5. Kickstarter, all projects funded by category*

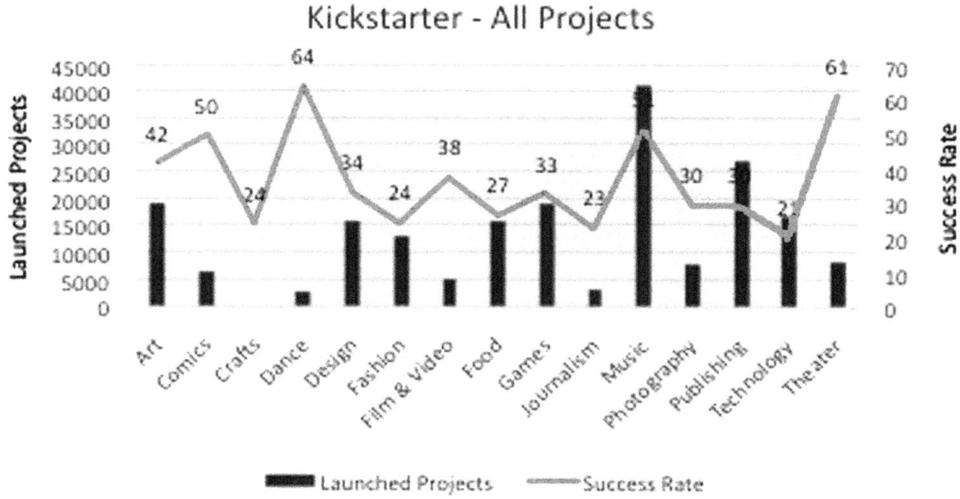


serial which was featured between 2004 and 2007. Rob Thomas who is the creator and producer of the program wanted to show Warner Brothers, owner of the property rights, the intensity of interest of the fans towards the program (Bennett, Chin, & Jones, 2014). Although the funding goal was $2 million, after the 30-day campaign period, $5.7 million have been gathered. Another successful project has been conducted on Kickstarter by musician Amanda Palmer in 2012. Although her funding goal was $100,000, she had reached about $1.2 million from almost 25,000 funders (Davidson & Poor, 2015). In some cases, a project which is launched in horizontal platforms, may transfer to another crowdfunding platform or to a specific independent platform created for the project itself after it becomes very successful. Star Citizen, the most successful funded crowdfunding project of all times, is an example of such project. Star Citizen, which is a video game project was first launched on Kickstarter with a $500,000 funding goal and raised $6,238,563. After its success, Chris Roberts, the owner of the project has created his own website on November 2012. Two years later, the project was funded with a total amount of $65,300,000 and was registered as a Guinness World Record (http://www.guinnessworldrecords.com/). As of August 25, 2015 over $87 billion has been collected from more than 967,000 funders. (https://robertsspaceindustries.com/).

Although horizontal platforms have extensive market, vertical crowdfunding platforms may offer actors some additional advantages about their projects. For instance, Myfootballclub gathers 25 pounds from their funders annually to finance a football team and also to vote for deciding which uniforms the players wear to, who will be the coach and which players the team will transfer through the platform (Martínez-Canas et al., 2012). Innovestment, as an equity-based unique crowdfunding platform in Germany, enables creators and funders to determine the price of the shares by using comparative bidding method (Bretschneider et al., 2014). The platform focuses technology related projects and the lowest funding amount is 1,000 €. The funders give their offers to identify the number of shares they want to buy with a certain amount of money. During the campaign period, the funders may increase their bids and at the end the highest offers get the shares (Bretschneider et al., 2014). In addition, although horizontal crowdfunding platforms have much more volume compared to vertical ones, vertical crowdfunding platforms might cause to increase the success rate of the project because of its specific market focus (Mollick,

2014; Crowdfunding Final Report, 2014). While the vast majority of the civic crowdfunding projects are conducted on horizontal type platforms, the number of vertical civic crowdfunding platforms are increased (Davies, 2014). The vertical civic crowdfunding platforms have unique characteristics such as involving agencies like municipalities or organizations instead of crowd are allowed to launch projects on the platforms of Neighbor.ly and Spacehive for example (Davies, 2015). Stiver et al. (2015) state that civic crowdfunding platforms have some different implementation to increase success rate of the projects thanks to their unique nature of the project scope. For instance, the civic crowdfunding projects are offered to fund gradual "phased" by the vertical crowdfunding platform of Citizinvestor to rise the possibility of finishing the campaign successfully. Spacehive, another crowdfunding platform related to civic projects, accepts "in-kind" aids and count them as financially for project funding goal. Moreover, vertical type crowdfunding projects have become more attractive than horizontal ones because of their lower fees. For example, while Kickstarter charges 5% of project total fee for successfully completed projects, the rate of fee in Spacehive is 3.7% (Stiver et al., 2015).

Another important factor to take into account about platform selection related to the increasing the success rate of the project is geographical criterion (Mollick, 2014; Crowdfunding Final Report, 2014). Location-centered campaign should be launched on a geographically suitable platform. Moreover, it is revealed that the geography of the creator might play a role on the success of the project (Mollick, 2014). Stiver et al. (2015) indicate that all vertical civic crowdfunding platforms are local and also the civic model projects on the horizontal crowdfunding platforms are generally supported by local funders.

## The Time Frame of Projects

Similar to other financial institutions, how long it takes to access the required capital is a notable topic for any crowdfunding system. Figure 6 illustrates the time period by models in 2012 (crowdsourcing.org cited in Isler, 2014). Whereas nonfinancial return crowdfunding models including donation-based and reward-based have similar duration from the beginning to the end of the campaigns which is about 10 weeks, financial return crowdfunding models that are lending-based and equity-based have shorter time frame for project campaign, like 4.8 and 8.2 weeks, respectively.

Especially in lending-based crowdfunding projects, almost one-month long is very fast compared to other models to actualize an idea. Moreover, the average crowdfunding project campaign takes around two months. On the other hand, it is reported that before the project is launched, approximately three to six months are required for identifying characteristics of the project including deciding the model of the project and the platform, funding goal, rewards/values promised and making business plan and preparing campaign materials like making a video (Hui et al, 2014).

Number of funders and the amount of funding of the projects vary during the campaign period. It is found that funders tend to make more contribution to the campaign both at the beginning and shortly before the end of the project rather than at the middle period (Marom & Sade, 2014). One possible reason of the increase in the number of funders and pledged amount in the first and last weeks of the campaign is the contribution of close social circles of the creators (Kuppuswamy & Bayus, 2014). In 2011, Kickstarter shortened the maximum campaign period from 3 to 2 months due to the decrease in quantity of the supporters in the mid-stage of the funding cycle (Kuppuswamy & Bayus, 2014). Figure 7 shows the average number of funders acquired per week for both successful and failed projects for the 3-year period starting from 2009 on Kickstarter.

*Figure 6. Time frame of projects by models in 2012*
*Crowdsourcing.org cited in Isler, 2014, p. 12.*

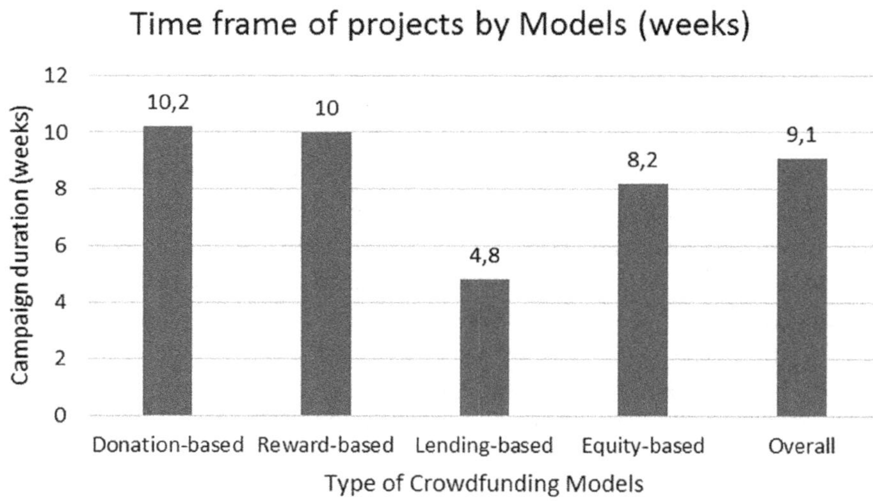


*Figure 7. Acquired funders by week for successful and failed projects*
*Marom & Sade, 2014.*

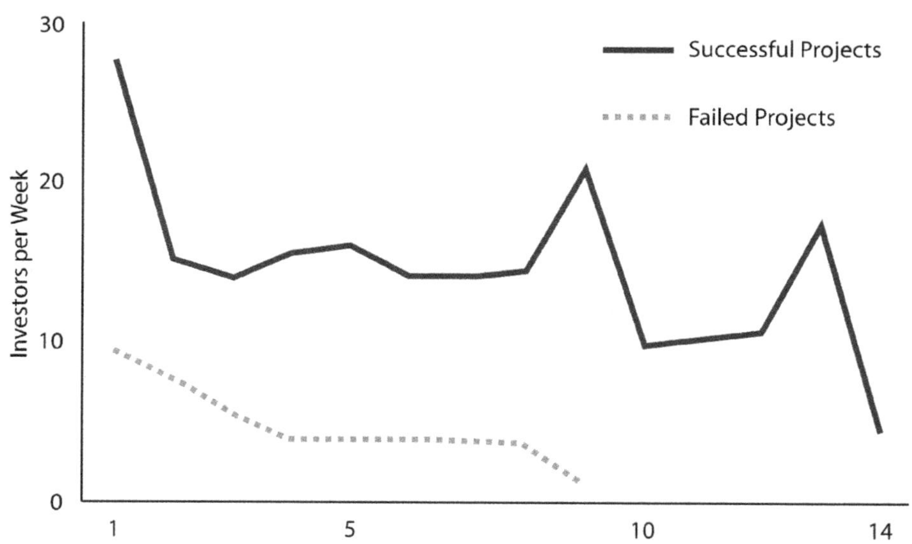


In line with the previous literature, the number of supporters reach its peak at the beginning of the campaign for both successful and failed projects. It can be said that the quantity of the failed campaign contributors is approximately one third of the ones in successful projects. In addition, Kuppuswamy and Bayus (2014) state that trends of contributions are directly proportionate to updates during the campaign period. Moreover, it is critical to know in which period projects attract more funders during the funding cycle, since in some cases creators of successful campaigns are benefiting from crowdfunding system more than one time for the same project (Buysere et al., 2012). This valuable information might be used in the second round of funding.

## THE CROWDFUNDING MARKET

According to the Massolution's 2015 CF (Crowdfunding) Industry Report, the global crowdfunding markets have increased from $6.1 billion in 2013 to $16.2 billion by 167% growth in 2014. It is forecasted that total global funding volumes would reach to $34.4 billion in 2015 (Massolution's 2015 CF Industry Report, 2015). The crowdfunding market is growing considerably in recent years and is expected to grow further as new ways of doing business along with state-of-the art projects arise in digital economy and new means are required to support such innovative projects. The crowdfunding system which is ground-breaking itself will continue to realize its own potential as well as to offer improvements through its differing models and platforms.

### General Extent of the Market

The crowdfunding mechanism has become an alternative funding system to banking industry, venture capital and angel investments. Also it has a potential to encourage innovation, entrepreneurship, and job creation by facilitating capital access and leveraging social network (Milosevic, 2015). Therefore, crowdfunding mechanism offers social and economic development by influencing employment and geographical capital market.

#### Employment Created

Crowdfunding system is an enormous opportunity based on its low-risk nature for start-ups and SMEs to provide capital. There are approximately 23 million SMEs in Europe and about 10 million of them have trouble to find money from traditional financial sector for their business and try to use their social relations to gather capital (Milosevic, 2015). As reported by European Commission, SMEs which constitute 99% of the business, create around 80% of new jobs and also according to Kauffman Foundation, all new jobs have been created in the USA by the young companies' age of those below 5 years old for almost 30 years (Buysere et al., 2012). Crowdfunding, as an alternative financing method, offers chances to fulfill the gaps in the financial sector and to enable start-ups and SME's to cope with capital access problem in a feasible manner. In 2012, after the 2008 financial crisis-period, SMEs were funded about $100 million via donation and reward-based models, almost $650 million via lending-based model and around $40 million via equity-based model of crowdfunding activities (Meric et al., 2015). In line with the objective of The Entrepreneurship 2020 Action Plan, the crowdfunding system stimulates entrepreneur activities to rise area of employment in Europe (Crowdfunding Final Report, 2014). For example, it is revealed that almost 7500 new jobs were created in Spain through about 2800 crowdfunding projects in 2012 (Unleashing the potential of Crowdfunding in the European Union, 2014). Moreover, The Digital Single Market strategy aims to regulate using the digital technology online activities for individuals and business across Europe until the end of 2016. The Digital Single Market paves the way for improvement in the level of employment for young companies and SMEs in the market with 500 million people and it will provide € 415 billion contribution to European economy in every year (http://ec.europa.eu/digital-agenda/en/digital-single-market).

Users of crowdfunding system are heterogeneous (Boh & Goh, 2014) and they are typically not professional investors (Ordanini et al., 2011; Macht & Weatherston, 2014). Approximately 31% of the funders support more than one project on Kickstarter as of August 24, 2015 (www.kickstarter.com). As reported by the research of Massolution (2012), only 10% of the funders and 30% of the creators involve more than two crowdfunding projects (Buysere et al., 2012). It is expected that in the near future, crowdfunding platforms attract professional investors such as angel investors (Ordanini et al., 2011; Macht & Weatherston, 2014). There is a few research about the typology of the crowdfunding actors. Gerber et al. (2013) state that users of the crowdfunding system are generally less than 35 years old, childless and with a college educated and earn at least $30,000 in a year. Another study reveals that members of the crowdfunding platforms are married and middle-aged and have a high education degree and medium or high socio economic status (Schweinbacher & Larralde, 2010). However, actors of the civic crowdfunding platforms have unique characteristics that might be municipalities, government entities, private or non-profit organizations, and institutions like universities (Davies, 2015).

## Geographical Distribution

Although crowdfunding market contributes economic growth in the worldwide year by year, it is much more widespread in North America and Europe compared to other regions (Massolution's 2013 CF Industry Report, 2013). Whereas crowdfunding market size in North America rose $1.6 billion in growth of 105%, European crowdfunding volume increased $945 million by 65% growth rate in 2012 (Massolution's 2013 CF Industry Report, 2013). USA has the largest crowdfunding market size in every year compared to the other regions. In Europe, as the second largest crowdfunding market, approximately 470,000 projects were successfully funded in 2012 (Unleashing the potential of Crowdfunding in the European Union, 2014).

According to the Massolution's 2015 Crowdfunding Report (Massolution's 2015 CF Industry Report, 2015) in 2014, North America keeps going on its regional market leadership by the funding volume of $9.46 billion with the 145% growth rate. While, Asia rose to the second rank in the regional market list with the 320% growth rate by the amount of $3.4 billion and Europe became the third one and only reached $3.26 billion funding size with the 141% growth rate. Regional crowdfunding market growth rates of South America, Oceania and Africa are 167%, 59% and 101%, respectively (Massolution's 2015 CF Industry Report, 2015). As shown in Table 4 the number of crowdfunding platforms are parallel to the market shares of various countries.

Concerning the geographic dispersion, although there are many platforms which serve locally, popular platforms such as RocketHub and IndieGoGo operate globally. On the other hand, while projects on Kickstarter cannot be initiated by the US residents due to tax procedure, there is no residence requirement for funders (Kuppuswamy & Bayus, 2014). In 2014, the vast majority of the funders (72%) on Kickstarter were from the US (www.kickstarter.com). On the other hand, civic crowdfunding projects differentiate from other type of projects due to its location-based characteristic. Although the civic campaigns generally focus on urban locations, the platforms are established in developed areas (Davies, 2015).

*Table 4. The number of crowdfunding platforms in selected countries in 2013*

| Country | Platforms | Country | Platforms |
|---|---|---|---|
| United States | 344 | Brazil | 17 |
| United Kingdom | 87 | Australia | 12 |
| France | 53 | India | 10 |
| Netherlands | 34 | South Africa | 4 |
| Canada | 34 | Russia | 4 |
| Germany | 26 | | |

Crowdfunding's Potential for the Developing World, 2013.

## Extent of Market by Crowdfunding Models

The crowdfunding market size is rapidly increasing by models with different growth rates year by year. Also, the structure of the models evolves and new types of crowdfunding models have been recently generated. For example, a comparison of the 2012 and 2014 is made in Figure 8 based on the crowdfunding market size and the ratio of percentage change is given (Massolution's 2013 CF Industry Report, 2013; Massolution 2015 CF Report, 2015).

In 2012, donation and reward-based crowdfunding have reached $1.4 billion with an 85% increase, lending-based crowdfunding which had the largest growing rate by 111%, has risen over $1.2 billion and equity-based crowdfunding relatively had minor market size as $116 million by growing 30% (Massolution's 2013 CF Industry Report, 2013). On the other hand, the dominant crowdfunding volume in 2014 comprises of lending based-model with the amount of $11.08 billion –with a growth of 223% (Massolution 2015 CF Report, 2015). The report presents the growth rate and market size information of equity-based crowdfunding into two subcategories including equity-based crowdfunding and royalty-

*Figure 8. Extent of market by crowdfunding models*
*Massolution's 2013 CF Industry Report, 2013; Massolution 2015 CF Report, 2015.*

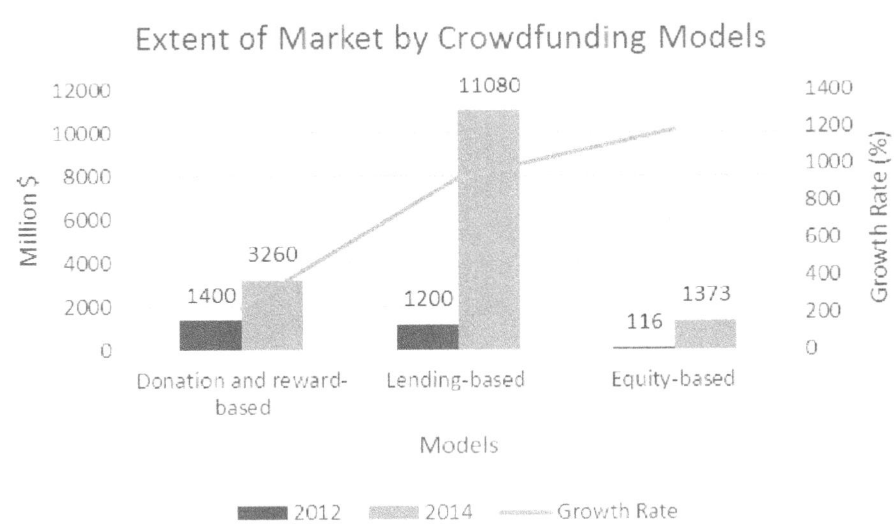

based crowdfunding. Royalty-based crowdfunding, which offers funders to receive a portion of revenue from the enterprises, is commonly used for business expansion projects like recording a CD through Sellaband.com (Beaulieu et al., 2015). Whereas equity-based crowdfunding relatively has a larger market size with $1.1 billion by 182% in 2014, royalty-based crowdfunding is rapidly rising to $273 million with an increase of 336%. In addition, hybrid-based crowdfunding in which multiple models are combined (Buysere et al., 2012) reaches $487 million by 290% increase (Massolution 2015 CF Report, 2015).

It is notable that between the years of 2008 and 2012, lending-based market has increased two times annually while the average annual growth rate of market size of this model in the US and UK is 79.1% and 99.5%, respectively. (Kirby & Worner, 2014). Moreover, the distribution of financial crowdfunding market including lending and equity-based by country in 2013 is presented in Figure 9.

The figure reveals that the US, the UK and China totally constitute 96% of the market where the US has approximately half of the market at 51%, followed by China with a 28% of market volume and the UK, as the third one, with 17% of market size.

## Extent of Market by Crowdfunding Platforms

According to the Massolution's 2015 CF (Crowdfunding) Industry Report, 1250 crowdfunding platforms are operating in the world and it is expected that global funding volumes would be $34.4 billion in 2015 (Massolution's 2015 CF Industry Report, 2015). One of the main reasons of this growth in the number of platforms could be attributed to the powerful touch of such settings to the crowd. Although the outcome is related with raising most money or creating a viable business, in most cases these platforms work as a perfect medium of validation of demand with their flexible and innovative characteristics. For instance, in 2014, $529 million in total was pledged on Kickstarter, the largest crowdfunding platform in the world, which is equivalent of 3.2% of total amount of crowdfunding market volume. Platforms may perform within their essential structure as explained at the very beginning of the chapter, or hybrid types can also be offered, but all serve to increase the market size of crowdfunding activities through enabling financial or nonfinancial support for various projects. For instance, some of the major lending-

*Figure 9. Financial crowdfunding market by country in 2013*
*Kirby & Worner, 2014.*

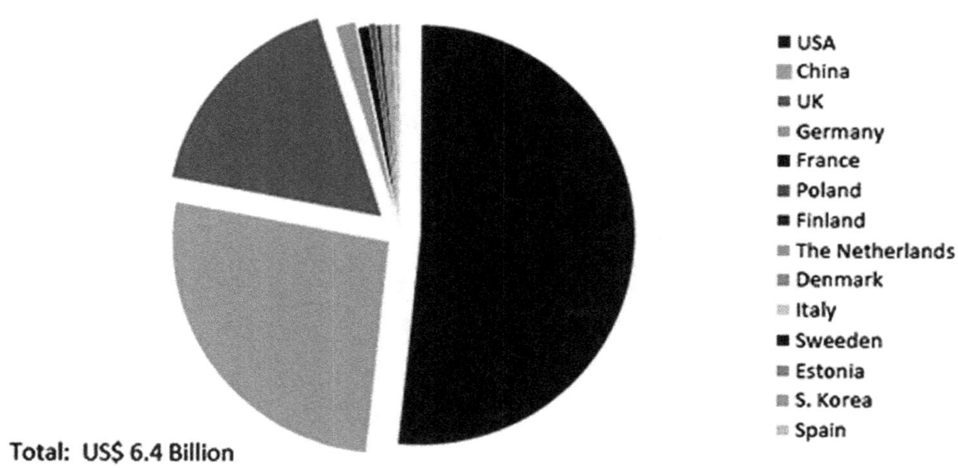

based and equity-based platforms and their characteristics including inception date, market size (as of September 30, 2013) and country of domicile are shown in Table 5 (Kirby & Worner, 2014). According to the table, it is stated that Prosper and Lending Club, two leading lending-based crowdfunding platforms dominate the US market. In the UK, Zopa is the largest lending-based platform which hold 50% of the market volume and Funding Circle and RateSetter are other major players. Also, CreditEase, the leading lending-based platform in China, have the $1.6 billion market size which constitute 27% of the market volume in the world (Kirby & Worner, 2014).

In addition, Figure 10 represents the top equity-based crowdfunding platforms (indicated country of domicile) listed based on the total estimated funding amount (million Euro) by the end of 2013. It can be said that UK has the vast majority of the equity-based crowdfunding market volume at 41%, followed by Germany and France, with 26% and 20% respectively.

*Table 5. The major lending-based and equity-based platforms and their characteristics*

| Company | Date of Inception | Market Size ($ Millions) | Country of Domicile | Crowdfunding Model |
|---|---|---|---|---|
| Lending Club | 2007 | 2,563 | USA | Lending-based |
| CreditEase | 2006 | 1,600 | China | Lending-based |
| Zopa | 2006 | 618 | UK | Lending-based |
| Prosper | 2006 | 612 | USA | Lending-based |
| Funding Circle | 2010 | 251 | UK | Lending-based |
| RateSetter | 2010 | 183 | UK | Lending-based |
| Auxmoney | 2007 | 87 | Germany | Lending-based |
| Pre d'Union | 2009 | 57 | France | Lending-based |
| AngelCrunch | 2011 | 40 | China | Equity-based |

Kirby & Worner, 2014.

*Figure 10. Estimated amount of equity raise by end of 2013 (in million Euro)*
*Crowdfunding Final Report, 2014.*

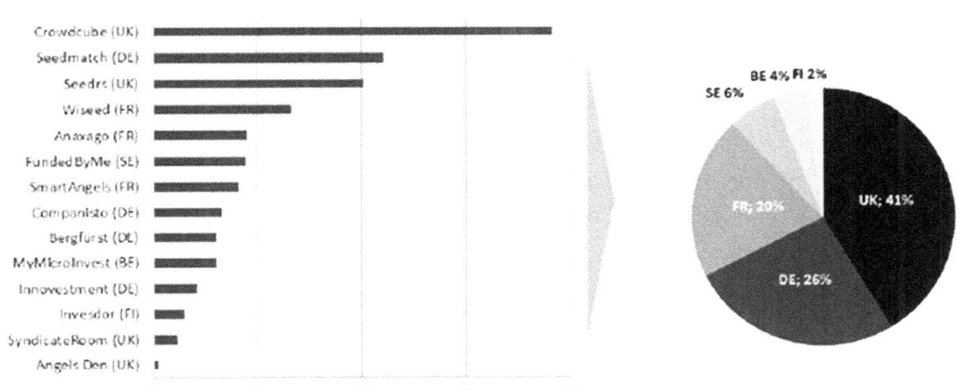

The significant growth in the number of crowdfunding platforms and market size between the years of 2010 and 2014 is illustrated in Table 6.

For the last five years, the quantity of the crowdfunding platforms and the total crowdfunding market volume has increased with an accelerating performance in a sustainable manner and it seems that this trend will continue in the future due to evolving financial system and legal regulations around the world.

## FUTURE TRENDS OF CROWDFUNDING

The crowdfunding market is considerably growing along with offering diversifying platforms with rich examples of projects through its dynamic platforms. This growth is especially apparent in the past few years and the trends are valuable to analyze in order to shed light to future developments in the field since a large worldwide potential for crowdinvesting is expected to materialize in the next decade (Crowdfunding Final Report, 2014).

### General Market: Trends

Global crowdfunding has witnessed accelerated growth in 2014, expanding by 167% to reach $16.2 billion raised, up from $6.1 billion in 2013 which was largely due to faster growth than predicted in Asia. In 2015, the industry is set to more than double once again, on its way to raising $34.4 billion (Massolution's 2015 CF Industry Report, 2015). In terms of the general market data, North America still accounts for the largest market, growing by 145% and raising a total of $9.46 billion, but 2014 saw Asia surpassing Europe. With exponential growth in Asia, Massolution forecasts this lead will increase significantly in 2015 increasing to over $4 billion. It is predicted that Europe's 20.1% of market share in 2014 will decline slightly in 2015 when Europe will account for 18.8% of the worldwide market (Massolution's 2015 CF Industry Report, 2015). Another observable trend is about surprising growth in the P2P lending market in Asia, caused largely by Chinese market. On the other hand, the top five crowdfunding platforms in North America increased their market share during the period 2011-14 while in Europe the top five lost 30% of their market share, which shows that the markets in North America and Europe develop in different ways (Massolution's 2015 CF Industry Report, 2015).

*Table 6. Growing size and platforms of crowdfunding market*

| Year | Platforms | Market Size (Billion $) |
|---|---|---|
| 2010 | 283 | 0.8 |
| 2011 | 434 | 1.5 |
| 2012 | 536 | 2.7 |
| 2013 | 630 | 6.1 |
| 2014 | 1250 | 16.2 |

Isler, 2014; Massolution's 2015 CF Industry Report, 2015.

Thus the strong growth in 2014 was due in part to the rise of Asia as a major crowdfunding region. Although this makes the region slightly ahead of Europe as the second-biggest region by crowdfunding volume, Europe is still powerful in equity-based crowdfunding and has a very brilliant past record by generating 55% of all successful crowdfunding campaigns (Hallow, 2013).

The crowdfunding market is still in progress in developing countries, but there is a real potential as it is shown in Figure 11. The estimated 344 million households in the developing world are potential customers of small crowdfund investments. Along with their income of at least US$10,000 a year, and at least three months of savings it can be predicted that they have the ability to use up to US$96 billion a year by 2025 in crowdfunding investments. The figure clearly presents that the greatest potential is in China, which accounts for up to US$50 billion of that figure, followed by the rest of East Asia, Central Europe, Latin America/the Caribbean, and the MENA region (Crowdfunding's Potential for the Developing World, 2013).

In relation with above explanations, and with the support of related regulation and other enabling factors, the total market potential by 2025 is estimated to be up to US$90-96 billion per year. Although China is seen to have significant potential, there are critical issues with intellectual property, user expectations, and limitations on securities markets that may make China's potential size much smaller (Crowdfunding's Potential for the Developing World, 2013).

The future trends indicate that the crowdfunding activities will target toward cross border markets. However, basic accepted standards are needed to make this development healthy in international markets. (Kirby & Worner, 2014). The completion of the Digital Single Market (DSM) as identified among the 10 political priorities of European Commission could have important impact in the evolution of the market around the world.

*Figure 11. Market potential for crowdfunding activities*
*Crowdfunding's Potential for the Developing World, 2013.*

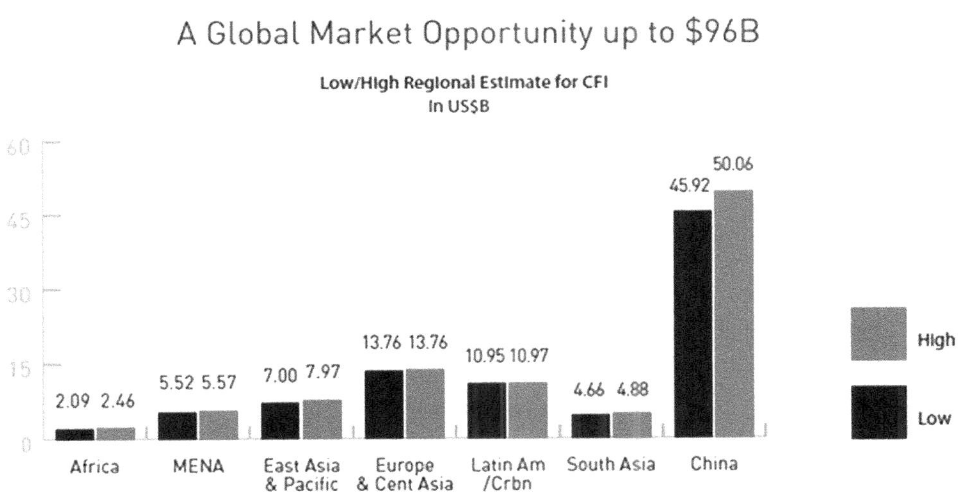

## Models: Trends

The crowdfunding market is expected to grow across all models over the next decade: The World Bank predicts that the total crowdinvesting market potential will reach EUR 70 billion per year by 2025, excluding North America. This simply shows that crowdfunding models can be used easily to support start-ups and established businesses and can become an alternative solution for early stage financing for several projects (Crowdfunding Final Report, 2014).

In terms of general data, most of the crowdfunding projects were using reward-based model as of 2012, although equity-based platforms were representing the fastest-growing category (http://www.crowdsourcing.org/) (Mitra, 2012). It is evident that crowdfunding originated as a platform to generate funds in the creative and social sectors in exchange for nonmonetary rewards (Hemer, 2011).

Although the reward-based model maintains its popularity, the equity-based model is currently expected to gain power and bypass the reward-based one especially after the introduction of the JOBS Act. The main motivation to incline towards equity-based platforms is based on the assumption that expectancy of investors to get better financial rewards from their investment (Cholakova & Clarysse, 2015). Therefore, the famous Kickstarter using reward model, aspire for the recent legalization of equity crowdfunding, with many of its features written into the JOBS Act (Mollick, 2014). In fact, by the JOBS Act, equity-based crowdfunding has become a legal alternative for early-stage equity investments and an important tool for raising seed capital (Cholakova & Clarysse, 2015).

Equity crowdfunding may be one of the most favorable ways to increase small business growth. The cross-border development of this platform is subject to harmonization of regulations. The failure of a cross-border platform could raise insecurity that is why this will need to be addressed in the future if this industry is to be able to grow into a viable source of investment in international aspect. (Kirby & Worner, 2014)

On the other hand, when 2014 data show that rewards- and equity-based campaigns typically get the most popularity, however lending-based crowdfunding model dominates the industry; $11.08 billion dollars are collected using this model at this period (Massolution's 2015 CF Industry Report, 2015). The strong growth of lending-based model has been observed in Asia especially in P2P and P2B lending markets (Massolution's 2015 CF Industry Report, 2015).

Finally, civic crowdfunding's potential seems underdeveloped, however there is a rising trend in the number of projects on civic crowdfunding platforms. For instance, Indiegogo has hosted several projects under "community" label, fitting with the civic crowdfunding concept. Kickstarter has also hosted civic projects, such as the LowLine, a project aiming to turn an abandoned trolley terminal in New York into a public park. The UK-based Spacehive has funded £1 million of civic projects since going live in 2012. The US-based ioby has successfully funded 282 civic crowdfunding projects and raised US$732,011 since its beta launch in 2009 (Stiver et al., 2015).

National regulatory frameworks will undoubtedly affect market development of crowdfunding models. Currently, Northern and Western Europe own the most refined regulations. To be able to extend internationally, different national regulatory frameworks should be analyzed to see if crowdfunding models are applicable in each market they aim to operate in (Crowdfunding Final Report, 2014).

## Platforms: Trends

The increasing global interest in crowdfunding is evident along with the development and spread of its platforms. The penetration and acceptance of some crowdfunding platforms are related with the social, cultural and economic progress in different countries. For instance, Finland's Joukkoenkeli manages co-creation between individuals, public officials, private companies, and NGOs, whereas the recently launched platform Crowdera supports both non-profit organizations and individuals with civic projects around the globe (Stiver et al., 2015). iFundIT was able to organize a crowdfunding trial for a 5500-person within IBM (Muller, Geyer, Soule, & Wafer, 2014).

In recent years, another development is about the emergence of angel investor networks and the potential of equity-based crowdfunding platforms functioning as a means to involve angels in a more systematic way -e.g. Angels Den (UK)-. Another example showing the potential of crowdfunding can be found in Belgium where business angel organizations serve as the main investors in equity-based crowdfunding campaigns on MyMicroInvest. Other business angel groups utilize equity-based crowdfunding platforms to manage closed rounds for their members in the process of evaluating and deciding to invest for a project. Therefore, equity-based platforms may serve as an important source for early stage financing especially for SMEs and in some countries the authorities have implemented measures to support related operations which have certainly a considerable impact on the rise of equity-based crowdfunding (Crowdfunding Final Report, 2014).

Research reveal that the need for EU action on financial return crowdfunding is important. Market access and investor protection appear to be equally critical; for instance 51% of the participants of a survey reported a need to ensure appropriate investor protection for lending and securities crowdfunding. Respondents also emphasized the risks for contributors such as loss of invested capital, risk of fraud in lending-based platforms, as well as the need for transparency on platform charges, interest rates, and on the expected yields. The survey report presents that 81% of non-financial returns-based platforms already operate cross-border, while 14% would like to do so in the future; 44% of all financial returns-based platforms declare the lack of information about applicable rules preventing them from operating in more than one country; 27% of them refer to the high costs of getting an authorization in another member state (Unleashing the potential of Crowdfunding in the European Union, 2014).

It is critical to inform contributors of investment and credit risks and provide advice on investment diversification in order to make the full usage of crowdfunding platforms since potential contributors might not get involved in crowdfunding activity because of such concerns over risks. The main issues appear as different investment limits in different countries, the shareholder voting rights, the rules concerning transparency, investment suitability, client money protection and investment advice (Unleashing the potential of Crowdfunding in the European Union, 2014).

## CONCLUSION

Although crowdfunding market is very recent thus juvenile to establish its own dynamics, many transformations seem to be on the way. At first, crowdfunding system will be more internationalized because of the rising number of global crowdfunding platforms and new legislative arrangements to minimize geographic barriers. Globalism enables platforms to increase their market size and to have more potential actors with differing roles, including creators/initiators/borrower/investees and funders/supporters/lenders/investors and less operational costs. On the other hand, legislative structures of the countries vary in Europe, therefore, preparing common laws might cause problems among the members of the Union. National platforms represent localization which is crucial for domestic social connection and identification to the campaign. In recent time, European equity-based crowdfunding settings have expanded geographically their service areas to across all Europe (e.g.UK-based crowdfunding platform Seedrs started to operate in other European countries.). From the funders' perspective, regional expansion empowers investors to connect with more qualified enterprises. Likely, lending-based approach has similar tendency, for instance, FundingCircle from the UK, has started to operate in the US (Crowdfunding Final Report, 2014). Moreover, not only in Europe but also in the other regions efforts are made to arrange laws for equity-based crowdfunding. Although the US market is aiming to complete legislations at the federal level, all states have published their own regulations due to the latency. Equity-based platforms are operated in Australia, Japan, and Korea thanks to the existing legislative framework, but China has not a particular arrangement concerning this issue.

Crowdfunding will not only draw individual entrepreneurs but also business firms, NGOs, universities, local and central governments. Crowdfunding activities inside the company have potential to become widespread because of their advantages. This kind of approach expands the level of creation inside the company and fosters technology-related improvement. As a result, creative projects can be produced via this intermediary settings and workers may donate fund if they desire to support the projects and this climate. IBM, which is the most successful company used this system, has given a quota to each workers to support their favorite projects. The decision making mechanism has changed from upper levels to lower levels in the organizational hierarchy and this leads to the increase in efficiency of the business activities. Also, collectivistic approach has been adopted and new business networks have risen between the departments.

Along with companies, it becomes ordinary to see universities and not-for-profit organizations launching committed crowdfunding platforms on present third-party platforms or even their own website. The so-called 'do-It-yourself crowdfunding' shows the company's commitment to social responsibility, controlling the online traffic from the corporate website and improve the performance of crowdfunding campaigns by this way, and most importantly functioning as an outstanding marketing means. Universities engage in crowdfunding activities, not because they target direct commercial purposes, but because they aim to raise funds for their own research projects and to help their associate companies to grow by attracting funding. (Crowdfunding Final Report, 2014).

An increase in the amount of vertical platforms will be observed. For all types of crowdfunding models, it is expected that the smaller platforms will become much more specialized and dedicated in order to distinguish among all types of prominent mainstream platforms. The most effective ways for platforms to differentiate themselves could be driven by a very strong local or geographical focus, or by enhancing their offering by including crowdsourcing of nonfinancial elements specific to their market. Seed&Spark (US) is such a specialized platform which focuses entirely on the independent movie industry; it presents not only the donation funding instrument but also the opportunity to lend movie production materials to the film maker (Crowdfunding Final Report, 2014).

On the other hand, it may be expected that in the mid-term the crowdfunding industry can be consolidated especially based on the equity and lending-based platforms, but with the platforms carrying the most interesting campaigns and investors, driving competitors with smaller platforms out of the market (Crowdfunding Final Report, 2014).

Conventional or long-established sectors will open up to crowdfunding and disintermediation of existing sectors will be observed. These trends can be expected especially for the businesses supported by equity-based models. The crowdfunding campaigns will be diversifying more and more into traditional market sectors, beyond investment in technology start-ups. Secondly, a great potential lies is in the disintermediation of current industries. For instance, the real estate industry is witnessing a boost in campaigns in recent years; campaigns were delivered through both specific platforms and generalist platforms as well. Large real estate agencies or developers have decided to engage in campaign management with their own crowdfunding models. Industry experts claim that the insurance industry will be the following to generate new market segments (Crowdfunding Final Report, 2014).

Another prevailing trend will be closely related with the economic development level of the communities as well as the social context they are tied within. The entrepreneurship is encouraged in many institutions as it is seen as an engine of economic growth.) The dynamics of crowdsourcing could help to find ways to create both new ventures in cultures where cooperation and teamwork are embedded in social structure (Beugre & Das, 2013). Specific crowdfunding models or the hybrid ones could contribute to economic development by increasing entrepreneurial activities initiated and managed by the crowd.

## REFERENCES

*About MyFootballClub*. (n.d.). Retrieved August 20, 2015 from http://www.myfootballclub.co.uk/about-myfootballclub

Ahlers, G. K. C., Cumming, D. J., Guenther, C., & Schweizer, D. (2012). *Signaling in equity crowdfunding*. Retrieved May 15, 2015, from http://ssrn.com/abstract=2161587

Belleflamme, P., Lambert, T., & Schwienbacher, A. (2014). Crowdfunding: Tapping the right crowd. *Journal of Business Venturing*, *29*(5), 585–609. doi:10.1016/j.jbusvent.2013.07.003

Bennett, L., Chin, B., & Jones, B. (2014). Crowdfunding. *New Media & Society*, 1–8.

Beugré, C. D., & Das, N. (2013). Limited capital and new venture creation in emerging economies: A model of crowd-capitalism. *SAM Advanced Management Journal.*, *78*(3), 21–27.

Bradford, C. S. (2012). Crowdfunding and the Federal Securities Laws. *Columbia Business Law Review*, *1*, 1–150.

Braet, O., Spek, S., & Pauwels, C. (2013). Crowdfunding the movies: A Business Analysis of Crowdfinanced Moviemaking in Small Geographical Markets. *Journal of Media Business Studies*, *10*(1), 1–23. doi:10.1080/16522354.2013.11073557

Bretschneider, U., Knaub, K., & Wieck, E. (2014). Motivations for crowdfunding: What drives the crowd to invest in start-ups?*Twenty Second European Conference on Information Systems.*

Burtch, G., Ghose, A., & Wattal, S. (2012). *An empirical examination of the antecedents and consequences of investment patterns in crowd-funded Markets*. Retrieved May 15, 2015, from http://www.krannert.purdue.edu/academics/mis/workshop/2012/crowdfunding%20journalism.pdf

Buysere, K. D., Gajda, O., Kleverlaan, R., & Marom, D. (2012). *A framework for European crowdfunding*. Retrieved May 14, 2015, from www.crowdfundingframework.eu

Cholakova, M., & Clarysse, B. (2015, January). Does the possibility to make equity investments in crowdfunding projects crowd out reward-based investments?. *Entrepreneurship Theory and Practice,* 145-172.

Creative Funding for Creative Media. (2014). *Future Media Lab. & European Crowdfunding Network.* Retrieved June 14, 2015, from http://www.eurocrowd.org/wp-content/blogs.dir/12/files/2014/01/FML-ECN-EMMA-crowdfunding-guide-final.pdf

*Crowdfunding Industry Report*. (2013). Retrieved May 14, 2015, http://www.buzzacott.co.uk/getattachment/a4e47539-1038-40c9-b90d-8709e3f4e4f/crowdfunding-industry-report-1.pdf.aspx

Crowdfunding's Potential for the Developing World. (2013). infoDev, Finance and Private Sector Development Department. Washington, DC: World Bank.

Davidson, R., & Poor, N. (2015). The barriers facing artists' use of crowdfunding platforms: Personality, emotional labor, and going to the well one too many times. *New Media & Society*, *17*(2), 289–307. doi:10.1177/1461444814558916

Davies, R. (2014). *Civic crowdfunding as a marketplace for participation in urban development*. Center for Work, Technology and Organizations, Stanford University. Retrieved June 18, 2015, from http://ipp.oii.ox.ac.uk/sites/ipp/files/documents/IPP2014_Davies.pdf

Davies, R. (2015). Three provocations for civic crowdfunding. *Information Communication and Society*, *18*(3), 342–355. doi:10.1080/1369118X.2014.989878

*Fan Funded Music*. (n.d.). Retrieved July 12, 2015, from https://www.sellaband.com/en/pages/about_us

Galuszka, P., & Bystov, V. (2014). The rise of fanvestors: A study of a crowdfunding community. *First Monday*, *19*(5). doi:10.5210/fm.v19i5.4117

Gerber, E. M., Hui, J. S., & Kuo, P. Y. (2013). *Crowdfunding: Why people are motivated to post and fund Projects on crowdfunding platforms*. Retrieved May 16, 2015, from http://www.juliehui.org/wp-content/uploads/2013/04/CSCW_Crowdfunding_Final.pdf

*Guinness World Records 2015 : Most Crowdfunded Projects*. (n.d.). Retrieved August 25, 2015, from http://www.guinnessworldrecords.com/2015/preview/files/assets/basic-html/page17.html

Hemer, J. (2011). *A snapshot on crowdfunding*. Working papers firms and region, No. R2/2011. Retrieved May 16, 2015, from http://hdl.handle.net/10419/52302

Hollow, M. (2013). Crowdfunding and civic society in Europe: A Profitable partnership? *Open Citizenship*, *4*, 68–73.

Hui, J. S., Gerber, E. M., & Greenberg, M. D. (2012). Easy money? The demands of crowdfunding work. *Northwestern University Segal Design Institute Technical Report*, *12*, 4.

Hui, J. S., Greenberg, M. D., & Gerber, E. M. (2014). Understanding the role of community in crowdfunding work. In *Proceedings of the 17th ACM conference on Computer supported cooperative work & social computing*, (pp. 62-74). ACM.

Isler, S. T. (2014). Modern finansmanda yeni bir dönem: Kitlesel fonlama – Crowdfunding. *Gösterge*. Retrieved May 10, 2015, from http://www.tspb.org.tr/tr/Portals/0/aim__gosterge_guz_2014_02.pdf

*Kickstarter 2014 : By The Numbers*. (n.d.). Retrieved August 25, 2015, from https://www.kickstarter.com/year/2014/data?ref=yir2014

Kirby, E., & Worner, S. (2014). *Crowdfunding: An infant industry growing fast*. Staff Working Paper of the International Organisation of Securities Commissions Research Department. Retrieved May 14, 2015, from http://www.memofin.fr/uploads/library/pdf/Crowd-funding-An-Infant-Industry-Growing-Fast[1].pdf

Kuppuswamy, V., & Bayus, B. L. (2014). *Crowdfunding Creative Ideas: The Dynamics of Project Backers in Kickstarter*. UNC Kenan-Flagler Research Paper. Retrieved June 27, 2015, from http://funginstitute.berkeley.edu/wp-content/uploads/2013/11/Crowdfunding_Creative_Ideas.pdf

Kuti, M., & Madarász, G. (2014). Crowdfunding. *Public Finance Quarterly*, *3*, 355–366.

Lin, Y., Boh, W. F., & Goh, K. H. (2014). How different are crowdfunders? Examining archetypes of crowdfunders and their choice of projects. *Organizational Communication & Information Systems*, *1*, 13309.

Macht, S. A., & Weatherston, J. (2014). The benefits of online crowdfunding for fund-seeking business ventures. *Strategic Change*, *23*(1-2), 1–14. doi:10.1002/jsc.1955

Marom, D., & Sade, O. (2014). *Are the life and death of an early stage venture indeed in the power of the tongue? Lessons from online crowdfunding pitches*. Retrieved May 28, 2015, from http://www.researchgate.net/publication/256060198

Martinez-Canas, R., Ruiz-Palomino, P., & Pozo-Rubio, R. (2012). crowdfunding and social networks in the music industry: Implications for entrepreneurship. *International Business & Economics Research Journal*, *11*(13), 1471.

*Massolution's 2013 CF Industry Report*. (2013). Retrieved May 15, 2015, from http://www.compromisoempresarial.com/wp-content/uploads/137356857-Massolution-2013CF-Excerpt-Revised-04182.pdf

*Massolution's 2015 CF Industry Report*. (2015). Retrieved May 10, 2015, from http://www.crowdsourcing.org/editorial/global-crowdfunding-market-to-reach-344b-in-2015-predicts-massolutions-2015cf-industry-report/45376

Meric, J., Bouiass, K., & Maque, I. (2015). More than three's a crowd... in the best interest of companies!: Crowdfunding as zeitgeist or ideology? *Society and Business Review*, *10*(1), 23–39. doi:10.1108/SBR-09-2014-0042

Milosevic, M. (2015). *Crowdfunding as an innovation booster factor and its dependence on information systems*. Retrieved June 10, 2015, from https://hal-paris1.archives-ouvertes.fr/hal-01158238/document

Mitra, D. (2012). The role of crowdfunding in entrepreneurial finance. *Delhi Business Review*, *13*(2), 67–72.

Mollick, E. (2014). The dynamics of crowdfunding: An exploratory study. *Journal of Business Venturing*, *29*(1), 1–16. doi:10.1016/j.jbusvent.2013.06.005

Muller, M., Geyer, W., Soule, T., & Wafer, J. (2014). Geographical and organizational commonalities in enterprise crowdfunding. CSCW Technology and Information Workers.

Ordanini, A., Miceli, L., Pizzetti, M., & Parasuraman, A. (2011). Crowdfunding: Transforming customers into investors through innovative service platforms. *Journal of Service Management*, *22*(4), 443–470. doi:10.1108/09564231111155079

Pyburn, G. (2014). *Crowdfunding (Equity-Based) In Asia-Pacific: Legal and Regulatory Ambiguities without an Analogous us Jobs Act Exemption*. Academic Press.

Report, C. F. Regulatory Framework in EU Member States and Perspectives for the EU. (2014). *European Commission*. Retrieved July 10, 2015, from http://capacity4dev.ec.europa.eu/crowdfunding/document/crowdfunding-regulatory-framework-eu-member-states-and-perspectives-eu-en-version

Review of Crowdfunding Regulation. (2014). *Interpretations of existing regulation concerning crowdfunding in Europe, North America and Israel.* European Crowdfunding Network. Retrieved May 25, 2015, from http://www.eurocrowd.org/files/2014/12/ECN-Review-of-Crowdfunding-Regulation-2014.pdf

Safner, R. (2015). *Crowdfunding expressive works, crowding out copyright?*. Job Market Paper.

Schwienbacher, A., & Larralde, B. (2010). Crowdfunding of small entrepreneurial ventures. In *Handbook of Entrepreneurial Finance*. Oxford University Press.

*Star Citizen: Stats*. (n.d.). Retrieved August 25, 2015, from https://robertsspaceindustries.com/funding-goals

Stiver, A., Barroca, L., Minocha, S., Richards, M., & Roberts, D. (2015). Civic crowdfunding research: Challenges, opportunities, and future agenda. *New Media & Society*, *17*(2), 249–271. doi:10.1177/1461444814558914

*The whole truth about MegaTotal.pl*. (n.d.). Retrieved July 12, 2015, from http://www.megatotal.pl/pl/page/about

*The year in Kickstarter 2014*. (n.d.). Retrieved August 25, 2015 from https://www.kickstarter.com/year/2014?ref=2014datayir#stats-4

*Trampoline Systems*. (n.d.). Retrieved June 15, 2015, from http://crowdfunding.trampolinesystems.com/index.html

*Unleashing the potential of Crowdfunding in the European Union*. (2014). European Commission. Retrieved June 9, 2015, from http://ec.europa.eu/internal_market/finances/docs/crowdfunding/140327-communication_en.pdf

Van Veelen, T. (2015). *Civic crowdfunding: A financial kickstart to urban area development? An analysis of the institutional structure in which civic crowdfunding in urban area development is embedded.* Retrieved June 27, 2015, from http://repository.tudelft.nl/view/ir/uuid%3Abf8caf4f-79a8-4257-ac88-948518f5cc36/

Zheng, H., Li, D., Wua, J., & Xu, Y. (2014). The role of multidimensional social capital in crowdfunding: A comparative study in China and US. *Information & Management*, *51*(4), 488–496. doi:10.1016/j.im.2014.03.003

# Chapter 7
# Horizontal Crowdfunding Platforms:
## Socio-Economic Impact and Exogenous/ Endogenous System of Rules

**Tomaso Greco**
*Università degli Studi dell'Insubria, Italy*

## ABSTRACT

*In recent years arose a multitude of crowdfunding platforms of different nature and with different goals. In this chapter some peculiar aspects of crowdfunding platforms will be examined, both in terms of their socio-economic impact and in terms of the exogenous and endogenous system of rules with which they interact. Crowdfunding is not governed exclusively by the rules of law and the regulations of the platforms that host it. Every campaign involves the construction of a community, with its specific rules, that are in part pre-established by the campaign creator and in part the result of the interaction of the community. However, these rules do not live in a closed set, building a case history to which the future campaigns will not necessarily have to conform, but with high probability they will use as a parameter. Although it is not possible to identify the rules that are common to all the crowdfunding campaigns, could be observed an evolution mostly harmonic between different campaigns with respect to the methods, the mechanisms of involvement and, of course, the rules.*

## INTRODUCTION

In recent years we have witnessed the spread of crowdfunding platforms of different nature and with different goals. In this chapter we will examine some peculiar aspects of crowdfunding platforms, both in terms of their socio-economic impact and in terms of the exogenous and endogenous system of rules with which they interact.

DOI: 10.4018/978-1-5225-0568-6.ch007

## A Major Distinction between Crowdfunding Platforms

Firstly, we can make a macro distinction with respect to the goals of the platforms, identifying three main categories: equity crowdfunding, donation crowdfunding and reward crowdfunding.

The first one is aimed at collecting investments around a business plan, recognizing to the (micro) investor a participation in the company. They have proven to be particularly useful for business in the start-up phase, a phase in which it is of vital importance to raise the necessary funds to cover the initial expenses. Furthermore, the road of equity crowdfunding does not necessarily represent an alternative road to traditional methods of raising venture capital. It rather belongs to an earlier stage, leaving the door open to further rounds of financing when the company is "more mature" and, therefore, more attractive for the investors (whether they are business angels or private equity funds, etc.).

However, we will not dwell here in a specific manner on the system of rules of equity crowdfunding, for at least three different reasons. The first reason is the lesser extent - both in absolute terms and in geographical terms - of equity crowdfunding with respect to other forms of crowdfunding, while the second reason is related to the intimate relationship between the rules of the crowdfunding platforms and the set of rules and structures that govern the corporate participation in the different Countries, and the third one is that several countries – as, for example, the United States - are structuring rules and regulations specifically for equity crowdfunding and investor protection, a regulatory process of great interest, but which will necessarily need some time to produce results and to influence the large-scale spreading of this funding model. Here - postponing to a more specific discussion the analysis of critical aspects and perspectives of equity crowdfunding - we will just hope for a system of shared rules regarding equity crowdfunding that allows to collect micro loans from different parts of the globe, without creating many small or large regional sandboxes that would weaken the characteristic of globality (to which we will return later) of the funds collected through the network, and the solidaristic-interested nature (we will also return to this important aspect) of crowdfunding.

## Donation Crowdfunding Platforms and Reward Crowdfunding Platforms: Focus and Funding Models

Therefore, only the donation crowdfunding platforms and the reward crowdfunding platforms will be at the center of our investigation. More than of platforms it would be correct to speak of crowdfunding methods because it often happens that the same platform hosts both the donation campaigns and the reward campaigns, perhaps applying different commissions to the campaign creators (as for example the case of Indiegogo that through Indiegogo Life applies a commission of 0% to the personal fundraising campaigns and of 5% to the normal reward campaigns).

However, as above-mentioned, we will analyze these methods in relation to the attitude of the platforms to create communities and to the system of rules that they apply. In fact, despite the most obvious distinction between the two methods is tautological, that is the reward campaigns offer a reward, usually equated to the value of the offer, while the donation campaigns offer rewards that are purely symbolic - and, therefore, not comparable to the value of the offer - or they do not offer material rewards, limiting to consider as reward the same implementation of the project and, of course, the moral gratification for having participated in the campaign, the relationship between reward and donation will be the subject of a further in-depth analysis.

In fact, we cannot just accept a schematic and rigid separation between the two methods. An aggregate of realities, which are very different between them, hides behind the expression donation crowdfunding: civic crowdfunding or crowdfunding based on geographical and emotional proximity, philanthropic crowdfunding and crowdfunding as a substitute for welfare. These three methods have in common, in addition to the information technology technicalities, a redistribution mechanism of individual wealth on a voluntary basis. But there are also several points of difference. Civic crowdfunding or crowdfunding based on geographical and emotional proximity, which can be an instrument activated by the public administration in order to carry out a project to promote the area and/or the fabric of the city, but which can also be seen as the initiative of one or more individuals aimed at overcoming the shortcomings (no matter whether political or economic) of the administration or at acting as a complementary activity with respect to the administration in a spirit of subsidiarity, necessarily brings with it the above-mentioned element of interested solidarity that both the proponent (campaign creator) and the backers have in common. This is the case, for example, if we consider the expansion or maintenance of a museum structure (it is possible to make the example, for all cases, of the use of crowdfunding by the Louvre in Paris for the conservation of its "Winged Victory of Samothrace" and for the acquisition of Lucas Cranach's "The Three Graces"), the implementation of a bike sharing service (this is the case of Kansas City) or of public works (as in the case of the +Poll campaign, an initiative to bring a floating swimming pool to the East River, on the Manhattan/Brooklyn banks, in New York City).

The case of those who finance a hospital in a distant area of the world (philanthropic crowdfunding) or who help to finance medical care or studies of an unknown person (crowdfunding as a substitute for welfare) is different. In these last two assumptions, in fact, the impact in the sphere of the campaign creator and the impact in the sphere of the backer do not coincide at all, the first one having a direct-material interest and the latter an ethical interest. Not even the objection that the backers might help to have a beneficial effect of psychological and/or social type, thing which is common to all types of philanthropy, including also the traditional ones, is sufficient to close the significant gap asymmetry of the interests of campaign creator and backers.

Therefore, this premise that for reasons of space fails to consider other experiences of sure interest and spread as the lending crowdfunding and the loan crowdfunding, as well as other forms of micro-patronage, is necessary to define the perimeter of the discourse relating to an in-depth analysis: reward crowdfunding platforms and, to some extent, donation crowdfunding platforms that have the characteristics (at least potentially) of the global fundraising and a solidaristic-interested nature prevailing in the proposed campaigns. To these requirements it should be added the horizontality of the platform, that is the availability of the same to host campaigns aimed at financing products or activities of different nature (we will return to the vertical and specialized ones in another chapter of the same volume). This additional requirement is aimed at identifying and describing the most possible homogeneous set of rules that is applied,

The homogeneity of the rules needs inevitably not to consider jointly, in this comparative analysis, platforms that apparently provide the same service, but that basically aggregate communities and sets which are profoundly different between them.

This does not mean that the generalist platforms do not aggregate communities, quite the opposite (approximately 12% of the campaign creators of Kickstarter launched more than a campaign - Kickstarter data March 24, 2015 - and 29% of the backers finance more than one project, for a total of 60% of the funds raised - Kickstarter data October 23, 2013), but it rather means that some communities know interactive methods that are very different with respect to the vertical and territorial platforms. But, as it has been said, we will return to this interesting aspect later.

A main difference between the philanthropic oriented campaigns and the campaigns designed to achieve a service or product is the nature generally keep-it-all of the first ones and all-or-nothing of the second ones (some reward platforms, as Kickstarter, do not allow keep-it-all campaigns, while where both types of campaigns are permitted, as Indiegogo, relatively to the reward campaigns the percentage is 95% of all-or-nothing, against only 5% of keep-it-all. Data referred to the period 2008-2013). This aspect indicates a real disparity, which is not limited, as above-mentioned, to the different spirit that animates the campaign creator and backers, but it reaches the deep relational mechanisms underlying the campaign itself and, more specifically, it helps to determine the set of rules of the campaign.

## THE SYSTEM OF RULES OF THE CROWDFUNDING PLATFORMS

In fact, the cardinal points of a campaign are: project, campaign creator's profile, duration and economic objective to be achieved. In the campaigns that we will consider these points represent a rigid set, a parallelepiped with pre-defined angles, precisely because of the first rule (which is, therefore, the real challenge of the campaign): that is to collect a certain amount of money in a given time frame. If the pre-determined figure is not reached, the all-or-nothing mechanism means that none of the offers is actually collected, due to the rule, fixed on the basis of a more or less arbitrary evaluation of the campaign creator on the minimum amount of funds required to implement the project, an arbitrary evaluation but that becomes a shared rule through the participation of backers to the campaign. Mr. Blue (the campaign creator) can assume that he can build the time machine if he collects $100,000, but there is nobody who certifies that this is the case. It may not be feasible at all, or feasible only with figures much higher or lower, or feasible at the current level of science and technology, but not with the skills of Mr. Blue. Mr. White, and Miss Orange and the other backers, participating in the campaign of Mr. Blue, decided not only to declare their interest in the project (and to participate economically to its implementation), but also to accept the premise on which Mr. Blue has based the project itself: what for the rest of the world can remain assumptions, becomes a real contract between the backers and the campaign creator. But this contract is subject to the achievement of the pre-determined economic objective. If it will not be possible to reach $100,000, the money of the backers will not be withdrawn and Mr. Blue will not receive it, thus dissolving the moral and legal obligation to build the time machine.

In particular, from the legal point of view and not only, the rules involved in crowdfunding of reward type are those that regulate generally pre-sale contracts: a deposit is paid for the purchase of a product not yet implemented, the contract will be finalized upon the occurrence of a pre-determined condition (the achievement of the economic objective according to schedule) and this will implicate the obligation for the backers to make the promised payment (but here there is the intervention of the main online payment methods, adopted by almost all platforms, which will make automatic and immediate this step at the positive end of the crowdfunding campaign) and, the most important thing, they will allow the campaign creator to implement the work.

In this way, both the campaign creator and the backers have to observe at least three distinct sets of rules: the exogenous one that regulates the agreements subject to a suspensive condition, and those that are endogenous to the crowdfunding platform, between which we can distinguish a set of general rules, valid for all campaigns hosted on a certain platform, and a set of specific rules, valid for the single campaign.

They are all three sets of pre-determined rules, in the sense that even before the campaign starts they were explicit in their basic parts and, in good logic, will not change during the campaign. Relatively to the specific set of rules, for example, it is not possible to lengthen the duration of the campaign (and neither turn an all-or-nothing campaign into a keep-it-all campaign), nor to modify the economic objective nor to diminish the scope of the project.

It must be said that with respect to these parameters the platforms apply rules that can be defined passive at the beginning and rigid during the running phase of the campaign.

Generally, in fact, to accept or not a campaign the platforms do not evaluate neither the feasibility of the project, nor the attitude of the campaign creator to implement it, nor whether if the target set for the implementation is reasonable. They just make sure that the limiting criteria are respected, within which the campaign creator has great autonomy. But this autonomy is rigidly limited at the moment when the campaign becomes public and enters the fund collection phase.

In fact, the "forbidden campaigns" or not accepted campaigns are those that fall in a general definition with large approximation for Indiegogo:

*Campaign Owners are not permitted to create a Campaign to raise funds for illegal activities, to cause harm to people or property, or to scam others. If the Campaign is claiming to do the impossible or it's just plain phony, don't post it. Users must comply with all applicable laws and regulations in connection with their Campaigns, including offering Perks and using Contributions. Campaign Owners shall not make any false or misleading statements in connection with their Campaigns. (Indiegogo Terms of Use, 2015)*

while Kickstarter, after specifying "Kickstarter does not allow projects to fundraise for charity or offer financial incentives" (Kickstarter Creator Questions, 2015) prepares a list of activities and materials that are not allowed on the platform:

*Item claiming to cure, treat, or prevent an illness or condition [..], contests, coupons, gambling, and raffles, energy food and drinks, offensive material, political fundraising, pornographic material, resale, drugs, nicotine, tobacco, vaporizers and related paraphernalia, weapons, replicas of weapons, and weapon accessories. (Kickstarter Prohibited Items, 2015)*

To these it adds some rewards that may not be offered, even if they were related to accepted campaigns:

*... genetically modified organism, alcohol, financial, money-processing, or credit services; financial intermediaries or cash-equivalent instruments; travel services (e.g., vacation packages); phone services (e.g., prepaid phone services, 900 numbers); and business marketing services. (Kickstarter Prohibited Items)*

But it should be said that Indiegogo also integrates the above-mentioned rules in relation to the allowed campaigns with rules on rewards, and in particular:

*... any form of security (as such term is defined in the Securities Act of 1933), any form of financial incentive or participation in any profit sharing, any alcoholic consumer products, any controlled substance or drug paraphernalia, any weapons, ammunition and related accessories, any form of lottery or gambling, any form of air transportation, any items promoting hate, discrimination, personal injury,*

*death, damage, or destruction to property; or any items (a) prohibited by applicable law to possess or distribute, (b) that would violate applicable law if distributed, or (c) that would result in infringement or violation of another person's rights if distributed. (Indiegogo Terms of Use).*

It should be noted that these lists can largely overlap each other while setting the boundaries of the field inside which "everything that is not forbidden is allowed", both platforms apply forms of control and both make use also of users' reports to find eventual violations of rules.

It is true that the major platforms promote certain campaigns as "featured", ensuring them so not only more visibility, but in general a feeling of trustworthiness of potential backers. This combination of images between the one of the platform and the one of the campaign, that is generally related also to the dissemination activities of the campaign through the social channels of the platform (Facebook, Twitter, etc.), follows rules but not necessarily coded, which can be related to several factors: the innovativeness of the proposal, the reputation of the campaign creator, the result that the proposal is having in terms of money and supporters, but that Is difficult to translate into an assumption of responsibility by the platform towards the project. For the platform of a "featured" campaign that although reaching the goal does not keep its promises (product not made or different from the description, or poor workmanship), the only negative effect could be a reputational damage for the platform itself.

Therefore, the generalist platforms essentially tend to present themselves as a passive tool, which hosts campaigns that fall in parameters not particularly strict, and which monitor some aspects relating to the conduct of the campaign, but not on what happens after the campaign, that is when the platform/tool is no longer the instrument for collecting pre-orders in exchange for promises of money. Moreover, the fee retained by the platform is quite low (both Kickstarter and Indiegogo provide a fee of 5% for campaigns in the United States) and this hardly could bind the platform to an action of control and verification that goes beyond the methods of implementation of the campaign or beyond a closed group of possible actions (e.g., violation of a copyright by the campaign creator), as specified in the Terms of Use of Indiegogo

*Indiegogo makes no representations about the quality, safety, morality or legality of any Campaign, Perk or Contribution or the truth or accuracy of User Content (as defined below) posted on the Services. Indiegogo does not represent that Campaign Owners will deliver Perks or that Contributions will be used as described in the Campaign. Users use the Services at their own risk.*

which adds that in case of disputes between campaign owners and contributors "Indiegogo is under no obligation to become involved in disputes between Campaign Owners and Contributors, or Users and any third party." This approach is common to Kickstarter, a platform that provides in its Terms of Use:

*Kickstarter isn't liable for any damages or losses related to your use of the Services. We don't become involved in disputes between users, or between users and any third party relating to the use of the Services. We don't oversee the performance or punctuality of projects, and we don't endorse any content users submit to the Site. When you use the Services, you release Kickstarter from claims, damages, and demands of every kind — known or unknown, suspected or unsuspected, disclosed or undisclosed — arising out of or in any way related to such disputes and the Services. All content you access through the Services is at your own risk. You're solely responsible for any resulting damage or loss to any party.*

However, it is useful to note how the situation is evolving fast, especially in light of some negative experiences at the limit of the scam perpetrated through crowdfunding, some episodes that, if replicated on a large scale, could put at risk the element of trust that underlies the crowdfunding campaigns. As we have already seen, the relationship of trust is essentially always between campaign creator and supporters, not between supporters and platform. But this is not necessarily the perception of general public and, in fact, to support a project it is still necessary to connect to a particular platform, to create an account (becoming a user of the platform) and to make a payment - whether without immediate effects - on a page of the platform, accepting the rules of the platform itself. Therefore, it seems obvious that, quite rightly, the backer will put his trust also in the control activities carried out by the chosen platform.

Beside the four basic elements of a crowdfunding campaign (project, campaign creator's profile, duration and economic objective to be achieved), there are, however, other elements deserving to be taken into account. They differ from the previous ones because of their less characterizing and less rigid nature with respect to the campaign. For example, they can be inserted in an ongoing campaign, or can be changed, or can have a merely indicative function, or can be the result of a confrontation between campaign creator and backers.

Below we will try to make a (not exhaustive) list of these elements. The rewards are the first of them. The so-called reward levels correspond to the amount committed by the backers. They are essentially items which provide that a certain financial commitment of the backer matches a certain type of product or a particular service offered by the author. They can be primary rewards, that is the product that the campaign proposes to make, or secondary, that is gadgets related to the product, or opportunities to meet with the author, or a mix of primary and secondary rewards. To exemplify, in a campaign aimed at producing a music album, the CD of the album will be a primary reward, a message of thanks by the musician a secondary reward, the CD with a thank-you note by the musician will be a mixture of primary and secondary rewards.

However, the author may decide to include new rewards in the ongoing campaign, giving or not the possibility to those who have already supported the campaign to opt for a new reward, maybe paying the difference. It is not said that a reward for the same offered product/service is related to the same figure during the entire duration of the campaign. Returning to the example of the CD, the singer may decide to make available the CD for $15, but only for the first 100 supporters, then he may make available the CD for $20 only for a further 100 supporters, and then he may decide to make it available to anyone for $25 (we could make many examples in this sense, now it is a fairly widespread practice; not being able to list all of them here, I will only indicate as an example the campaign for LightBlue Bean+ on Kickstarter). In fact, limiting the availability of a level of a product related to the campaign, or the availability of that product to a given price, could be an important instrument in the hands of the campaign creator to attract early backers, supporters that already in the first hours of the campaign make their promise of payment. It should be remembered how it is useful so that a campaign is perceived as successful, to give the impression of being able to easily reach the goal. The backers, although the all-or-nothing mechanism protects them against the risk related to the failure of the campaign, prefer to contribute when it is reasonable that the goal will actually be achieved. Through this mechanism, vice versa, there is the dual effect of attracting early backers with a promise of economic savings (or now or more expensive) and more supporters having shown them that the campaign has a good response from the public.

But what are the rules that the campaign creator has to follow? Substantially two, the first one is a common-sense rule, that is, not to change a reward already entered (if the campaign creator says that there are 100 CD at $15, it would be good that he did not subsequently diminish the availability of the same level of price to 90 or he did not increase it to 110), while the second rule is that he does not betray the trust of the early adopters, making available later the same reward at a lower price, maybe at the end of the campaign to raise additional funds. We will return later to the fiduciary relationship between backers and campaign creator.

Another element of great interest is the inclusion of overgoals, that is further objectives with respect to the achieving of goals that may be announced during the campaign and that involve, if they will be reached, a further development of the project. The overgoals respond to a double need. On the one hand, they protect the campaign creator against controversies about how the funds raised in excess will be used, since it is true that if there is a surplus of collected funds this is due to a large number of orders, implicating, consequently, higher expenses. But it is equally true that the campaign creator must necessarily get some margin on each order to cover the costs of the project (it could be just the fee of the platform) and we could hypothesize that the costs for the implementation of the single reward will significantly decrease.

On the other hand, they maintain a high interest for the campaign in the backers and public, giving a further incentive to involve more the potential supporters and to create a larger audience.

While overgoals generally involve all the backers who supported a project, intermediate goals are less used, but no less important, being rewards that the campaign creator decides to grant to the supporters precisely when some intermediate objectives related to the funds that have hitherto been collected or to the relationship between the money raised and the time elapsed since the start of the campaign have been achieved. Usually these intermediate goals are not pre-announced. They do not add anything to the project (which at that time has not yet been financed) but integrate the description of the project, showing specific features (in the case of technology projects) or parts of the concept not made public until then.

The campaign creator can decide these elements (rewards, overgoals, etc.) unilaterally or together with the backers who, until then, have decided to believe in the project. This is the case, for example, of the campaign of Jide to finance Remix Mini, a computer with small dimensions and an affordable price, the first campaign created by Chinese campaign creators up to exceed one million dollars on Kickstarter. To decide the development of the project and some level of rewards the campaign creators have decided to make some real surveys between the backers. In other words, they have introduced explicitly some elements of democracy and bilateralism in the construction and development of their crowdfunding campaign. They were surely not the first ones to have done this, but can be considered a significant example for the size of the campaign, the number of people involved and the consultation methods that have been chosen.

In fact, the construction of the community is a central moment of a crowdfunding campaign. And it proves to be the crucial moment in our investigation on the setting of a system of internal rules in a crowdfunding campaign.

To simplify, it can be argued that the crowdfunding campaign is proposed to an indeterminate plurality of subjects, without any linkage with it strictly speaking, and that this plurality decides to participate under certain conditions. These conditions cannot be contracted because they have been pre-determined by the campaign creator. In this sense, the birthright of the community and its boundaries, are created by the campaign creator clarifying what the project is, what its goal implicates and what not until to achieve the economic objective. The campaign creator accepts as limit to his power to establish rules and conditions, the limits of the current legislation and the ones of the platform. But within those limits he may provide the rules and conditions that he considers the most useful, correct and appropriate. We will define this moment "unidirectional".

When a person decides to participate in the campaign, he joined the community of the campaign. Bragging rights. In part these rights will result from a self-restraint by the campaign creator, that is strictly part of the contractual synallagma which was originally proposed, but to some extent the subscriber will acquire a right to participate, to be informed, to communicate with the campaign creator, to ask questions, to give advice, to interact, using different forms and methods, with other backers that are part of the community. It should be noted how this process goes to build a set of other rules, expressed or implied, that together with the aforementioned rules define the overall framework of the life of the community built around the campaign. This moment can be defined as bidirectional (with respect to the relationship between the community and the author), and multi-directional (with respect to the relationship between the members of the community).

The person who chooses to support a campaign acquires the status of backer, that is of active member of a community of purpose, a community of purpose that has its first regulatory element in the unidirectional determinations established by the campaign creator, but then lives a democratic participatory life. We could dare a comparison that seems fitting to reflect the position of those who choose to enter the community. When a person gets the status of a citizen of a Country, with all probability that country has its own pre-existing Constitutional framework which that person has not helped to build. From then on, contributing to the civil and democratic life of that Country, the citizen will not only (or not exclusively) be a passive party in the process to establish the regulations and the system of rules of the country, that is he will get an active role. This does not means that the active role will be exercised in its entirety, for example, relying on the rights of which the person is a bearer, exercising the right to vote or participate in the political life, but it is a potential of which the citizen is the bearer. Similarly, the backer will simply wait that the project is finalized and his reward is packed, shipped and delivered, as if he had participated in a unique buying group. But he may also decide to participate without that nobody can stop him, at least until he respects the system of rules that was established.

However, this does not mean that all backers enjoy the same rights; if we look diachronically the evolution of a crowdfunding campaign we could note that a decision taken by the campaign creator and the backers a certain day affects those who decide to become backers the next day. However, those who decide to become backers do it accepting the conditions of the moment, without that this represents an imbalance. After all, returning to the example of the citizenship and democracy, two people that become citizens of a Country within a few days, the first one on the day before an important referendum and the second one on the day after the referendum, will be in a condition of inequality: the first one will participate in the referendum while the second one will not. But it is also true that they have both to adapt their future expectations (and their behavior as a consequence) to the result of the referendum and that, if the first one has had an active role and the second one has had exclusively a passive role, it is also true that the second person chooses to become a citizen already knowing the outcome of the referendum, while the first one, whether it is the preference for which he expressed his vote, must comply to what has been decided by the majority.

Therefore, it is evident that, considering a crowdfunding campaign as a process to establish a regulatory system and to define the structures of a community, the first one who participated has more chances to influence the choices of the community and the campaign creator.

But even in the case that the community is built around a democratic participation - and not forced - of the backers, it is equally clear that the expectations of each backer and more generally of all the backers are placed in the campaign creator and in his determinations, rather than in the community itself. In other words, the issue is to understand whether the campaign creator can be seen as a "primos inter pares" within the community, or whether he must be considered as an absolute monarch, in conditions to assign more or less margins for participation to the supporters of his project, but this margin of participation is only granted in order to meet a need of the sovereign, that is to motivate the backers to seek other backers to achieve the economic objective in time.

Moreover, even the economic objective, once reached, marks an asymmetry between those who receive money (the campaign creator) and those who pays money (the backers).

Giving a clear answer to this dilemma would be to deny the premise, that is how each crowdfunding campaign builds a peculiar set of rules and a community whose composition, interests and participatory attitude may vary significantly.

In fact, nothing prevents the campaign creator to minimize the interaction with the backers, limiting it to merely occasional communications on the progress of the project. Neither the issue of the participation and construction of a community in a communicative technicality, in an instrument that the campaign creator can use to better reach the goal can be solved. It would not be simply restrictive, but it would be, in my view, an incorrect evaluation. Of course, as we know, a significant part of the marketing tries, with always more sophisticated and effective techniques to create an involvement among the intended audience of the campaign, transforming it from an audience into real testimonials and to feed a word-of-mouth using gaming mechanisms and social networks, uniquely designed to spread the pre-determined advertising message.

If the assumption of the communicative technicality could be accepted, a crowdfunding campaign would be classified simply as a campaign of marketing and communication and, as such, it should be evaluated through the social and economic sciences that deal specifically with marketing and communication.

However, I think that it can be found at least a strong element of difference, so strong and central to become, in the light of a careful analysis, an element of divergence between a marketing campaign and a crowdfunding campaign.

The concept of moral property, as a legal and economic element, differs from the concept of ownership in the strict sense. The moral property of a crowdfunding campaign, or rather the result promised in the event of reaching the goal, does not belong to the campaign creator, but to the community constituted by the campaign creator and the backers. Without the backers the product, the specific product developed according with the proposed methods, could not exist. This creates a deep relationship that appears not only in the active involvement of the backers during the campaign (of which it is a prerequisite). The linkage is the moral reason with which the community identifies itself and at the same time is the condition of formal and substantial legitimacy of the participation to the crowdfunding campaign. A potential supporter might expect that other people finance the product so that he will enjoy it after the purchase, once available through the traditional channels. Or, if the product is not financed, he might expect to turn to other products to meet his own specific needs. He decides, instead, to get involved, to participate, not buying, or rather, not simply pre-buying a tangible good, but registering in his name a fraction of the moral property of that product.

The element of the sharing of the moral property, which is accepted by the campaign creator even before he can describe the product, the methods of implementation and before to be able to open to the public his campaign, distinguishes the crowdfunding and its inherent need to build a community. After all, it is based on an exchange. The campaign creator, owner of the potential property of the product/idea, provides the moral property in exchange for the feasibility of the same. He shares something to get something else in a mechanism of mutual exchange and mutual enrichment that creates relationships at different levels, some destined to manifest themselves in the material exchange as money for goods, others destined to precede this exchange and to survive to it once it has been perfected.

Crowdfunding shows all the elements that belong to the sharing economy: an element of exchange, an element of reciprocity and the construction of relationships of mutual satisfaction.

Therefore, it is not correct to say that a crowdfunding campaign is animated by two separate interests, that is to raise money to implement the work and to receive a copy of it, but rather these two interests, which constitute the premise of the campaign, merge together into one common interest on the basis of that element of interested solidarity of which we talked about at the beginning. This interest is shared by the entire community.

While keeping in fact a distinction of roles, both campaign creator and backers have an active function, because it is mainly the activity of the latter that determines the success of a campaign. In fact, we focused until now on the activity of the backers within the platform and in the page of the campaign, but it is only one aspect of the activity of the backers.

Backers are called to share the news of the campaign, its evolution and its results. The campaign creator can certainly push them in this sense, but the solicitation is implicit and inherent to the crowdfunding, so much so that almost all platforms integrate sharing tools, which are mainly designed for the backers in the hosted campaigns. In fact, it is assumed that the campaign creator is not able to reach a large audience alone - that is if he had the economic means and/or media he would not undertake, with all probability, a crowdfunding campaign - and it is, therefore, up to the backers to share the message of the campaign, either through a message addressed generically to their contacts on the social networks, either through more specific business relationships with other people potentially able to sensitize a wider audience (e.g., influencers, journalists, bloggers, etc.).

We will not dwell here on the means of dissemination, although they are interesting and innovative, but we will focus on the mechanism of involvement in the relationship of trust existing between campaign creator and backers. As we have already seen, the relationship of trust is a relationship aimed at establishing the rules of a community and, as such, it has a value merely within the same community. When a person, who has become a backer adhering to that relationship and has contributed to its subsequent construction and explicitness, suggests a crowdfunding campaign to others, he actually suggests adhering to that relationship, becoming its guarantor. We return to the campaign for the time machine of Mr. Blue, the campaign creator. Mr. Yellow decides to adhere to the system of rules proposed by Mr. Blue, becoming in this way a backer. Shortly after joining it, Mr. Yellow suggests publicly that the time machines, once they are ready, will be shipped by airmail, an aspect that has never been specified before in the campaign. Many of the backers adhere enthusiastically and Mr. Blue takes note of this request, perceived as a need by the community, updating the description of the campaign. Since then, the shipping by airmail has become an integral part of the rules of the campaign as well as of the expectations of the community. However, when Mr. Yellow shares the campaign, he does not share his specific contribution to the campaign, neither in economic terms, nor in terms of the proposal, but the entire set of rules as they have been established since that time. So, Mr. Yellow adhered to the proposal of Mr. Blue, after receiving a message from Mr. Yellow, Miss Red decides to become a backer giving her trust to Mr. Yellow, who she already knows to have a good reputation, even before than to the community or to Mr. Blue, who is a totally unknown stranger for her.

In this way, therefore, reproposing the original relationship of trust, Mr. Yellow is not only a substitute for Mr. Blue, because he represents the entire community.

It could be objected that the mechanism is not so innovative, it is the traditional word-of-mouth that always, or almost always, determines the fortunes of restaurants, museums, gyms, hotels, etc. However, we can identify two elements in my opinion highly innovative. The first one is that when Mr. Yellow suggests participating in the campaign, he has not had the opportunity to try the product of Mr. Blue, therefore the relationship is based only on trust and what happens in the moment when Miss Red is also involved reproposes precisely the relationship of trust. The second element, in my view highly peculiar of crowdfunding, is the attitude of this mutual trust to create value. The community does not die with the end of the crowdfunding campaign and we have seen more than one case in which the community at the end of the campaign is transferred to other portals (social networks, etc.), but it has continued to be active, representing on the one hand a source of strength for future works of the campaign creator and on the other hand an aggregate and active group of individuals joined by an experience as well as by a strong common interest. To remain in this example, it is certainly possible that those who financed the time machine of Mr. Blue are interested then, a few years later, to the illustrated guide for travelers in the time that Mr. Purple proposes to implement, if he reaches his economic objective.

That is to say that aggregating people concretely motivated around an interest, and for it willing to invest, makes easier to find in the future funds for activities that are related to that interest, but we will return to this in detail in the chapter on specialized (or vertical) crowdfunding platforms.

The elements that I would like to highlight in this reconstruction are that crowdfunding is not governed exclusively by the rules of law and the regulations of the platforms that host it.

Every campaign involves the construction of a community, with its specific rules, that are in part pre-established by the campaign creator and in part the result of the interaction of the community. However, these rules do not live in a closed set, building a case history to which the future campaigns will not necessarily have to conform, but with high probability they will use as a parameter. Although it is not possible to identify the rules that are common to all the crowdfunding campaigns, we can observe an evolution mostly harmonic between different campaigns with respect to the methods, the mechanisms of involvement and, of course, the rules. The lowest common denominator of the crowdfunding campaigns is the sharing of resources aimed at achieving a common interest as the sum of two different interests; the pursuit of this common interest can be defined as the moral property of the work or product. This moral property does not lose its importance with the subsequent commercialization of the product in the same way that the community and the sense of belonging/identification do not disappear. In this manner the crowdfunding campaign creates value also for third parties, not because, and not only because, there is the implementation of something that did not exist before and that can meet the needs and requirements which had not been satisfied until then (crowdfunding is not, in fact, the only existing system to finance projects), but because of its nature, of its being able to unify communities that previously did not exist, communities that could be facilitators, in their area, of future projects, which may, but not necessarily must, pass through new crowdfunding campaigns.

It also represents a democratic, transparent and accessible validation system for the proposals of the campaign creators, through a system that rewards the usefulness and interest in a project and that, in theory at least, puts in a position to collect important figures campaign creators little known, but with a good idea - and an educational and professional background - which demonstrates that they are able to implement it, or at least to create a trust relationship in this sense.

## NOTE

To refer to systems of specific rules, we have taken into account the internal regulations of Kickstarter and Indiegogo platforms, both because they are globally widespread and because the level of complexity of the issues faced by the two platforms has allowed them to develop a comprehensive system of rules that is functional to their activities and has become a reference point for all the reward crowdfunding platforms, at least as regards the policy.

## REFERENCES

*Indiegogo Terms and Rules*. (n.d.). Retrieved August, 2015 from http://www.indiegogo.com

*Kickstarter Terms and Rules*. (n.d.). Retrieved August, 2015 from http://www.kickstarter.com

# Chapter 8
# Vertical and Community-Based Crowdfunding Platforms

**Tomaso Greco**
*Università degli Studi dell'Insubria, Italy*

## ABSTRACT

*If the inherent nature of the horizontal crowdfunding platforms makes the relationship between campaign creator and backers basically without intermediation of the platform, in the case of vertical platforms there are different and sometimes complex levels of intermediation. The vertical crowdfunding platforms can be considered more than a subset of the horizontal crowdfunding, as an independent system of relationships and rules which has in common with the horizontal crowdfunding a large part of the fundraising methods, but that is also profoundly different with respect to the aggregative procedures and the system of rules.*

## INTRODUCTION

Beside the above-mentioned horizontal platforms, a large number of specialized and local civic crowdfunding platforms have spread in recent years.

The specialized crowdfunding platform has in common with the generalist platforms the method and, in some ways, the approach to funding, even if there are more or less differences - depending on the platform that we are considering - with regard to the development of internal communities and the system rules which determines their daily lives.

We have seen the emergence of crowdfunding platforms exclusively dedicated to publishing, culinary art, music, wine, culture, cinema and many other specific interests. They are, in almost all cases, platforms of reward type, aimed at the implementation of the project, project of which some examples/copies will be sent to the supporters as a reward.

Not all platforms have a global vocation; in fact, the most of them combine the criterion of specialization with the one of territoriality. The common feature, in any case, is the verticality.

In fact, if the inherent nature of the horizontal platforms makes the relationship between campaign creator and backers basically without intermediation of the platform, in the case of vertical platforms

DOI: 10.4018/978-1-5225-0568-6.ch008

there are different and sometimes complex levels of intermediation. It could be argued that this distinction is not necessarily due to organizational reasons: the staff of a horizontal crowdfunding platform cannot certainly be omniscient, while it is likely that the staff of a vertical crowdfunding platform can boast strong and specific skills on the specific area of interest of the platform.

In reality, this argument would be quite insufficient to explain the real reason for the need to create platforms that are strongly vertical. Moreover, this need finds a correspondence in the choice of many campaign creators to rely on a vertical platform rather than on a horizontal platform.

For a correct examination we will have to compare the three different types of methods of fundraising through crowdfunding: with horizontal platforms, with vertical platforms and with self-hosted campaigns, that is campaigns hosted directly on a pre-existing website of the campaign creator or on a website created for that purpose.

In the horizontal and vertical platforms the campaign creator will count on no entry costs or almost non-existent fees while for a self-hosted campaign he will need to acquire the necessary technology, to enter into agreements with one or more payment systems (especially if the campaign involves the all-or-nothing method, a method according to which the payments of the backers are effectively collected only if the campaign reaches its pre-fixed goal), and also to establish a system of rules relating to the campaign, the payment management and the privacy of the backers. The economic advantage, on the other hand, is that on a platform (either horizontal or vertical) the campaign creator will have to pay a fee on the total amount of the campaign, while on his website, as it is reasonable to expect, there will be no fees to pay.

But the choice of the campaign creator to rely on a platform rather than another, or to prefer a self-hosted campaign is certainly not only due to an economic aspect.

Otherwise, there would be a race to the bottom of the fees of the different platforms, a race that could not obviously have a different ending than the survival of a single platform: the one with the lowest fees. However, the offer of crowdfunding platforms is going in a different direction, and the reasons for the effective failure of the "war of the fees" is, in my opinion, due to the different needs of the campaign creators, related to the different characteristics of the campaign creators themselves and their projects.

Which characteristics, in fact, orient the choice of a campaign creator? We start from the lowest common denominator: to fund a project. However, to do it, there are also other channels, while the peculiarities of crowdfunding, as we have seen in the chapter on horizontal platforms, is to build a community around a project and, through the community, to give the maximum possible visibility to the project, developing its potential in a context of constructive dialogue with the backers.

In other words, we need to recall the four basic elements of a crowdfunding campaign: project, campaign creator's profile, duration and economic objective to be achieved. On the basis of these elements it is established or, rather, it is possible to establish a relationship of trust between the potential supporter and the campaign creator. However, to establish the initial relationship of trust, it is necessary and preliminary an element of validation of the project. This does not mean the attitude of the project itself to be valid, but that it appears to the potential donor as valid. This process can take place, in the horizontal crowdfunding platforms and the self-hosted crowdfunding platforms, through three different elements: the potential backer may have the technical knowledge to define the proposed project as valid, the potential backer may deem necessary to fund that particular project, the profile and reputation of the campaign creator provide sufficient guarantees to the project.

It is clear that the relationship of trust is based on the unknown, on the result of a promise, result that could be postponed at the end of the campaign and, in most cases, even several months after the end of the campaign. However, these three elements are generally sufficient, considered individually and even more combined with each other, to justify the participation in a campaign, validating at least the premises of the project.

But this does not implicate that all good projects can meet at least one of the three requirements. Vertical crowdfunding platforms compensate this lack, firstly - as we will see - aggregating communities with similar interests and, therefore, that potentially have strong technical knowledge in the specific area of the project, on the other hand - and above all - providing an independent validation element, express or implied.

In fact the most of the platforms of vertical crowdfunding evaluate project (and campaign creator's profile) even before that the campaign begins, in some cases granting or not hospitality to the project, in other cases ensuring or not the visibility on their platform. It is a rigid, or less rigid, barrier at the entry, which responds to two distinct needs: on one hand that potential supporters find a good quality and reliable proposal because it was already validated by the platform, on the other hand the same platform needs not to saturate its public with proposals of poor quality or that will hardly reach the pre-fixed goal (it must be remembered that the platforms, especially those that use the all-or-nothing model, receive their fee only upon successful completion of the campaign).

This selection at the entry, which is a first validation parameter of the project, is an intermediation element of the relationship between campaign creators and backers and, as we will see, it will not be the only one.

A feature common to different vertical crowdfunding platforms is precisely to provide services, boasting a specific knowledge of crowdfunding campaigns for a specific sector. These services can sometimes be optional, but most of the time they are inherent to the offer of the platform.

A service of great importance for the present discussion is the so-called campaign strategy service, that is the preparation of explanatory content of the project, of its methods and of the rules of the campaign. In this way, the platform meets two needs: the author's own need and the own need of the community of subscribers of the platform.

In fact, it is not at all certain that an author with a good idea is also able to communicate and build a series of promises around that idea for the potential supporters. For example, we can consider a writer who has a good project for a book, or at least a project for a book judged valid by the platform. It is not at all certain that he knows how to prepare his own crowdfunding campaign, nor it will be enough just to know the general rules of the platform for doing it. In fact, he could make a series of errors, in good faith, with the effect to ward off potential supporters, rather than to involve them in the project. He could, for example, omit an even hypothetical delivery date, or fix it as too remote in time, or not be available for dialogue and interaction with the backers or even not explain in sufficient detail the key elements of his manuscript, as well as he could not highlight the elements of his professional background and culture that make him the right person to write and publish that particular book. In meeting these needs the platform on the one hand will try to study the best strategy for that specific campaign, and on the other hand it will apply a set of rules and methods which, always through the campaign strategy service, have already been seen and experienced in other campaigns of the same platform. This aspect ensures strong homogeneity to the campaigns present on the vertical crowdfunding platforms and marks a profound difference with respect to the horizontal platforms.

Therefore, in the horizontal platforms, as we have seen, each campaign develops a set of rules, while respecting the general rules of the platform, and around these rules aggregates its own community of backers, it matters little, in our reasoning, that all the backers contributing to the various campaigns of a horizontal platform will form the user base, in fact user base and community are two very different concepts. Users of a horizontal platform may be very heterogeneous among them with respect to their interests: it is not at all certain that persons who have funded a new model of backpack may then be interested in an innovative coffee machine. Indeed, there are the same opportunities as for anyone else. Nor those who have supported the campaign for the backpack will expect to find the same rules and procedures at the time to support the campaign for the coffee machine. For those familiar with social networks, it is easy to make a comparison with Facebook groups. The platform (Facebook) provides a number of possibilities and provides a set of rules. The groups hosted on Facebook, within the rules of the platform, may provide specific rules designed to lead at the best the discussions on their specific theme. Each group may, however, adopt special rules, while sharing the same frame.

Unlike the case of vertical platforms, where the supporter of a campaign may theoretically be interested in supporting any proposed campaign platform, and, therefore, if it is true that each campaign builds its own specific community, it is equally true that all communities are, in fact, a subset of the community of the platform. In other words, the community that aggregates a single campaign will be partly made up of people who have already funded other campaigns on the platform and partly of new supporters who will actually populate also the general community of the platform. Therefore, there is not a rigid boundary between the communities of the platform and the community of individual campaigns; in this way the individual campaigns will have an incentive, when not explicitly forced, to adopt rules that overlap each other, in which each member of the platform can easily recognize himself and that generate the same expectations. This does not exclude margins of customization of rules, but such customization must not affect the overall system of rules of the platform, or the system of rules and procedures that generally the campaigns of the platform adopt. We return to the example of the writer: who decides to fund the campaign for his book on a vertical platform will expect, if he has already funded a book earlier on that platform, the same methods of interaction with the author. Methods which may vary, mainly in the sense of being strengthened only if the specificity of the text requests to do it and the exceptional nature of that greater interaction is such that it is clear to all supporters and does not modify their expectations with respect to the platform.

To conclude this part of the reasoning, it seems a correct simplification to argue that in the horizontal crowdfunding the relationship of trust is mainly between campaign creator and backer, while in the vertical crowdfunding it is, instead, characterized as a trilateral relationship of trust between platform, campaign creator and backer. For this reason, the horizontal platforms will tend to give a good dose of freedom to the campaign creator in identifying the rules and procedures of his campaign, while the vertical platforms will aim as much as possible at maintaining homogeneous procedures and rules, not being satisfied only to define the scope of the rules, but building together with the campaign creator - or even replacing the campaign creator - the profile of the campaign. It is, therefore, a relationship explicitly intermingled, not only as regards the methods. Backer and campaign creator, in fact, expect that also the final product is the result of a relationship between platform and campaign creator, more or less invasive depending on the number and the type of services that the platform offers to the campaign creator.

In other words, a project that is funded through a successful crowdfunding campaign on a horizontal platform will have taken a path of validation from the bottom, having found a certain number of backers willing to commit a certain amount of money. A project that is funded through a vertical crowdfunding platform, vice versa, will have obtained a double validation process: from the top with respect to the platform, from the bottom with respect to the backers.

But also in the specific relationship between campaign creator and backer it is possible to assist, in vertical crowdfunding, to a situation of co-creation that goes well beyond what has already been observed about the instruments of participation and internal democracy of horizontal crowdfunding campaigns.

In fact, being able to turn to a less heterogeneous audience and, of course, more vertical, the campaign creator cannot only record its desires and draw important information on how to enhance the project, but he can draw from the individual skills and experiences that will characterize a vertical crowdfunding campaign as a true moment of collective intelligence. In fact, the campaign creator can often find, among his backers, those willing to make available skills aimed at achieving the common goal of making feasible the final product. In this way, returning to the example of the book, our writer can find people willing to proofread, to carry out a thorough editing, to make the cover or the website of the author. Or, more generally, the writer could find some people willing to give advice not generically motivated by their personal taste or their desires, but rather on the basis of a specific knowledge of the target market.

It should be noted another important difference between the horizontal and vertical platforms precisely with respect to the target market. In fact the latter will, in all probability, be most frequented by market operators (and consumers), with the result that there will almost automatically be the passage of the project from a prototype for a community of supporters/early adopters, to the distribution in the target market.

We will explore more in depth the rules and procedures of the different platforms of vertical crowdfunding in the following paragraphs. As mentioned above, there are considerable differences between them, due to the specific field under consideration, their territoriality, some peculiar choices of their founders and, more generally, due to the fact that if the crowdfunding is still a young and, in some ways, unexplored field, the vertical crowdfunding being one of its expression will still need some time to reach full maturity.

## CROWDFUNDING PLATFORMS FOR MUSICIANS

The most common vertical platforms include platforms to finance independent music, that is the musical productions that rather than use traditional methods to raise money, prefer to rely on fundraising from the bottom. It should be noted that a certain part of the music production in crowdfunding does not pass through the vertical platforms, and we will examine some platforms merely for illustrative purpose, aimed at comparing the rules and operational methods of the different platforms. In fact, it is not possible and it is not our aim, to provide an exhaustive list neither for music platforms nor for other platforms.

In particular, we will compare the models of PledgeMusic, Sellaband and MusicRaiser, platforms, that differ profoundly, despite having in common the crowdfunding and the verticality with respect to the music.

### PledgeMusic

PledgeMusic is a London-based platform, but it also has offices in New York and Boston, exclusively dedicated to music, although in its broadest sense (in fact, you can also finance concert DVDs, books etc., if they are related to music). It is a reward platform, in the sense that it accepts campaigns that recognize a reward to the supporters (basically the musical album that has been funded). It accepts campaign creators (musicians) regardless of whether they are signed record label artists or completely independent. It provides both of all-or-nothing crowdfunding methods, for the projects yet to be implemented and funded, and methods of keep-it-all campaigns, called pre-order, which allow artists to make their albums available for purchase before they are available on the market. This second model, which actually is not intended as a real crowdfunding, however, allows to international artists to participate in the platform, raising their profile and, ultimately, helping indirectly emerging artists. The fee, in both the cases, is 15%.

The platform does not provide a selecting mechanism at the entry ("You must evaluate, and bear all risks associated with, the use of any Content or any reliance on said Content, and in no circumstances will PledgeMusic be liable in any way for any Content or for any loss or damage of any kind incurred as a result of the use of any Content made available via the Service. You acknowledge that PledgeMusic does not pre-screen or approve Content, but that PledgeMusic shall have the right (but not the obligation) in its sole discretion to refuse, delete or move any Content that is available via the Service for any reason." fromTerms and Conditions, PledgeMuisc, 2015), nor provides special services to the artists, however, it advises to involve the public as much as possible.

In doing so the platform does not assume responsibility for the music produced by the artists nor with respect to the contractual obligation contracted between artists and audience ("PledgeMusic is under no obligation to become involved in disputes between Pledgers and Artists, or between Pledgers and any third party. In the event of a dispute, you release PledgeMusic, its officers, employees, agents and successors from claims, damages, or demands of any kind, known or unknown, suspected or unsuspected, disclosed or undisclosed, arising out of or in any way related to such disputes and our service. Artists are wholly responsible for fulfilling obligations both implied and stated in any campaign they create." from Terms and Conditions). However, the platform maintains a form of control on the effective delivery of the rewards to the backers, as we can read in the Payment Processing and Payout Details "Automated payments are executed as follows:

1. 75% of the target 15% PM (PledgeMusic nd) commission is released the Thursday after the funding finalizes. This equals 60% of the target amount (and accounts for all of PM's commissioned funds on the target amount).
2. 25% of the target is released the Thursday after the download portion of the campaign is released on PM. If PM is not sending this release, or there is no digital release product, a bespoke payment request needs to be justified to accounting in order for these funds to be allocated.
3. All remaining funds above the target amount 15% PM commission are released the Thursday (post a 2 week holding period) after all items have been marked as shipped and all outstanding questions have been answered and closed."

This expectation appears aimed at ensuring the credibility of the platforms and to protect against the eventual delays in the implementation and shipping of the work.

The management of the rights of the artists is also interesting because, even if specifying that these rights remain available to the artists, the platform requires in Terms and Conditions:

*Artist shall fully own all the intellectual property to all publishings and recordings they create through or result from their Campaign. Artist grants to PledgeMusic a non-exclusive, worldwide, royalty free license to reproduce, digitally distribute, and publicly perform the recordings (as well as any underly-*

*ing musical compositions) created or contributed by the Artist as part of the Campaign via any means developed, owned, or controlled by PledgeMusic or its partners solely for the purposes of promoting the Artist's Campaign or PledgeMusic. Artist also grants to PledgeMusic a non- exclusive, worldwide license to (i) to reproduce, display and distribute any artwork, photographs supplied by the Artist, liner notes, metadata, track data lyrics and editorial content relating to the recordings, and (ii) to use the name (including professional name(s)), likeness, performances, photographs, and biographical material of each performer, producer, and songwriter featured on a recording, in connection with promoting the Artist's Campaign or PledgeMusic. Notwithstanding the foregoing, Artist acknowledges and agrees that PledgeMusic's business model may involve free promotional downloads and/or streams of recordings. (PledgeMusic, n.d.)*

It is interesting to note that the types of behavior and interaction within the campaign are also closely regulated, providing a true "Code of Conduct", which is not reported only for reasons of space, tracing in a large extent, the rules of conduct and etiquette that every user, and in any case the users of a crowdfunding platform, should follow on the global network.

## SellaBand

The mechanism proposed by Sellaband has many features in common and some element of distinction. Since from the introduction page of the site it is clear that Sellaband relies on the value of the community (the backers are called "Believers" in the front end and "project sponsors" in the policy) and on the experience of the platform in the sector.

However, the platform does not make quality control on the proposed campaigns, but it establishes a minimum funding target to the campaigns of € 3,000 and a maximum funding target of € 250,000. Like PledgeMusic, Sellaband also accepts both artists signed to record labels and independent artists. Sellaband accepts campaigns aimed at three separate targets "for music recordings, for promotion/marketing, for live concerts/ tours/festivals" (Sellaband, 2015) provided that they are always closely related to the world of music.

It is very interesting the role of Sellaband, which acts as "interface between you, the artist, and your investors, who are hereinafter referred to as project sponsors, and as the administrator for the invested money" (Project Financing Agreement Sellaband) and as such assumes a role which hardly could be called passive or mere tool. The maximum limit provided for campaigns is 12 months and is announced when the artist activates the project, even if "SellaBand can also determine final dates for achieving certain percentages, so-called milestones for the project financing goal", if these interim targets are not satisfied the project is de-activated (a kind of all-or-nothing made even more severe with regard to timing, however, this expectation is mitigated by the fact that the artist is allowed to increase or reduce the goal of the campaign before the goal is reached). Another peculiarity is that in the event of the successful conclusion of the campaign, the platform will retain 15% as fee, but will not directly transfer the rest to the author, in fact SellaBand "will rather pay the invoices authorized by the artist on his or her behalf for the experts in order to conclude the music project". It is an element of protection of the backer and also of the credibility of the platform, because if it is true that the artist can choose providers/technicians which he better prefers to finish his project, it is also true that the payment is handled by the platform. Consequently, the system is difficult to circumvent and the risks of a lost production of the album are low.

As regards the rights management, Sellaband specifies:

*The artist may design his or her career freely and, according to this agreement, has the right to complete contracts with record companies, managers, agents or publishing houses. The artist retains 100% of his or her entire music rights*

and also that

*SellaBand will not assert any claims to the copyright, master right or other rights in connection with the music project.*

The aspect related to rewards is particularly interesting, partly because the artist is obliged to include in the rewards:

*A music download in connection with the music project at minimum. The artist may offer other benefits in the form of benefits in kind, which may also include CDs, T-shirts, concert tickets, etc. All offers are effective only after achieving the project financing goal and completing the music project", but also because it is provided a kind of economic reward for backers, through a mechanism configured as follows: "The artist may distribute offers to potential project sponsors about the purchase of project shares in the music project. This offer may contain a financial component and benefits in the form of benefits in kind. The financial component corresponds to a percentage of the possible profits which are generated in connection with the music project. This percentage share must be determined before activation of the music project. The artist has the right to increase this percentage share before the project financing goal is achieved [...] If applicable, the project sponsors' participation in the profits runs throughout the entire duration of the project. For music recordings, this period is 5 (five) years and begins on the day of the publication of the music recording. The artist undertakes to distribute all net profits that are generated in connection with the project [...] If the model for the participation in profits is used, the parties are obligated to provide each other with quarterly reports showing all gross and net profits in connection with the commercial exploitation of the music project. The parties have the right to mutually review the accounts and documents associated with the gross and net profits of the music project. (Sellaband, n.d.)*

It is a model rather complex, involving the backers also from the economic point of view, in the destiny of the work, maintaining an active linkage that is not only moral but also related to material interests, even after the campaign for the implementation of the project has been funded. It could be drawn a parallel with the equity crowdfunding campaigns, in which the supporters become business partners, but I think it is more correct in this case to frame the backers as digital agents; in fact, they not only make possible the implementation of the musical project, with direct, contributions, but in all likelihood they will also determine its spreading and popularity, having the dual interest to spread something that they have helped to fund and with an economic return which could be potentially higher than the amount that they paid.

Speaking of backers and their offerings, it should be noted that the all-or-nothing mechanism operates differently in comparison to other platforms, meaning that if it is an all-or-nothing standard for the artist (if he does not reach the goal by a certain date he does not receive anything), it is not possible to say the same for the backers; in fact, it is provided that "if the project financing goal is not achieved by the final date, all project shares are transferred into the project credit of the project (sponsors of the music project". In other words, funding is not lost, but is transferred to the user's credit, which will then in turn transfer it to another project that he likes.

As regards the services offered by the platform, it must be highlighted that it offers a support service - through a helpdesk - which offers to directly bring on the market the music funded through the platform ("if you want to we can arrange the global online distribution of your music via our partners like Amazon in the UK, USA and Germany, BOL.com in the Netherlands and The Orchard who make sure the music is available in all download stores like iTunes and eMusic") without making this road mandatory for the artist and offers a network of selected partners for several other services related to music production.

## Musicraiser

The Musicraiser platform divides the offered services into three main categories: crowdfunding ("The Crowdfunding Service consents the Creator to collect from Users, so called raisers, the required amount of money to realize its Project; the Users, in fact, by disbursing an amount corresponding to the economic value of the chosen Reward will consent the Creator to reach the targeted objective"), wishow ("The Company offers, through the Website, the possibility to organize a music event, a festival or a concert, if the minimum amount of pre-sales is placed to Users. The event may be organized by artists, agencies, promoters, etc.") and pre-order ("The Pre-Order Services may allow the Creator

1. To sell his music album even f still in production (so called album pre-order);
2. To find alternative channels to sell tickets for an event, concert or music venue (so called ticket pre-order) (Musicraiser, 2015).

While crowdfunding and pre-order methods are essentially modeled on what has been seen about PledgeMusic (with the first one offering an all-or-nothing mechanism and the second one a keep-it-all), the wishow offers a hybrid model, offering a keep-it-all crowdfunding designed exclusively for the events, where the pre-fixed goal is not expressed in economic terms, but in number of tickets sold

The Creator must pre-determine the minimum number of tickets to be sold to organize the event. Once the threshold is reached, the tickets for the event may still be sold through the Website until the event date. In case of failure to reach the minimum number of tickets purchased by the Users, the Project will be deemed not realized.

As regards the projects, which are of course eligible only if strictly related to music, the platform reserves a controlling syndicate ("The Project and its profile, the Reward and its corresponding economic value may be subject to Company's assessment and control through discussions and/or verification of personal documentation") and, with regard to the events to be promoted through the wishow, the campaign creator will have to demonstrate to the platform that he received confirmation by the involved artists ("In case of Wishow Services the Creator must deliver to the Company satisfactory evidence of the relevant artist's acceptance to perform the live event on the chosen date.").

The duration of the campaigns is exclusively established by the platform ("Project duration is decided by the Company only") and the artist will have to complete the project within one year from the end of the campaign ("The Creator must realize the Project and deliver the Rewards within 1 year from the end of the relevant campaign"). The all-or-noting mechanism is of classic type in the sense that the money is automatically returned to the supporters in the event of failure to reach the goal, unlike what has been observed with regard to SellaBand.

The most interesting aspects is related to the rewards, which are determined by the campaign creator, after approval by the platform ("established by the Creators, subject to the Company's prior approval"), not so much because their costs and the responsibility for their implementation are up to the author ("The Creator is fully responsible for the delivery of Rewards to the Users and shall bear any related costs"), a feature common to other platforms so far analyzed, as well as the rule of modest economic value of rewards ("Rewards are strictly non-monetary in nature, with little economic value and can no longer be modified once the Project is uploaded on the Website"), generally a characteristic inherent to the platforms of donation type, while the reward platforms usually provide rewards with an economic value quite similar to the amount committed by the backer.

Therefore, this is a platform that adopts an average criterion between the total opening to every proposal and a strict form of control, that is maintaining an approving judgment on the proposed campaigns, but at the same time not making selections and not taking responsibility for the campaigns

*The Company makes no representations or warranties, express or implied, with respect to accuracy, completeness, authenticity, legality or decency of the contents and materials uploaded on the Website by the Users. [..] The Company is not responsible for missed delivery of the Rewards by the Creator, events cancellation or failure to deliver tickets in case such activities are performed by third parties.*

Rights management is transparent and limited to the offered service, advertising and communication:

*Musicraiser has no intellectual property rights on the Projects uploaded on the Website. However, the Creator grants the Company and others acting on its behalf, the right to use the contents, brands, logos or similar intellectual property […] The above rights, limited to the Project uploaded on the Website, are irrevocable, royalty free, transferable and with no territory limits. As effect, the Company shall use the materials uploaded and published by the Creator on the Website with the purpose to distribute them, create derivative works, visualize related products without sort of limitation.*

The platform provides a fee of 15%, promises to invest in promotion of the hosted campaigns, offers a service to the artists/campaign creators for the creation of the campaign and the distribution of music funded through the Tunecore service.

As in the case of SellaBand, there is a specific network of partners for the music sector.

## CROWDFUNDING PLATFORMS FOR WRITERS

Publishing is another area of great interest for vertical crowdfunding. We will here consider only two platforms, both in English, which have in common the verticality, but differ profoundly for goals and methods: Pubslush and Kickstarter.

### Pubslush

Pubslush is proposed as an open platform, in the sense that it does not make a selection at the entry (it only requires that proposals are "edited for spelling and grammar, and formatted properly" and that they do not impose unrealistic rewards) (Pubslush, 2015), specifically for writers who want to fund their own book through crowdfunding or who want to sell copies of their book before it is published through a pre-order mechanism.

On all aspects, two in particular appear to be of greater interest. The first one is the length of the crowdfunding campaigns, which have a deadline between 15 and 45 days (therefore, these campaigns are very short), the second one is the fee of 4%, which is lower not only than the standards of vertical platforms, but also than the ones of horizontal platforms - it should be noted that the fees of Kickstarter and Indiegogo are 5%).

Against a fee so accessible, the platform offers a series of packages upon payment, including promotional activities and publications, ranging from 50 to 1000 dollars, aimed at the success of the campaign.

It is a large list of activities, made possible by the verticality of the platform. However, it should be noted that in the keep-it-all crowdfunding campaigns there are immediate costs which have to be paid by the author, without being sure of achieving the goal (the risk, in this case, is asymmetrical between author and platform).

With respect to the copyright, Pubslush requires the writer a license limited to the needs of the campaign and platform

*With respect to all Content, by submitting Content to Pubslush, you hereby grant to Pubslush, a worldwide, non-exclusive, royalty-free, sub- licensable, perpetual, irrevocable and transferable license to use, reproduce, adapt, publish, translate, distribute, prepare derivative works of, display, and perform the Content in connection with the Service and the business of Pubslush, and of its successors and assigns.*

Pubslush boasts a rather large community, made up of writers, readers, publishers and partners in the publishing world.

## Unbound

A different model is proposed by Unbound, a platform that accepts only crowdfunding campaigns aimed at the implementation, preparation and marketing of a book, strictly unpublished ("We cannot publish books which have been previously published or have been available to buy previously, even if the book is now out of stock or (in the case of self- publishing) you have stopped selling this yourself.") (Unbound, 2015). Firstly, the model differs because Unbound makes a selection of the proposed texts, through a selection process lasting about 6-8 weeks and it should be said that to participate in the selection the author must provide detailed information about himself and his work but the latter does not necessarily have to be already completed ("We do accept submissions for books which aren't yet written, but we would need to see evidence of the quality of your writing."). After passing the selection, the author will receive support for the creation of the campaign in all its aspects ("We need to get print estimates, work out the funding target for the project, sign a finished contract, shoot and edit the video and come up with a marketing plan before a project goes live").

The campaign is designed to cover the costs "Our editorial team review all projects to determine the cost they will need to raise in order for Unbound to publish them. This varies depending on the type of project. The average cost for text only books is around £15,000 whereas illustrated, photographic books cost considerably more. The costing includes everything from copy editing, editorial, production, marketing and distribution into the trade") of the activities necessary for the publication, and once covered these costs the platform will share profits with the author ("Unbound shares 50% of all profits (after costs) with the author").

Consequently, it can be said that on a campaign with a standard duration of three months (but which can vary depending on the project's needs), the platform withholds 100% on the economic objective, holding then 50% on the amounts subsequently collected (moreover, the platform when asked whether there are provided advance payments for the author, has specified

*The target of the project goes towards the producing of the book, so no advance is given as this makes the funding target higher. After the project reaches 100% and the book is published the profits are split 50/50 with the author and Unbound. So, although there is no advance authors do get all of the benefits of having their book published by a publisher and get more of the profits than with a traditional publisher.*

It should be considered that the goal proposed to the public, particularly important considering the all-or-nothing model adopted by the platform, is expressed as "pledges" regardless of whether there are offers from 10 to 20 pounds, or 1000, and then not in relation to a figure to reach. If on the one hand this allows to evaluate the actual potential of readers of a book (and it will not be published a book with a few - but generous - supporters), on the other hand this could discourage supporters very motivated to donate more because to donate a larger amount would not achieve any practical benefit for the book and the writer.

The role of Unbound thus seems to be a hybrid between the role of a crowdfunding platform and a traditional publishing house ("We act as your publisher") drawing up a specific publisher-author contract, by reason of which the platform generally holds the publishing rights of the volume that has been funded ("When a project is published Unbound usually own the worldwide or English language rights, but this can vary on a project by project basis as this can be negotiated in the contract."), for which it will provide all activities necessary for its publication ("We edit and proof all of our titles using in house or freelance editors and proofreaders. This is arranged by the managing editor and production manager. We organise for the cover art and design work to be completed along with the typesetting and internal design work. We collaborate with our authors on the production side of their books, and we would certainly work with you on some design aspects.") and will ensure the dissemination of the book by the distribution into bookstores ("We use a third party sales team to sell trade editions of Unbound books into bookshops") and advertising ("We help with traditional publicity, but encourage authors to help with this where they can"), always maintaining a high level of involvement of the author and sharing 50% of profits with him.

The all-or-nothing model has some points in common with the one proposed by SellaBand in the music sector; in fact, if a book is not funded, the credits are returned to the availability of the backer and the backer can use them to support other books, or may request a refund to the platform ("If a project fails to reach its target of readers all credits will automatically be returned to the user's account. The user can then allocate those credits to other book projects or they can contact Unbound for a full refund."). The rewards are shipped directly by the platform, which updates the backer on the timing of preparation and shipping ("After a book reaches 100% it normally takes around 6-8 months from the manuscript hand-in date to go through all of the editorial and production processes. We'll send you updates throughout this time.") and deals with the reimbursement if it was not able to send the reward ("If a project reaches its reader target, but for whatever reason, Unbound fails to deliver the associated rewards to the user, then Unbound will refund these credits so they can be applied to another project.").

Although they both are dedicated to the publishing world, and mainly (but not only) to the emerging writers, Pubslush and Unbound differ to a greater extent than we have seen with regard to platforms for music. It could be argued that, beside the crowdfunding mechanism which represents the common phase in Pubslush and Unbound (albeit with significant differences related to the method), the services offered by the two platforms differ so much to have two different purposes: Pubslush for the publishing world, Unbound for the traditional circulation of books. Therefore, we should focus on the goal that motivates a writer to choose a crowdfunding platform (in primis) and to prefer a vertical platform to the inevitably more popular horizontal platforms. In fact, the author will not want to collect money, since the amount collected by the campaign should be functional to the creation and promotion of the submitted book and it should not - or at least not only - be seen as an economic compensation for the author, but in order to provide the best service to his own work.

That is to say, the entry into the target publishing market whether it is self-publishing or traditional publishing.

## THE PATREON CASE

In conclusion of this brief and illustrative overview of some types of vertical platforms, I decided to consider Patreon, a platform dedicated to "content creators" (ranging from movies to writing, from comics to educational content), that more than a real crowdfunding platform could be called a form of micro-patronage.

It is, in fact, a mechanism that allows supporters to offer a monthly amount (which is a kind of subscription) to allow the content creator to continue his activity being financially supported..

In a platform like Patreon, the value of the community - and of every community built around a project and its creator - is even stronger and more central than what has been observed previously, for two distinct reasons: the pledge was renewed several times and it is, in most of projects, difficult to distinguish between campaign creator and project. This centrality is also recognizable in the rules of the platform.

In fact, the provided guidelines - in addition, of course, to ensure the authenticity of what is proposed by the creators - ("Do not put up fake pages or collect money for things you're not actually doing. You also can't use Patreon as a prank or to fund non-activity."), inhibit the use of the platform for certain categories of proposals ("We don't allow funds to be collected for organizations that promote, forums that distribute, or anything else that primarily facilitates harmful or dangerous activities. For example, an organization that promotes sexual abuse, intellectual property violations, weapons, commercial spamming, self harm, drug manufacturing techniques, or property crimes would be prohibited from receiving funds through Patreon") and even of individuals because of their conduct outside the platform ("We impose restrictions not only on the types of media and projects that can be funded on Patreon, but also on which people can and cannot receive funds through Patreon. After creating a Patreon page, any Creator caught in the act or convicted of making credible violent threats, committing violent crimes, malicious doxing, coordinating nonviolent harm such as fraud, or encouraging others to do the aforementioned harmful activities may be banned from using Patreon.

People with a dangerous criminal history or known affiliation with violent or dangerous groups including terrorist or cyberterrorist organizations, organized criminal groups, and violent hate groups, cannot receive funds through Patreon, no matter the purpose or apparent intention of their Patreon page.

Similarly, anyone who has ever been convicted of child sex abuse, fraud, or money laundering is not permitted to collect funds through Patreon.")(Patreon, 2015).

These are limitations that rigidly regulate the entry into the community, limitations that can be defined more subjective than related to the proposals and their contents. From this point of view Patreon itself as platform/tool, specifying that

*atreon.com does not take part in the interaction between Users. Patreon.com does not have control over the quality, timing, legality, failure to provide, or any other aspect whatsoever of any ratings provided by Users, Content provided by Users, nor of the integrity, responsibility or any of the actions or omissions whatsoever of any Users.", and also "The Service is only a venue for connecting Users. Because Patreon.com is not involved in the actual contact between Users or in the completion of the Content Creation, in the event that you have a dispute with one or more Users, you release Patreon.com (and our officers, directors, agents, investors, subsidiaries, and employees) from any and all claims, demands, or damages (actual or consequential) of every kind and nature, known and unknown, suspected and unsuspected, disclosed and undisclosed, arising out of or in any way connected with such disputes.*

The platform holds 5% of the amount collected as fee, the goals of the campaigns, considering the model of Patreon, are entirely optional and indicative, although the authors are encouraged to include them.

It is particularly interesting to see whether, and to what extent, Patreon becomes a system used on a large scale, not so much to fund projects, but to ensure an income - and especially the main source of income - for creators of cultural and informational content.

## CONCLUSION

For reasons of space, we analyzed only some vertical platforms. However, in my opinion, this short comparison allows to identify the rapid evolution from a sociological point of view, from the point of view of the crowdfunding models and from the point of view of the system of rules.

In fact, it could be observed that, at least from the point of view of music crowdfunding, there are several points of contact between the platforms, without that this has undermined the research to offer a service as close as possible to the needs of the artist through elements of strong originality.

Even where we have observed the strongest differences, that is in the platforms for writers, it should be noted a common element in the value of the community and the creation of an ecosystem of relationships with subjects belonging to the target market.

Then, the vertical platforms can be considered more than a subset of the horizontal crowdfunding, as an independent system of relationships and rules which has in common with the horizontal crowdfunding a large part of the fundraising methods, but that is also profoundly different with respect to the aggregative procedures and the system of rules.

At the same time, it is likely that a larger spreading of horizontal crowdfunding could be useful for the world of vertical crowdfunding, also to allow more people to assimilate the economic and relational dynamics that are, as we have seen, a common and shared basis between the two types.

## REFERENCES

*Musicraiser Terms and Rules*. (n.d.). Retrieved August, 2015 from http://www.musicraiser.com

*Patreon Terms and Rules*. (n.d.). Retrieved August, 2015 from http://www.patreon.com

*PledgeMusic Terms and Rules*. (n.d.). Retrieved August, 2015 from http://www.pledgemusic.com

*Pubslush Terms and Rules*. (n.d.). Retrieved August, 2015 from http://www.pubslush.com

*Sellaband Terms and Rules*. (n.d.). Retrieved August, 2015 from http://www.sellaband.com

*Unbound Terms and Rules*. (n.d.). Retrieved August, 2015 from http://www.unbound.co.uk

# Chapter 9
# The Crowdfunding Communities and the Value of Identification for Sustainability of Co-Creation

**Melek Demiray**
*Istanbul Technical University, Turkey*

**Yonca Aslanbay**
*Istanbul Bilgi University, Turkey*

## ABSTRACT

*For a sustainable future, the actors of the market need a change in spirit that will elevate the life-chances of people by being an "empowered part" of the economic system. A participative, open knowledge economic and market system that will ensure the material well-being, the social visibility, the happiness and the consciousness of the individual is essential. Crowdfunding, as a recent online social community network market model is to be a new socio-technical system of co-creation through self "making and funding". The aim of this study is to discuss the role of identification in co-creation for sustainability of newly rising crowdfunding communities. In line with this objective, the following two aspects are clarified: the key characteristics of crowdfunding platforms as online communities and the role of identification for co-creation in online crowdfunding communities.*

## INTRODUCTION

The crowdfunding as a civic co-creation of innovation network system is a promising paradigm for a sustainable future, thus promotion of it requires a theoretical understanding of its organisational challenges. The motivation to participate in such civic networks derives from a number of sources; the sense of responsibility, the satisfaction of participating for a common purpose; the identification of a public problem, and the belief that the involvement will make a difference (Carpini, 2000). Studies claim that being a part of a project could be one of the main motivations in crowdfunding community

DOI: 10.4018/978-1-5225-0568-6.ch009

(Ordanini, Miceli, Pizzetti, & Parasuraman, 2011; Gerber, Hui, & Kuo, 2013), that can be the driver of the sustainability of this funding system. The aim of this chapter is to discuss the role of identification for sustainability of crowdfunding communities. In line with this objective, the following two aspects are clarified: the key characteristics of crowdfunding platforms as civic online communities and the role of identification for co-creation of innovation in online crowdfunding communities.

In this theoretical paper, social identity provides a conceptual framework for discussing identification within crowdfunding communities. The authors employ Ellemers, Kortekaas, and Ouwerkerk (1999)'s social identity framework, three components of social identity including cognitive, evaluative and emotional. Through analysis of articles / literature, this theoretical framework will be used to discuss how components of social identity affect actors' in-group favouritisms and behaviour in a crowdfunding community.

Aiming to understand *the role of identification for co-creation of innovation in crowdfunding communities*, this study is in the quest of answering the below questions;

- What kind of social identity concept is salient in a crowdfunding community?
- What is the nature of identification for creators and funders associated with different crowdfunding models?

The study lastly discusses whether it is required to strengthen identification in the community to pursue a sustainable co-creation through crowdfunding system?

## CO-CREATION OF INNOVATION THROUGH ONLINE CROWDFUNDING COMMUNITIES

Crowdfunding platforms are recently launched innovation basis that are likely to give way to significant changes upon the individuals' roles in markets in the near future. This development will not only allow individuals to be creators in the innovation systems along with big corporations but will also let them design and even manufacture at home, or own laboratories and small scale production places. Individuals will also be active investors by using their small amounts of capital to be a part of new project or venture. Generation Y and Z, who seem to be the major actors of this system, are expected to build up a new social output that they themselves will innovate, contribute, produce, share, consume and be influenced by. This is a new socio-technical system of co-creation of innovation through self "making and funding".

Through mass production period under Fordism in the first half of the twentieth century, the individuals as consumers implicitly have accepted to be passive receivers of standardized goods supplied to the market. In mid 1980s the CAD / CAM processes brought flexibility to production runs, hence customers gained control for customization of the products they are offered. Meanwhile, the diffusion of the Internet into homes for personal use widened the communication channels between individuals and producers, which built up the basis for open co-creation of innovation platforms. Through these communication bases backed up by digital information systems, the consumers were able to put their ideas at the centre of innovation process of producers. In these platforms, technology made ideas easier to visualize by the sharing of designs and creating physical representations of concepts. This improved the efficiency of communication among suppliers, designers, engineers, producers, and most importantly prospective customers demanding products.

Starting from the 1990s, more consumer-oriented customization was offered by producers, employing industrial rapid prototyping technologies of 3D printing machines that could manufacture products with customer specified features. Prototyping elevated the customer presence all through producers' decision-making process (Dimitrov, Schreve, & de Beer, 2006). The new system was a variant of mass customization. It opened up the opportunity for consumers to interactively personalize products in such a way that motivated users for not only sharing their new product ideas but being active designers of objects. Thus consumer's voice was brought into the design and production process more. The consumer evolved to be an effective 'prosumer', an autonomous co-innovator balancing the role of being both the producer and the consumer (Kotler, 1986; Ritzer & Jurgenson, 2010; Cova & Cova, 2012). This had implications for the design and development of products in the pre- and post-production stages, the production duration of the goods themselves and the distribution process which gave way to the manufacturing of objects that had exactly the features an individual user needed (Bradshaw, Bowyer, & Haufe, 2010; Morlock, 2010). Although the individuals became considerably empowered stakeholders of co-creation of innovation with manufacturers recently, they want to be more participatory in the market system (Wright, 2005). Today they do not only demand to collaborate in the customization but seek out increasingly individualized experiences. The development of technology further enables individuals around the world to interact with each other. The technology is just a tool, individuals demand to design for the process, figure out which process is best, actualize the process (Betts, 2010), even invest on innovative projects guarantying a value in return.

The interactions of people on the Web have evolved to form online communities where members communicate to have a conversation, share ideas or resources or acquire information (Hui, Greenberg, & Gerber, 2014; Kraut & Renick, 2012). Online communities allow individuals to have a global connection regardless of time and place through a platform where a shared focus is created to collect and glue members together who have common interest and needs. Building and managing online communities include launching the community, engaging the users, establishing the commitment, stimulating the participation and regulating sustainability of the community (Kraut & Renick, 2012). In these virtual settings, community can be enhanced via members' contributions such as generating the content, providing the feedbacks and involving the activities.

Crowdfunding platform is seen as an online community (Gerber et al., 2013; Gerber & Hui, 2013; Hui et al., 2014; Lin, Boh, & Goh, 2014) since it connects the members with each other to enable interaction about specific areas of interest. Lambert and Schwienbacher (2010, p. 6) describe crowdfunding as: "An open call, essentially through the Internet, for the provision of financial resources either in form of donations or in exchange for some form of reward and/or voting rights in order to support initiatives for specific purposes." (cited in Macht & Weatherston, 2014). Crowdfunding is a funding system, which enables ideas come to fruition via collecting money from a great number of individuals in the society, the 'crowd'. Crowdfunding concept is derived from the expansive concept of crowdsourcing in which 'crowd' is used to collect ideas and get feedbacks, comments and suggestions to improve business practices (Bellaflamme et al., 2013; Macht & Weatherston, 2014). Crowdfunding describes a collaborative and pro-active behavior, encouraging and harvesting the creativity and commitment of all its members to put forward new ideas for the design and development or improvement of products or processes. Gerber et al., (2013) state that due to the community based structure, crowdfunding platforms have long-term relationships rather than short term relationships that other online funding mechanisms hold. Also due to the funding system involved, crowdfunding settings have long-term interaction in contrast to virtual discussion settings (Gerber et al., 2013). However, these platforms have their unique characteristics as they involve monetary or reward returns and risks for community members.

In these networks there are different actors; creators who initiate a project, funders who support financially project(s), observers who may become further creators and/or funders and the managers of platforms who put users together that desire launch a new venture and/or support new ideas/projects under the crowdfunding system (Gerber et al., 2013; Gerber & Hui, 2013; Ordanini et al., 2011). As stated by Kozinets (1999) online communities have their own shared language to allow members to understand each other and provide a sense of common identity. Shared language which means build and use common vocabulary in the community helps exchange information, comments easily while also strengthening the communication between members who have similar mind and interest of area. Chiu, Hsu, and Wang (2006) state that common language encourages actors in the community to participate effectively knowledge transfer activities and improve the quality of transferred knowledge. Similar to the online community, crowdfunding communities have their own shared language such as actors who initiate a project are called as "creators", "creatives", "designers", "inventors", or "activists" and actors who support financially projects are called as "backers", "fuelers", or "funders" (Lambert & Schwienbacher, 2010; Gerber et al., 2013).

Today there are five types of crowd-funding models that the members are engaged. These are

1. Donation-based (without any reward or financial return),
2. Civic crowdfunding (without economic revenue in projects for producing public assets (Davies, 2015)
3. Reward-based (with non-monetary reward such as products/services related supported project including preordering),
4. Lending-based (with financial refund including interest) and
5. Equity-based (with monetary return like equity or dividends in the project) (Buysere, Gajda, Kleverlaan, & Marom, 2012; Bretschneider, Knaub, & Wieck, 2014).

The design of websites in crowdfunding platforms encourages co-creation of members via allowing project creators to build a profile, add pictures and/or videos and express why they want to collect the money under any of these models. Through these websites with the social networking perspective, project owners willingly share information with intended funders (Kirby & Worner, 2014). Potential funders also actively contribute in crowdfunding websites to provide feedback, comments and recommendations about a new venture. Indeed, crowdfunding communities move ahead of traditional online communities in terms of participation since they allow their members being more proactive by choosing new initiatives to encourage and fund those (Ordanini et al., 2011).

For a sustainable economic system, the actors of the market need a change in spirit that will elevate the life-chances of people for not just being a value initiating contributor, but be an "empowered part" of the economic system. A participative, open knowledge economic and market system that will ensure the material well-being, the social visibility, the happiness and the consciousness of the individual is essential. While a crowdfunding has the characteristics of conventional crowdsourcing and social networking concepts, it has some exclusive features such as members' incentive and financing roles for supporting of projects. Social and financial motivations are critical to increase interactions among members (Sakamoto & Nakajima, 2013). The crowdfunding system requires collaborative networked communities, those will recast the meanings of both production and innovation as well as will blur the producer and the consumer dichotomy further. Actors use crowdfunding platforms as a co-creation mechanism. Relationships in these virtual settings are supportive rather than competitive. These communities are online and bring high potential as drivers of co-creation, allowing network actors to access to new knowledge, sharing risk and resources, joining complementary skills and capacities. As a result, crowdfunding platforms include the characteristics of online communities in which actors of a community exchange their knowledge, share their experiences, solve their problems, use common language and have a sense of common identity. Crowdfunding platforms have community-based experiences to create "community benefits" for actors (Ordanini et al., 2011).

As an eco-system, collaborative online crowdfunding networks back up innovation, co-create new sources of value, combine the resources and technologies, as well as create synergies (Camarinha-Matos & Afsarmanesh, 2006). In such a framework all institutional and individual actors will be broadly enfolded in a synergistic and efficient market system celebrating innovations. This prospect is not a vague optimism but is a quest of holistic democracy within its limitations, rooted in the relationship of mutual and bounded interdependence that exists among institutional and civic individual actors of the markets.

## THE ROLE OF IDENTIFICATION FOR SUSTAINABILITY OF CO-CREATION IN ONLINE CROWDFUNDING COMMUNITIES

In researches (e.g., Hars & Ou, 2002; Lakhani & Wolf, 2005; Bagozzi & Dholakia, 2006a) identification with the community is seen as a vital driver for participation to sustainable co-creative projects. Identification with a community is a key mechanism for in-group favoritism defined as tendency of favoring members of one's own group over other groups (Muller, Geyer, Soule, & Wafer, 2014). According to Mael and Ashforth (1992, p. 104) "identification is the perception of oneness with or belongingness to an organization (group), where the individuals defines him or herself in terms of the organization (group) which he or she is a member". The concept is accepted as an initial point for building strong relations among group members by social psychologists (Dovidio, Gaertner, Pearson, & Riek, 2005). Identification within the community brings about to encourage active participation not only in the individual level but also in the group level which is called "fully cooperative group action" (Bagozzi & Lee, 2002). Bagozzi and Dholakia (2006a) state that individual's identification in the co-creative communities supports maintaining of the membership, thus, sustainability of the creation. In the existing literature there is lack of research or discussion about the dynamics of identification in the crowdfunding communities. Like other types of communities (i.e. brand communities), crowdfunding actors share identification with certain project(s).

The foundations of identification are rooted in social identity theory. Tajfel (1972) first proposed social identity which refers to "the individual's knowledge that he belongs to certain social groups together with some emotional and value significance to him of this group membership" (as cite in Shen, 2012, p.4). According to social identity theory individuals categorize members from non-members in the particular social group where they benefit collectively from a certain characteristic. Therefore, the theory is used to explain the relationship between self-characteristic of individual and membership characteristic in the social group as an effect of individual's activity (Muller et al., 2014). Social identity can be regarded as a vital antecedent of group behaviour (Bagozzi & Dholakia, 2006b). For innovations spurring in a team environment, internal dynamics of the team and the resulting creative outcome can be effective through the whole dynamics of the team and the context (Amabile, 1988). If the group dynamics occur online, the value of the network connections depend more on others' participation (Kosta & Kahn, 2002). This study will focus on the identification of two major active participants of this connected system; creators and funders.

Based on the Tajfel's definition, Ellemers et al. (1999, p. 372) developed social identity concept and introduced three components which constitute one's social identity: "a cognitive component (a cognitive awareness of one's membership in a social group-self-categorization), an evaluative component (a positive and negative value connotation attached to this group membership-group self-esteem) and an emotional component (a sense of emotional involvement with the group-affective commitment)".

To look through the lens of concepts of social identity, the nature of this new phenomenon can be understood clearly by discussing similarities and discrepancies between crowdfunding platforms and other online communities. Social identity theory offers to argue the link between individual level of the community and social aspect of the membership (Muller et al., 2014). Using social identity which is seen as an intermediary concept provides us an insight for building a bridge between micro level and macro level approaches (Crane & Ruebottom, 2012) to sustainability. From a substantive viewpoint, it is important to refine how the micro level dynamics of human identification with an online community impact the macro level emergence of a new sustainable innovative platform. More specifically this study probes how the micro-level context of entrepreneurs "crowdfunding communities around a project" may transform the settled ways of creating new structures for innovation "start up eco-system", employing drastically new dynamics. Since even different models of funding offer different motivations and attract different online actors, can there be an identification within crowdfunding communities so that this type of funding will become a synergistic and sustainable holistic innovation system?

## Cognitive Component of Identification

Concerning the cognitive component of social identity, identification with a group includes self-categorization mechanism in which an individual defines and keeps a cognitive awareness of his/her membership in the group such as "I see myself as part of the community", he/she underlines perceived likeness in-group members and unlikeness out-group members such as "My group is an important reflection of who I am", and member's personal identity overlaps with their group-based identity such as "I identify with other members of my group" and "I like other members of my group" (Algesheimer, Dholakia, & Herrmann, 2005; Ellemers et al., 1999; Bagozzi & Dholakia, 2006b). Cognitive component of social identity has a strong influence on group behaviors of being part of a community, supporting ideas of likeminded members, communicating with the community, reciprocity among members and collaboration within the community which are critical for sustainability of co-creation in crowdfunding community.

### Being Part of a Community and Supporting Ideas of Likeminded Members

Despite the monetary based structure of the crowdfunding platforms, cognitive component of social identity has an impact on individuals' behavior related their project preferences and causes of their involvement (Gerber & Hui, 2013). Both creators and funders might be part of a community which is compatible with their identities. From the creators' perspective, they may engage in crowdfunding community after they observe that other likeminded members launch initiatives (Gerber et al., 2013). Accordingly, from the funders' perspective, their motivation is to involve in community with similar interest and also to spur causes which are associated with their personal believes (Gerber & Hui, 2013). For example, if a funder is interested in technological developments, he/she may identify with the technology-focused initiative and see himself/herself part of that community, support the project to make it possible and also even he/she may consider that funding of this kind of efforts are his/her responsibility. Moreover, such kind of identification is high in) civic crowdfunding projects where outcome is public assets since funders identify and support the initiatives they desire to participate, manage and control for generally the community benefits they belong in (Davies, 2015; Stiver, Barroca, Minocha, Richards, & Roberts, 2015).

## Communicating with the Community

Crowdfunding platforms offer an opportunity for creators to meet people whom identities overlap their self-identities. Creators connect with likeminded individuals and increase awareness about their projects in a particular time period of online interaction (Gerber et al., 2013). When creators launch initiatives in the crowdfunding platforms, they can express the uniqueness and benefits of their ideas with the help of technological possibilities of the setting, such as uploading videos about their projects and/or replying the questions of their potential funders.

In consistent with the self-categorization identification, a powerful communication occurs within all kind of crowdfunding communities. For instance, creators build social impact about the donation-based projects and evangelize them in the public. Additionally, one of main motivations for some of the creators especially in reward-based crowdfunding models may be to communicate socially individuals with similar interest and attract their attention rather than collect money (Gerber et al., 2013). Creators can reach their target and niche audience about their unusual works by sharing it in the crowdfunding platforms and confirm people the usefulness of their projects via dialogue on the platform. An anthropologist, for example, who has posted her project about searching old Roman skeletons on one of the crowdfunding platforms, has experienced rapidly and incredibly spread the word about her work. Many people in the world and even one of the documentary TV programmers have been interested in her initiative in which supporters desire to reach a scientific end and her work has become cover in some of the world's largest magazines (Hui, Gerber, & Greenberg, 2012). If this researcher used another alternative financing method such as applying angel investors or publishing her work in a scientific journal, a few people would be aware of her study. Strong communication and evangelism which provides continuous and increased interests about the works within the group generate from the cognitive awareness of identification in the crowdfunding communities.

In the crowdfunding platforms creators not only come together with their potential supporters but also meet other creators whose identities overlap. They feel that they belong in the same group and like being with each other, talking about the similar areas of interests (Gerber & Hui, 2013). Other creators also motivate them to take part in the same community, thus the sincere connection among them provide active participation and sustainability of co-creation in the community.

## Reciprocity among Members

Reciprocity is another group behavior that may results from cognitive component of identification since creators also become funders due to seeing themselves as a part of community and identifying themselves with other community members. Reciprocity can be identified that individuals have tendency or feel responsible to respond when they benefit or get something from others (Bretschneider et al., 2014). Gerber and Hui (2013) reveal that creators and supporters within the community feel a strong tie with other community members. This provides a powerful willingness to make communication and obligation to help them. In the crowdfunding context, reciprocity is carried out not only financially but also socially (Hui et al., 2014). In the financial aspect, reciprocity can be described that creators who reach their funding targets about their projects, may feel obliged to support financially other works in the same crowdfunding platforms (Bretschneider et al., 2014). Social-based reciprocity such as sharing experiences and/or helping people in skilled works generally reveals in the crowdfunding platforms. Some creators act as a mentor within the community through giving the advice in the community website, helping the people who get in contact and request support and/or trying to find ways to encourage others in the platform (Hui et al., 2014). In reward-based crowdfunding platforms, most project initiators support at least one creator or potential creator who connects with them (Hui et al., 2014). Creators like supporting other people as a consequence of cognitive component of social identity approach such as "I like other members of my group" (Ellemers et al., 1999) and they may learn and get benefit from the members they assist during the interaction (Hui et al., 2014).

Some skillful members also move their support along from financial and intellectual level to practice. For instance, if a creator needs help about a qualified work such as preparing video for a campaign material, a member within the community may allocate time to support the creator by providing required equipment like video and/or teaching her/him how to do it and/or helping the creator while making the video (Hui et al., 2014). This kind of free professional support is a result of in group favoritism and is very important for sustainability of co-creation in crowdfunding platforms.

All kind of crowdfunding platforms offer their members to be in a supportive atmosphere rather than competitive, one difference from the entrepreneurship in order to strengthen their identification level. Financial reciprocity approach is common in the equity-based crowdfunding models (Bretschneider et al., 2014). In the lending-based crowdfunding model, cognitive identification as a result reciprocity fosters community participation and creation activities (Allison, Davis, Short, & Webb, 2015). Hui et al. (2014) assert that if project initiators feel responsible for returning knowledge and economic sources to community, growth of the community will be achieved in the reward-based crowdfunding models.

## Collaboration within the Community

Although monetary support is in the center of the crowdfunding communities, the crowdfunding setting is more socially collaborative unlike the other creative communities (Gerber & Hui, 2013). Financial transaction and following the project provide long-term engagement with the crowdfunding projects and strengthen the co-creation activities between funders and creators (Gerber et al., 2013). Because of this kind of collaborative space, creators are encouraged to share their projects to benefit from interactivity and committed communication to improve their works.

Community-based approach in the crowdfunding platforms enables potential creators to motivate, facilitate and make improvement associated with their works by benefiting from shared experiences and given feedbacks in the community website. In the reward- based crowdfunding models, for example, creators may get inspired about their campaign details from other project creators with similar interest such as determining the funding target, marketing their campaign, and identifying project rewards (Hui et al., 2012). If creators learn from this platform they also share their experiences and comments within the community, hence, sustainability of the community is provided through identification.

Giving advices to other creators is indispensable part of especially reward-based crowdfunding models. In the community each creator always come up with requests from other creators who are novice or has failed about their works and he/she gives feedback them due to seeing himself/herself as a part of the community (Gerber & Hui, 2013). Creators motivate each other to provide qualified ideas and share some tips learned during their campaigns (Hui et al., 2014; Gerber & Hui, 2013). Creators believe that supporting to others even they do the similar things is a key mechanism to continue a sustainable system (Hui et al, 2014).

## Evaluative Component of Identification

Concerning the evaluative component of social identity, identification with a group includes group self-esteem dynamics which can be defined as an affirmative or an unfavorable meaning referred to group membership such as "I think my group has little to be proud of", "I feel good about my group", "I have little respect for my group", "I would rather not tell that I belong to this group" and "I am a valuable member of the group" (Ellemers et al., 1999; Bagozzi & Dholakia, 2006). Evaluative component of social identity has a strong impact on factors of appreciation of the structure of crowdfunding, rely on the community, sense of being a valuable member of the community which all have important roles for the sustainability of crowdfunding communities.

### Appreciation of the Structure of Crowdfunding and Rely on the Community

Both creators and funders admire collecting money system of all kind of crowdfunding models. Creators find satisfying to get fund from a great number of small funders instead of a big investor and they appreciate the process of this mechanism. Supporters also like the nature of crowdfunding platform and describe it as valuable since they consider that this kind of space offers an alternative system to innovative people to raise funds even in the idea generation phase (Gerber & Hui, 2013). In civic crowdfunding projects, for instance, both initiators and funders find this system is convenient since public makes possible the change they desire where the government has lack of enough budged (Davies, 2015). In the crowdfunding platforms, risk of the initiatives is lower than the risk of conventional business venture projects (Ordanini et al., 2011). Traditional entrepreneurial ventures involve only a few experienced individuals, whereas there is a large number of participants in the crowdfunding community who support the project and believe it will reach success.

Concerning the evaluative component of the social identity, building trust within the community is an important determinant of the degree of group self-esteem. Especially in the preordering concept of the reward-based crowdfunding models, funders rely on the community and contribute the projects even before the product comes out such as ordering a music album before recorded or a book before written (Gerber et al., 2013). Therefore, one of the main focus of creators in the reward-based and donation-based crowdfunding models is providing confidence such as answering the questions sincerely in the community website, sharing regularly updates and keeping in contact with their audience (Hui et al., 2014). Moreover, trust in project initiators is an important motivation for funders in the equity-based crowdfunding models (Cholakova & Clarysse, 2015; Ahlers, Cumming, Guenther, & Schweizer, 2012).

Not only funders but also creators seek a credible likeminded community (Gerber & Hui, 2013). Creators rely on the community members and when they design their project, they ask members for comments. Even creators fail in their project, they also respect the community to make improvement before launching a new one (Greenberg & Gerber, 2014). As a consequence of evaluative identification, both creator and funder members of crowdfunding models respect their groups and maintain their active participation in the community.

### Sense of Being Valuable Member of the Community

Consistent with the evaluative component of the identification, crowdfunding platforms enable creators and funders to feel as valuable members of the community. When creators are supported financially in the crowdfunding platforms by a large number of people rather than one investor, appreciated about their projects and skills through conversation in the crowdfunding community website and attract attention from other members related to their works, their self-confidence about their competence and their self-esteem increase (Gerber et al., 2013; Gerber & Hui, 2013). Creators appreciate being validated rather than collecting funds (Gerber & Hui, 2013). Many creators who conduct and/or complete successfully their projects report that because of their success some novel creators are inspired from them to launch their projects (Gerber et al., 2013). Online approval by the community motivate creators to complete their projects and/or encourage them to start a new one even if their first projects fail (Gerber & Hui, 2013). They perceive that they gain validation from community and identify themselves as an important member of this group they respect.

Creators respect the community associated with the evaluative component of the identification and they obtain comments about project improvements, quality of the works and design of the products (Hui et al., 2014). Crowdfunding platforms allow funders to be involved in initiatives in the level of strategic decision-making stage due to the high interaction in the online settings and also financial transaction basis (Bellaflamme et al., 2013). Projects are presented in the early period and funders may contribute in the managerial level. Funders are aware that they create value, so they are sensitive in selecting which project is supported through which crowdfunding model (Gerber et al., 2013). Especially in the reward-based and equity-based crowdfunding models supporters may decide the design of the products and make innovative comments related to direction of the projects (Bellaflamme et al., 2013). Moreover, in some cases of finished civic crowdfunding projects such as the Glyncoch Community Center, a public asset financed through the Spacehive platform in 2012, are managed by volunteer funders thus they feel valuable member of the community (Davies, 2015). As a result, funders follow the project closely and participate actively, thus involvement degree within the community increases.

Regarding incentive inclusive nature of the crowdfunding community, financial and/or non-financial transaction may cause a sense of valuable membership of the community. The funding amount given may reveal the given value to the initiative from both perspectives of creators and funders (Gerber et al., 2013). Funders feel as an important member of the community when they get material or immaterial incentives such as having the product first or a limited version of good or seeing her/his name in the list of supporters (Gerber et al., 2013). Gerber and Hui (2013) assert that funders motivate to get symbolic reward such as getting a postcard and seeing their photo on the community website no matter having it takes a long time in the reward-based crowdfunding models.

## Emotional Component of Identification

Concerning the emotional component of social identity, identification with a group includes a sense of emotional involvement which can be defined as affective commitment such as "I would like to continue working with my group", "I dislike being a member of my group", "I would rather belong to the other group" and "I am attached to the group" Algesheimer et al., 2005; Ellemers et al., 1999; Bagozzi & Dholakia, 2006b). In the context of social identification, emotional component of the social identity is a more powerful influencer rather than cognitive component (Ellemers et al., 1999). Loyalty and citizenship in the community are strengthened because of emotional component of social identity (Chiu et al., 2006). People desire to continue their dedicated connection with the community. Ellemers et al. (1999) state that the level of affirmative evaluation of the group depends on the level of emotional identification of the individual if a community committed with a powerful affective social identity is assessed positively. Emotional component of social identity has a strong impact on motivational causes such as philanthropy and dedication which have important roles for maintaining participation in the crowdfunding community.

## Philanthropy

In the crowdfunding platforms, from the emotional viewpoint, funders want to be involved and engaged in the community. The settings offer an opportunity to their members who may not be an initiator, to collaborate in the innovative and/or creative and/or well-intentioned project process (Gerber et al., 2013). Therefore, they want to build a strong relationship within the community (Belleflamme, Lambert, & Schwienbacher, 2014; Macht & Weatherston, 2014).

When considering the emotional component of identification, philanthropic causes can be regarded as one of the main motivations for supporting the crowdfunding projects. Funders like being a part of this kind of altruistic and synergic community to indicate their appreciations of the projects in all kinds of crowdfunding models. As expected, such a sense is dominant in the donation- based crowdfunding models that funders have positive feelings about the community and less doubts (Ahlers et al., 2012). Funders support the projects such as employed women in a non-developed country due to the affective attachment to the group. Funders want to involve the project for charity and altruism and support the work without expectation of material returns (Ahlers et al., 2012). Although, there is no responsibility to give rewards to donors, sometimes immaterial incentives are supplied such as a thank you message or writing the donors' names on the albums cover (Hemer, 2011). Lending-based crowdfunding models also include altruistic and charitable motivations and Galak, Small, and Stephen (2011) define it as "prosocial lending". Allison et al., (2015) reveal that intrinsic causes like willingness to help others are more influential than extrinsic causes like expected revenue. In order to achieve sustainability of the lending-based crowdfunding models, intrinsic motivational characteristics of the works are sharpened (Sakamoto & Nakajima, 2013). Surprisingly, in the equity-based crowdfunding models, funders are motivated to support the projects due to their desire to engage in the community (Ordanini et al., 2011; Cholakova & Clarysse, 2015).

## Dedication

Supporting a project in the crowdfunding platforms means that the funder believes in the creator and the idea and/or project within the community. Thus they give fund, contribute and spread the project to actualize the idea as a consequence of emotional attachment in all kind of crowdfunding models (Ordanini et al., 2011; Gerber et al., 2013). Sometimes, funders want to be part of the project regardless of its cause, when launched by their family members, friends or anyone they know because of the emotional component of the social identity (Gerber et al., 2013). Crowdfunding communities have charming atmospheres that members are very excited about the initiatives. For corporates, building such an affective commitment can take a very long time.

Projects about music and recording arts have a unique characteristic in terms of emotional identification in all kind of crowdfunding models (Ordanini et al., 2011; Mitra, 2012; Bellaflamme et al., 2013; Galuszka & Bystrov, 2014; Booth, 2015; Macht & Weatherston, 2014). Due to the powerful sense of identification level, supporters of this kind of initiatives are called as "fanvestors" (Galuszka & Bystrov, 2014) or "believers" (Bellaflamme et al., 2013). Main motivation of the funders is to support the artist or contact with him/her rather than getting revenue as in the equity-based crowdfunding models for the music industry (Ordanini et al., 2011; Galuszka & Bystrov, 2014). Funders who are very dedicated also give funds to musicians without expecting a return or with a reward such as a copy of the album (Macht & Weatherston, 2014; Mitra, 2012). Booth (2015) states that supporting an artist in the donation-based crowdfunding model offers a strong satisfaction for the funders due to their high level of affective commitment. Funders' motivation in the music industry proves that emotional identification has a critical driver to achieve a sustainable community in crowdfunding context.

## SOLUTIONS AND RECOMMENDATIONS

Through the process of co-creation by more personalized experiences of informed, networked, empowered and active individuals, the markets evolve to be platforms for conversation and interactions between individual consumers, consumer communities, and firms (Prahalad & Ramaswamy, 2004). In parallel, recently expanding crowdfunding as co-creation platforms transform the instrumentality of consumption for reflecting self-identities of individuals as described by Levy (1959) of postmodern era, towards more individual expression by self-innovation, self-production and self-investing for the products consumed. Meanwhile the networked individualism (Wellman, Quan-Haase, Boase, Chen, Diaz, & Miyata, 2003) observed in the new millennium is expected to arise. Compared to the individualistic values of modernity, the individuals are even more detached in their lives within their own innovation, production and investment places but at the same time are in continuous digital touch with other networks. These newly rising tribes of innovators, both creators and funders, will be able to diffuse to the markets and the networks will open up to global markets. There seems to be the start of a new era of co-creation with new driving forces, leaving both the modern and post-modern dynamics of the twentieth century behind.

A remarkable foreseen social contribution of crowdfunding is that it may affect the balance of power between individuals and institutions. This is a new socio-technical development leading to co-creation though more lateral power structure and a sustainable networked society through online participation, self-control, collaboration and sharing. The hierarchies of authority and knowledge in the markets may flatten through connected crowdfunding platforms, increasing the democratization of the manufacturing and market system. More importantly, triggering individual creativity and democratizing innovation, anyone may have the opportunity to be an important actor of the future market. Furthermore, they will share their original experiences online. This will contribute to a newly rising innovation crowd and the early millennium will witness a new base of knowledge, intellectual culture and social capital accumulation. These cyberspaces will turn out to be new virtual communities, network forums, and co-creation interfaces.

A novel form of civic eco-system is expected to rise and have significant development impacts in the global markets. This will be a system of creating in which individuals not only participate in, but are also afforded the digital tools to engage in the act of making and investing directly. Besides global markets, crowdfunding communities may also have significant impacts on local development models. Cities are now increasingly becoming more than just infrastructures and are building up networking platforms for citizens. These innovation networks may affect the urbanization models by offering accessibility to knowledge and funding within the same locality. These will be very important issues for the future sustainability of local employment and development, bringing new opportunities for the empowerment of individual citizens. On the other hand like any new development, this transformative innovation system might lead to an advancement only if it can offer societies clear, considerable and most importantly sustainable benefits over existing practices.

About all social relationships online start as weak ties. Online communities form a basis of large number of weak ties, often with individuals that do not meet over a relationship but because of being members of the same network. Weak ties are very common and valuable because of their flexibility and diversity that enable crowdsourcing. On the other hand for the sustainability, a community must strengthen the ties, inducing high levels of identification, commitment, trust, networking and conforming established norms. Weak ties can be observed in both communities and social networks, whereas strong ties are built up mostly in communities sharing meaningful commonalities of values and topics of interest. Wellman (1992; pp. 211-212) characterizes strong ties as "a sense of the relationship being intimate and special, with a voluntary investment in the tie", "an interest in being together as much as possible through frequent interactions over a long period" and "a sense of mutuality in the relationship, with the partner's needs known and supported". Weak ties are more instrumental than strong ties, providing informational resources rather than support and exchange of confidences (Wellman, 1992). Zhao, Wu, and Xu (2010) reveal that weak ties are in fact the key to information diffusion. Thus deleting weak ties impede the flow of information more than removing strong ties (Granovetter, 1973; De Meo, Ferrara, Fiumara, & Provetti, 2014). In crowdfunding communities, innovative ideas, prototypes or even finished products are discussed on the basis of a situated creativity, then further developed and funded, inspiring to more contribution and long-term attendance of diverse actors.

Propensity to innovate is both facilitated and restricted by existing social community structure. The tendency among entrepreneurs toward innovation is seen to be a function of the patterns of these social ties those entrepreneurs stick to. The relative value of strong and weak ties is a critical aspect of this system. Considering the focus of this study as online crowdfunding communities around a project, these communities need a sufficient number of strong ties to sustain engagement of both creators and funders. However, strong ties on their own are not enough for crowdsourcing and regeneration of novel ideas requiring contribution of refreshing diversity of weak ties over the crowdfunding platforms. This diversity can be considered as the triggering mechanism of the start-up eco system of co-creation through crowdfunding communities (Ruef, 2002). Thus in the micro basis, identification may both be an impetus or a constraint for an online innovation community on macro level. Although sticking is necessary for the flow of pointless information to innovative entrepreneurs, at the same time, entrepreneurs should not fall into the trap of assimilation that might be encouraged by social lodging (Ruef, 2002).

Crowdfunding is a rising market that is also affected by macro context. Concerning the recently evolved regulations in USA through the Jump-start our Business Startups (or JOBS) Act which regulates equity-based crowdfunding models as "non-accredited" investments and legitimize them for starts-ups (Ahlers et al., 2012) and coming legal systems in Europe through Digital Single Market (DSM) which aims to regulate and enhance right of digital skills for individuals and businesses (Digital Single Market, 2015), it is predicted that the crowdfunding mechanism will be affected in several ways in terms of identification. Due to the new legitimized nature of the equity-based and lending-based crowdfunding models, uncertainty is expected to decrease. In parallel, with the increased level of trust in the community, "evaluative identification" may rise. On the other hand, since the dominant crowdfunding category in the market will be expected to change from reward-based model to equity-based model (Cholakova & Clarysse, 2015), this alteration may modify the perception of the crowdfunding platforms from strong community approach to much more business investment approach. This will potentially cause a decrease in the level of "emotional identification". Lastly, new laws facilitate internationalisation of the crowdfunding platforms and thus the size of the community will become larger. Bagozzi and Dholakia (2006b) assert that small size of community have stronger identification than the larger one. The internalization of the crowdfunding platforms may cause decrease in the level of "cognitive identification" within the community. Therefore, aside from the context of being constructed within an innovative culture and be backed up by regulatory policies, the successful management of content that expands the identification of all contributing actors, namely creators and funders over five diverse crowdfunding models is crucial for sustainability of this system.

Managers of crowdfunding platforms must consider that these online communities have their own regulations and group norms and specific characteristics which may play role in community preference of members. For instance, Kickstarter users who are from the USA, has an All-or-Nothing funding structure in which creators have to reach their funding target, otherwise capitals are paid back thus funders and creators do not obtain any monetary support. RocketHub users who are from all around the world, apply the All & More funding method in which although creators do not access their funding target, they can get the fund (Gerber et al., 2013). In line with online communities in which monetary incentives are used to increase participation to community (Garnefeld, Iseke, & Krebs, 2012), some of crowdfunding platforms offer discount rates in submission fees for creators in the first few initiatives they launch, if creators access their funding target. In addition to monetary aspects, similar to the group norms of online communities, in the social networking aspect, crowdfunding platforms have some sanctions such as obligating the creators to share their experiences with other creators in the community website (Hui et al., 2014). The nature of the online platform with its content and the context will determine the identification mechanisms of actors thus weaknesses and strengths of ties among them. A tie can also be characterized by its position in the network. Social networks are composed of communities, connected among them by long range ties known as bridges and that bonding capital creates and sustains the connections that keep individual community groups consistent. Linking macro and micro approaches in sociology through "strength of weak ties" theory, Granovetter (1973) states that communities with 'bridging' social capital over weak ties across groups, along with 'bonding' social capital over strong ties within groups are the most effective in organizing for collective action. He also develops this notion that weak ties act as bridges between groups and are important for the diffusion of new information across the network, while strong ties are usually located at the interior of the groups. De Meo et al. (2014) classify weak ties as linking individuals belonging to different communities and strong ties as connecting users in the same community. They suggest individuals in online communities may create well-connected communities, while weak ties make way to cohesion, optimizing the coverage of information spread.

More substantial for managers is that this new production technology will bring ambiguity to the meaning of craft and entrepreneurship. Since the home is going to be the new production zone or workplace, new arrangements, including articles in the labor law for self-employment at home, how to be taxed, to get organized and issues concerning social security, in addition, branding issue will become increasingly complicated. The difficulty of controlling what is offered as innovation is a significant threat also. Just as personal computers enabled everybody to become creators of content, these platforms will enable everybody to become creators of anything, even dangerous, insecure or un-durable objects.

There must be strict mechanisms to audit whether a dangerous product like a gun or an unsafe product like a toy that will poison a child are not produced. Another legal problem is about the possible intellectual property and privacy violations. The easy sharing of innovative ideas and associated files by just e-mailing and scanning will create a vulnerable system in terms of protecting intellectual property rights. Also privacy and protection of personal data will become critical issues, challenging the existing regional and global regulatory systems. Differences in laws and regulations of countries will make it further difficult to overcome this threat. All these necessitate training and guiding mechanisms that the managers of the crowdfunding platforms must consider to encourage a sustainable participation and identification among creators and funders.

## CONCLUSION

As with any significant transformation, the probable impacts of crowdfunding is open to multidisciplinary contributions, criticisms and discussions. It is been a very short time span in history that these specific online community platforms have been launched. Still rather big companies carry out the requisite monetary and human capital, not only for actualizing internal research but also for reaching the outside ideas to leverage their innovation skills and expertise. Since the innovative ideas are very valuable for competitive markets, the free access to crowdsourcing platforms opens up a deep debate whether the creators of ideas will be unpaid research and development labors of big corporations open to exploitation in the capitalistic system (Arvidson, 2008; Ritzer & Jurgenson, 2010).

On the other hand, the significant contributions of crowdfunding systems for the moment as discussed above are too important to be ignored. This new value co-creating innovation system will inevitably have impacts in a range extending from individuals to local citizens, at the end to all sizes of corporatized market players. Progress is inevitable and human beings are eager to control their destiny. Crowdfunding of innovations is in its preliminary stages. Once more the human beings are the key for a good or an evil. Pessimism or fears about the unknown should not prevent this happening. What is known is that; the economy of abundance has been left behind. Under the current social conditions, namely the global economic recession and the unknown future it implies, this seems to be a more sustainable road that individuals would like to wander, identify them-selves and live the pleasure of being an active actor in a community of co-creation.

So managers of crowdfunding platforms, as stakeholders of these innovation systems, must contribute to their sustainability as much as creators and funders. In order to scale up the innovation communities, it is important for managers of these crowdfunding platforms to identify and motivate the people which do not only grow strong identification with their group, but who also have many weaker ties to other groups. Additionally managers must build up bridging mechanisms among the ones who have weaker ties. Because people with weaker ties tend to disconnect unless their identification is sufficiently strong based upon the perceived value of the reconnection. The most critical; identification as the key to sustainability must open up the ways to collaborations of the enrichment of heterogenous crowds thus must not end up with homogenous crowds inhibiting creativity.

## REFERENCES


Ahlers, G. K. C., Cumming, D. J., Guenther, C., & Schweizer, D. (2012). *Signaling in equity crowdfunding*. Retrieved May 15, 2015, from http://ssrn.com/abstract=2161587

Algesheimer, R., Dholakia, U. M., & Herrmann, A. (2005). The social influence of brand community: Evidence from European car clubs. *Journal of Marketing*, *69*(3), 19–34. doi:10.1509/jmkg.69.3.19.66363

Allison, T. H., Davis, B. C., Short, J. C., & Webb, J. W. (2015, January). Crowdfunding in a prosocial microlending environment: Examining the role of intrinsic versus extrinsic cues. *Entrepreneurship Theory and Practice,* 53-73.

Amabile, T. M. (1988). A model of creativity and innovation in organizations. *Research in Organizational Behavior*, *10*, 123–167.

Arvidsson, A. (2008). The ethical economy of customer co-production. *Journal of Macromarketing*, *28*(4), 326–338. doi:10.1177/0276146708326077

Bagozzi, R. P., & Dholakia, U. M. (2006a). Open source software user communities: A study of participation in Linux user groups. *Management Science*, *52*(7), 1099–1115. doi:10.1287/mnsc.1060.0545

Bagozzi, R. P., & Dholakia, U. M. (2006b). Antecedents and purchase consequences of customer participation in small group brand communities. *International Journal of Research in Marketing*, *23*(1), 45–61. doi:10.1016/j.ijresmar.2006.01.005

Bagozzi, R. P., & Lee, K. H. (2002). Multiple routes for social influence: The role of compliance, internalization, and social identity. *Social Psychology Quarterly*, *65*(3), 226–247. doi:10.2307/3090121

Belleflamme, P., Lambert, T., & Schwienbacher, A. (2014). Crowdfunding: Tapping the right crowd. *Journal of Business Venturing*, *29*(5), 585–609. doi:10.1016/j.jbusvent.2013.07.003

Betts, B. (2010). Bringing the factory home. *Engineering & Technology*, *5*(8), 56–58. doi:10.1049/et.2010.0813

Booth, P. (2015). Crowdfunding: A spimatic application of digital fandom. *New Media & Society*, *17*(2), 149–166. doi:10.1177/1461444814558907

Bradshaw, S., Bowyer, A., & Haufe, P. (2010). The intellectual property implications of low-cost 3D printing. *ScriptEd*, *7*(1), 5–31.

Bretschneider, U., Knaub, K., & Wieck, E. (2014). Motivations for crowdfunding: what drives the crowd to invest in start-ups?*Twenty Second European Conference on Information Systems.*

Buysere, K. D., Gajda, O., Kleverlaan, R., & Marom, D. (2012). *A framework for European crowdfunding*. Retrieved May 14, 2015, from www.crowdfundingframework.eu

Camarinha-Matos, L., & Afsarmanesh, H. (2006). Collaborative networks: value creation in a knowledge society. *PROLAMAT 2006 Conference Proceedings- Knowledge Enterprise: Intelligent Strategies in Product Design, Manufacturing, and Management*, (pp. 26-40). Springer.

Carpini, M. X. D. (2000). *Gen.Com: Youth civic engagement and the new information environment. Departmental Paper*. University of Pennsylvania, Annenberg School for Communication.

Chiu, C. M., Hsu, M. H., & Wang, E. T. G. (2006). Understanding knowledge sharing in virtual communities: An integration of social capital and social cognitive theories. *Decision Support Systems*, *42*(3), 1872–1888. doi:10.1016/j.dss.2006.04.001

Cholakova, M., & Clarysse, B. (2015, January). Does the possibility to make equity investments in crowdfunding projects crowd out reward-based investments?. *Entrepreneurship Theory and Practice*, 145-172.

Costa, D. L., & Kahn, M. E. (2002). *Civic engagement and community heterogeneity: An economist's perspective*. Presentation on Conference on Social Connectedness and Public Activism. Harvard University.

Cova, B., & Cova, V. (2012). On the road to prosumption: Marketing discourse and the development of consumer competencies. *Consumption Markets & Culture*, *15*(2), 149–168. doi:10.1080/10253866.2012.654956

Crane, A., & Ruebottom, T. (2011). Stakeholder theory and social identity: Rethinking stakeholder identification. *Journal of Business Ethics*, *102*(S1), 77–87. doi:10.1007/s10551-011-1191-4

Davies, R. (2015). Three provocations for civic crowdfunding. *Information Communication and Society*, *18*(3), 342–355. doi:10.1080/1369118X.2014.989878

De Meo, P., Ferrara, E., Fiumara, G., & Provetti, A. (2014). On Facebook, most ties are weak. *Communications of the ACM*, *57*(11), 78–84. doi:10.1145/2629438

*Digital Single Market*. (2015). Retrieved May 26, 2015, from http://ec.europa.eu/digital-agenda/en/digital-single-market

Dimitrov, D., Schreve, K., & de Beer, N. (2006). Advances in three dimensional printing– state of the art and future perspectives. *Rapid Prototyping Journal*, *12*(3), 136–147. doi:10.1108/13552540610670717

Dovidio, J. F., Gaertner, S. L., Pearson, A. R., & Riek, B. M. (2005). Social identities and social context: Social attitudes and personal well-being. *Social Identification in Groups: Advances in Group Processes*, *22*, 231–260. doi:10.1016/S0882-6145(05)22009-X

Ellemers, N., Kortekaas, P., & Ouwerkerk, J. W. (1999). Self-categorisation, commitment to the group and group self-esteem as related but distinct aspects of social identity. *European Journal of Social Psychology*, *29*(2-3), 371–389. doi:10.1002/(SICI)1099-0992(199903/05)29:2/3<371::AID-EJSP932>3.0.CO;2-U

European Commission Report. (2013). *3D printing*. Retrieved January 9, 2015, from http://ec.europa.eu/digital-agenda/futurium/en/content/3d-printing

Galak, J., Small, D., & Stephen, A. T. (2011). Microfinance decision making: A field study of prosocial lending. *JMR, Journal of Marketing Research*, *48*(SPL), 130–137. doi:10.1509/jmkr.48.SPL.S130

Galuszka, P., & Bystov, V. (2014). The rise of fanvestors: A study of a crowdfunding community. *First Monday*, *19*(5). doi:10.5210/fm.v19i5.4117

Garnefeld, I., Iseke, A., & Krebs, A. (2012). Explicit incentives in online communities: Boon or bane? *International Journal of Electronic Commerce*, *17*(1), 11–37. doi:10.2753/JEC1086-4415170101

Gerber, E. M., & Hui, J. S. (2013). Crowdfunding: Motivations and deterrents for participation. *ACM Transactions on Computer-Human Interaction. 20*(6). *Article*, *34*(December), 1–32.

Gerber, E. M., Hui, J. S., & Kuo, P. Y. (2013). *Crowdfunding: why People are motivated to post and fund Projects on crowdfunding platforms*. Retrieved May 16, 2015, from http://www.juliehui.org/wp-content/uploads/2013/04/CSCW_Crowdfunding_Final.pdf

Granovetter, M. S. (1973). The strength of weak ties. *American Journal of Sociology*, *78*(6), 1360–1380. doi:10.1086/225469

Greenberg, M. D., & Gerber, E. M. (2014). Learning to fail: Experiencing public failure online through crowdfunding. In *Proceedings of the SIGCHI Conference on Human Factors in Computing Systems*, (pp. 581-590).

Hars, A., & Ou, S. (2002). Working for free? Motivations for participating in open-source projects. *International Journal of Electronic Commerce*, *6*(3), 25–39.

Hemer, J. (2011). *A snapshot on crowdfunding*. Working papers firms and region, No. R2/2011. Retrieved May 16, 2015, from http://hdl.handle.net/10419/52302

Hui, J. S., Gerber, E. M., & Greenberg, M. D. (2012). Easy money? The demands of crowdfunding work. *Northwestern University Segal Design Institute Technical Report*, *12*, 4.

Hui, J. S., Greenberg, M. D., & Gerber, E. M. (2014). Understanding the role of community in crowdfunding work. In *Proceedings of the 17th ACM conference on Computer supported cooperative work & social computing*, (pp. 62-74). ACM.

Kirby, E., & Worner, S. (2014). *Crowdfunding: An infant industry growing fast.* Staff Working Paper of the International Organisation of Securities Commissions Research Department. Retrieved May 14, 2015, from http://www.memofin.fr/uploads/library/pdf/Crowd-funding-An-Infant-Industry-Growing-Fast[1].pdf

Kotler, P. (1986). The prosumer movement: A new challenge for marketers. *Advances in Consumer Research. Association for Consumer Research (U. S.)*, *13*(1), 510–513.

Kozinets, R. V. (1999). E-tribalized marketing?: The strategic implications of virtual communities of consumption. *European Management Journal*, *17*(3), 252–264. doi:10.1016/S0263-2373(99)00004-3

Kraut, R. E., & Resnick, P. (2012). *Building successful online communities: Evidence-based social design*. Cambridge, MA: MIT Press.

Lakhani, K. R., & Wolf, R. G. (2005). Why hackers do what they do: Understanding motivation and effort in free/open source software projects. In J. Feller, B. Fitzgergerald, S. Hissam, & K. R. Lakhani (Eds.), *Perspectives on Free and Open Source Software* (pp. 3–27). Cambridge, MA: MIT Press.

Lambert, T., & Schwienbacher, A. (2010). *An empirical analysis of crowdfunding*. Retrieved May 10, 2015, from http://www.crowdsourcing.org/document/an-empirical-analysis-of-crowdfunding-/2458

Levy, S. J. (1959). Symbols for sale. *Harvard Business Review*, *37*(4), 117–124.

Lin, Y., Boh, W. F., & Goh, K. H. (2014). How different are crowdfunders? Examining archetypes of crowdfunders and their choice of projects. *Organizational Communication & Information Systems*, *1*, 13309.

Macht, S. A., & Weatherston, J. (2014). The benefits of online crowdfunding for fund-seeking business ventures. *Strategic Change*, *23*(1-2), 1–14. doi:10.1002/jsc.1955

Mael, F. A., & Ashforth, B. E. (1992). Alumni and their alma mater: A partial test of the reformulated model of organizational identification. *Journal of Organizational Behavior*, *13*(2), 103–123. doi:10.1002/job.4030130202

*Massolution's 2015 CF Industry Report*. (2015). Retrieved May 10, 2015, from http://www.crowdsourcing.org/editorial/global-crowdfunding-market-to-reach-344b-in-2015-predicts-massolutions-2015cf-industry-report/45376

Mitra, D. (2012). The role of crowdfunding in entrepreneurial finance. *Delhi Business Review*, *13*(2), 67–72.

Morlock, G. S. (2010). Personal fabrication. *Communications of the ACM*, *53*(10), 14–15. doi:10.1145/1831407.1831414

Muller, M., Geyer, W., Soule, T., & Wafer, J. (2014). Geographical and organizational commonalities in enterprise crowdfunding. CSCW Technology and Information Workers.

Ordanini, A., Miceli, L., Pizzetti, M., & Parasuraman, A. (2011). Crowdfunding: Transforming customers into investors through innovative service platforms. *Journal of Service Management*, *22*(4), 443–470. doi:10.1108/09564231111155079

Prahalad, C. K., & Ramaswamy, V. (2004). Co-creation experiences: The next practice in value creation. *Journal of Interactive Marketing*, *18*(3), 5–14. doi:10.1002/dir.20015

Ritzer, G., & Jurgenson, N. (2010). Production, consumption, prosumption: The nature of capitalism in the age of the digital prosumer. *Journal of Consumer Culture*, *10*(1), 13–36. doi:10.1177/1469540509354673

Ruef, M. (2002). Strong ties, weak ties and islands: Structural and cultural predictors of organizational innovation. *Industrial and Corporate Change*, *11*(3), 427–449. doi:10.1093/icc/11.3.427

Sakamoto, M., & Nakajima, T. (2013). Micro-crowdfunding: Achieving a sustainable society through economic and social incentives in micro-level crowdfunding.*12th International Conference on Mobile and Ubiquitous Multimedia*. doi:10.1145/2541831.2541838

Shen, K. (2012). Identification vs. self-verification in virtual communities (VC): Theoretical gaps and design implications. In P. Maret & C. Morr (Eds.), *Virtual Community Building and the Information Society: Current and Future Directions* (pp. 208–236). IGI Global. doi:10.4018/978-1-60960-869-9.ch011

Stiver, A., Barroca, L., Minocha, S., Richards, M., & Roberts, D. (2015). Civic crowdfunding research: Challenges, opportunities, and future agenda. *New Media & Society*, *17*(2), 249–271. doi:10.1177/1461444814558914

Wellman, B. (1992). Which ties provide what kinds of support? *Advances in Group Processes*, *9*, 207–235.

Wellman, B. (2002). Little Boxes, Glocalization, and Networked Individualism. In M. Tanabe, P. van den Besselaar, & T. Ishida (Eds.), *Digital Cities* (pp. 10–25). Berlin: Springer-Verlag.

Wellman, B., Quan-Haase, A., Boase, J., Chen, K. H., Diaz, I., & Miyata, K. (2003). The social affordances of the Internet for networked individualism. *Journal of Computer-Mediated Communication*, *8*(3), 1–43.

Wright, P. K. (2005). Rapid prototyping consumer product. *Communications of the ACM*, *48*(6), 36–41. doi:10.1145/1064830.1064856

Zhao, J., Wu, J., & Xu, K. (2010). Weak ties: Subtle role of information diffusion in online social networks. *Physical Review*, *82*(1), 1–8. PMID:20866687

# Chapter 10
# Principles in Crowdfunding Benefits and Validation Tools:
## How to Ensure Campaign Success

**Paul Morrell Niederer**
*Crowdsourcing Week, Australia*

## ABSTRACT

*This chapter provides the key benefits of a crowdfunding campaign together with a rich variety of validation tools provided to enhance the chance of a crowdfunding campaigns success. Benefits cannot be achieved in a crowdfunding campaign if it is not successful. Thorough validation of the campaign upfront can heighten success. The author in an internationally respected crowdfunding authority who has managed over 300 crowdfunding campaigns through the platforms he has been involved with. The author lists 11 benefits and 11 validation tools together with his own road tested processes plus a plethora of links to tools to assist both the novice and experienced crowdfunder.*

## CROWDFUNDING BENEFITS AND VALIDATION TOOLS

This chapter takes a look at the *benefits of crowdfunding* and some of the many tools that can be used to check or validate your campaign before or after it goes live. It is often difficult for people to know if your campaign is successful or not. Validation tools can assist in this assessment.

With crowdfunding some of the benefits are obvious but some of them are often only visible to the campaign owner. Here is an overview of the benefits of crowdfunding and some of the tools that can be used to validate campaigns.

DOI: 10.4018/978-1-5225-0568-6.ch010

## Benefits

### The Benefit Is Money

"Show me the Money". The oft-quoted phrase from Tom Cruise is to most people the primary benefit of a crowdfunding campaign. Most people have a target amount of money they would like to receive and also a "beyond my wildest dreams" amount that they have seen evidenced in successful campaigns for coolers, watches and beehives. Some platforms are structured around keeping every dollar you raise while others have a minimum target you have to raise before you get any money. Before choosing the platform for your crowdfunding raise you should base your selection on these criteria if the "money" is important to you.

### The Benefit of Staying Compliant

Collecting money in collection boxes, tins or in your hand from supporters may be quick and easy but it is far from transparent and often not compliant. Most countries have regulations around contributions where the donor gets nothing in return. Usually those collecting need to have permission and identification.

With crowdfunding there is usually a benefit or reward transferred at an agreed future time to the contributor. By doing this on a crowdfunding platform the transaction is transparent, can be updated as delivery gets nearer, or delayed, and the financial part of the transaction is properly handled.

In the area of equity or debt crowdfunding compliance is even more important as there are strict rules about how these transactions are handled and recorded.

Since the beginning of time people have lent money and invested in others businesses and it is fair to say that many of these transactions have no loan agreement or share certificates and this aspect often sours relationships when things are go wrong. A property constructed compliant crowdfunding platform can alleviate this by transparently keeping the handling of contributions compliant.

### Reaching Contributors Far and Wide Is Very Beneficial

A platform provides the opportunity to gather investors, customers, contributors, supporters, followers, ambassadors, influencers and evangelists far and wide. This is because of the intersection of two major trends.

The two major trends and "technical disruption" and "cultural disruption".

Figure 1 is a modified version of a chart from Jonathan Sandlund's (2013) piece "The Rise of Meaningful Investing at TheCrowdCafe."

The first is the ability of the Internet to easily and efficiently handle financial transactions without people going to the bank, posting a letter or physically having to front people they know and don't know. The Internet and the growing number of Internet based financial services can handle transactions from a few cents to hundreds of thousands of dollars. It also provides much needed access to capital for those locally or in far off lands

The second trend is the growing demand for human beings to be more meaningfully involved in the things they do, the things they buy and the things they invest in. At the intersection of these trends is where crowdfunding platforms have evolved. If you are interested in finding people that are passionate about the fact that the bee population is declining and believe the solution is to have a beehive in your

*Figure 1.*

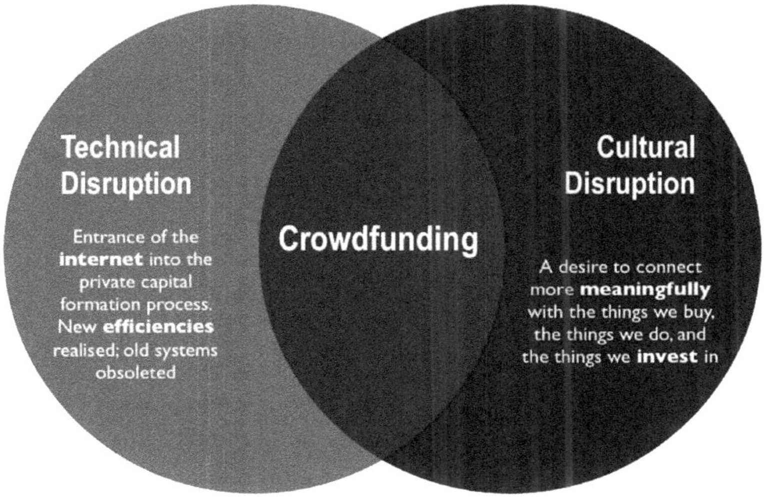


backyard then a properly constructed crowdfunding campaign will connect you with thousands of people worldwide that have this interest or desire to make a meaningful difference.

In most cases you can tell your audience or "your crowd" about this in a single click. Gone are the days of printing leaflets, flyers, posters and other material to catch people's attention about your opportunity. A good crowdfunding platform can communicate all they need to know in just a few clicks. Contributors or investors can also do due-diligence on you or your company or your campaign by entering a few simple phrases into Google and if there is something astray then it is usually recorded in discussion sections of the platform it the platform operator has not already taken the campaign down.

## Hedging Risks Is a Big Benefit

Not so many years ago if someone wanted to launch a new product they had to have it ready for sale before they started approaching buyers. They had to design it, source the parts, build a prototype, get someone to build it, manufacture a good supply and then start advertising or showing it to wholesalers. A good example is the book industry. Most budding authors had to agree to get 5,000 or 10,000 books printed before any book distributor would even look at carrying their titles. On a crowdfunding platform today a budding author can "pre-sell" copies of their book in a crowdfunding campaign and then get the exact right amount printed at an on-demand printer paying with the funds used from the crowdfunding campaign. Gone are the days of an author ending up with thousands of copies of their book in their garage.

This means that authors, investors, innovators, creators and artists can test the water effectively with their present and future audiences to see if there is enough demand for their next venture. This considerably lowers risk and is good for society, as only what is needed will be produced.

## It Allows the Benefit of Crowdsourcing of Creation and Innovation

A crowdfunding campaign can begin long before the page goes live to collect money. The product idea can be floated to the innovators crowd and as the campaign is built so are ideas as to what the suitable rewards and benefits of the campaign are. While a creator may have an idea of what their audience would like in exchange for their contribution they can often be surprised at which of the rewards end up being the most popular. A way to meet this head on is to involve "their crowd" in the creation of the campaign rewards. While the creator might think an extra copy of the book reward will be the number one reward, the supporter might actually prefer an audio version to listen to while they take long walks or drives.

In addition, the actual product or service can be crowdsourced by ideas from early followers and supporters. They may be able to assist you to fine-tune your product or service so that you can have a better chance of success.

## Crowdfunding Is Beneficial in Introducing Prospective Customers

In any marketing endeavour gathering a following is one of the surest ways to achieve sales success. One of the major benefits of a crowdfunding campaign is that it serves as a marketing tool and not only tests your marketing and prospective market audience but also builds a following that can have other products and services marketed to them in the future. This means the promotor get a lot more "free" promotion and often a very loyal following as they have shared in their journey and shared their glory and frustrations.

Popular, attractive, weird, interesting are things that get free PR. If you structure the messaging in your campaign properly you will gain extra customers form article in newspapers, blogs, on the radio and sometimes if you are lucky on TV.

These contacts can come from 1, 2 and 3 degrees out from you and your opportunity. They can come from friends, friends of friends and others who you have no direct contact with you at the moment. As your relationship develops with your "new" contacts some will cross the line and become your customers. Others may become your customers for a second product or service.

## One Benefit Is the Ease of Operations

Because a crowdfunding campaign can be run virtually from a bedroom there is the benefit of not having to get dressed up and visit an office or meet with contributors or investors. During the course of the fundraising campaign there is seamless transition from launch to close as the days pass by. The beauty of a crowdfunding campaign is that it follows a process and a sequence that moves forward in a manner that is very visual and transparent.

In the past a funding campaigner had to build a website, get it hosted, organise merchant facilities with the bank and go door to door to get sales. A crowdfunding platform organises most of these things for you and allows you to do most of it, with ease from the workplace you feel most comfortable with.

## The Benefits of Pre-Selling

Carrying stock, or guessing how many items to produce, is a real cost and hassle in business so it is refreshing that in crowdfunding the transactions are effectively "pre-selling". That means people come to a funding endeavours profile page on the website, order and this results in a "pre-sale". This gives the promotor the opportunity to place the production order once the campaign has closed as all contributors know from the words on the profile page that the promotor has allowed time to produce the rewards before the supporter receives them.

Pre-selling at this level was never possible before internet based funding platforms were invented. Pre-selling considerably reduces the risk for failure and is a great way to see if there really is a future for a product with customers the promotor doesn't even know yet.

## Being Free Is a Benefit

Most crowdfunding platforms do not charge any fees upfront. This means that any costs incurred are usually paid out of the funds people pay for the "pre-sale". That means costs come from their money not yours. This has benefits. You avoid incurring debts you may have needed to incur to fund stock and promotion. This means that because you haven't used up any of your capital in supporting your opportunity you are financially free to experiment with more campaigns. Just because the first one you didn't work it is not the end. Most successful entrepreneurs often say the thing they have in common with other successful entrepreneurs is that most of them have failed their way to success.

## A Beneficial Way to Develop Ideas for a New Product

Many large companies scour crowdfunding listings to get new ideas for products. If you have a general idea for a product you can also look through similar listings to fine tune what you are doing. The benefit of this is that you can check how similar products are travelling and see how the best of them are pitching and promoting their products.

Through this process you will also develop initial prototypes. For most platforms you need to have an actual prototype before you list your product. This is because the platform wants to be assured that you will be able to actually deliver the product once the time comes to deliver on your commitments made for the funds you have received. Having a launch goal for your product means you are then focussed on getting your prototype ready so that you qualify to list and no that what you describe can actually be produced. This is your proof of concept and you can stand tall and show people you can deliver what you have promised if they don't believe you can make one of those!

Whether you use crowdfunding to develop your products by looking at other people's products and progress or you use the crowd to assist you in building your product or service crowdfunding is a very beneficial way to develop new products.

## The Benefit of Moving Your Followers from a Transaction to a Relationship

One of the main benefits of a crowdfunding campaign is that the customers you gain through the platform are your own customers. That means you can communicate with them and build a solid relationship and eventually sell them your next product. This can either be direct or via another crowdfunding campaign.

Moving people from being a follower to a longer-term relationship through interactions and transactions is more of an art than a science. To get an initial understanding of whether your product, service or campaign will work you need to start small. The best place to start is by sharing your idea with friends, fans family and followers. When dealing with them, engage them, observe and record reactions, listen to questions being asked, and try to determine which parts of your idea isn't really resonating. The feedback from your friends, fans, family and followers should offer the insight you need do some further research and development.

One way to do this is to build a business plan. For those who aren't seeking investment nor plan on building a long-term business, developing a business plan may seem over-the-top. However, the process of creating a plan will force you to do focussed research, answer difficult questions about your idea, and help figure out your audience. A lot of this can be achieved by studying the steps involved in moving your followers from initial contact to a long-term relationship.

Important! The benefits of crowdfunding can only be obtained if you have a successful crowdfunding campaign. If you are in doubt here is a quick way to check. This is a model developed by the writer and used to pre-assess hundreds of raisings. The model is called "Fundability Circles".

Before venturing on to the second part of this chapter that covers Validation Tools, here is a good way to check if your funding campaign is worth proceeding with.

Basically you have 3 categories to assess your prospective funding campaign by:

1. The Story,
2. The Team,
3. The Supporters.

Each of the circles reflects a strong success pillar.

1. A Story told with clarity. It needs to be convincing, compelling and credible.
2. An enthusiastic, grounded Team. It needs to be a balanced (Steve Jobs and Steve Wozniak), passionate, capable and likeable team.
3. Supporters who believe in what you are doing. There needs to be lots of suitable people to share the story with and engage with.

Figure 2 was first published by Paul Niederer (2015) in a blog titled "A Simple Equity Crowdfunding Fundability Evaluation Tool Revisited" at paulniederer.com.

*Figure 2.*

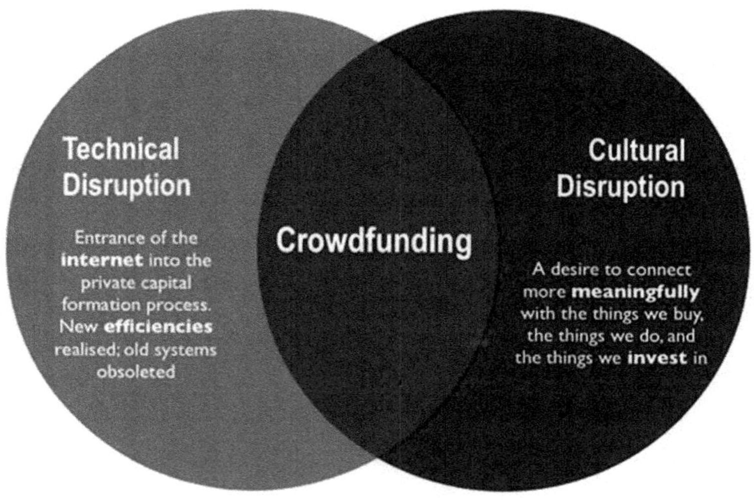

In addition, it is at the intersection of the circles that a simple check reinforces the overriding strength of Story, Team and Supporters.

1. **Relevance:** The story being pitched needs to be relevant to the supporters of the product, the service, the technology, the team, the geography or for whatever reason they are "supporting".
2. **Capability:** The team must be capable of implementing the story they are selling.
3. **Credibility:** The supporters need to see the team as credible and worthy of their trust.

You can work out your own scoring system depending on the weight you put on each part but a good guide is to rate each part out of ten so if the total is say 27, 28, 29 or 30 you know you have a good opportunity.

## Validation Tools

In the old days a company with an opportunity or a new product would design and build a prototype, manufacture it, create a professional marketing campaign to target an unproven concept to a group of people that may or may not be interested. The strategy here is often referred to as 'build it and they will come'.

You would imagine this only happens to small companies. Untrue. On the physical product side there are many examples in the corporate world. Some examples of this are Amazon's Fire product, Zune from Microsoft and the innovative Newton from Apple. These companies may or may not have used some validation tools. Perhaps there was validation through consumer focus groups or customer trials or other methods but none of these products had the advantage of the extensive validation tools afforded by the crowdfunding process.

What seem basic questions to crowd funders these days were probably not even considered by the big corporates. Would they have asked questions like: Is the campaign going to succeed at a crowd engagement level? Could it fail by getting no supporters? What is a success in terms of both exposure and funds pledged by supporters? How big a success is a success and would we be able to cope if we got al to more interest than we expected.? Take for example the recent "FlowHive" raise for your own personal beehive. "Flow Hive: Honey on Tap Directly From Your Beehive" They initially had a target of selling of around USD100,000 but pre-sold over 37,000 units for over USD12 million. I doubt you would ever find a marketing department in a large corporate that would overshoot their targets by that much. The usually considerably over estimate sales and are disappointed as a result.

Crowdfunding has become the ideal low-cost, low-risk way to validate consumer demand for a product or service. No up-front investments. No misleading consumer surveys. With crowdfunding backers vote with their money so it is a live visible survey or window into possible success or not.

However, you will not be the only one asking these questions about your crowdfunding campaign. A crowdfunding campaign is very visible to both the crowdfunding industry, journalists and public and amongst them there is a heavy interest in how campaigns fare.

Validation is a way of confirming that a campaign, service or product meets the needs of its potential users. For start-ups and campaign creators there are many tools to assist in the validation of success possibility. Here are some ways:

## The Campaign Itself as Validation

Instead of investing money and time into prototyping, manufacturing and marketing unproven concepts, both small and large opportunity creators are increasingly turning first to crowdfunding to validate demand and potential success. One of the best indicators of this is of course the campaign itself. It is out there in the open for all to see and people can see visually if you have met your goal, exceeded it magnificently or failed. Some campaigns fail the first time around and then with some re-focus and become very successful. This actually happened with the Coolest Cooler. How the campaign actually performs is where the rubber meets the road. You should have at least 40% of your contributions lined up before your campaign goes live so that it gathers the traction and momentum it needs to succeed. Many people make the mistake of "build it and they will come". They believe that just putting it up on a known platform will be enough. It never is. You need to work the campaign to get success. The outcome of the campaign is one of the best tools to validate how the campaign did.

## Validation by Feedback

Validation doesn't always come in numeric terms. Meaning it's not always about the numbers reaching your profile page or how many contribute. Sometimes you need to get back feedback from potential customers and backers as to what they really think of your product and campaign. This can assist you in reshaping your product range or campaign. Some people may say, "I love your goal but the rewards you are offering just don't do it for me," or "I don't believe the rewards don't offer the value I am seeking," or they may even suggest "if you fine-tuned it as follows it would have more value for me." Now that's pure crowdsourcing!

This feedback loop will continually improve the opportunity and gives stronger and stronger validation as improvements are implemented.

## Mentor Validation

There are many very knowledgeable and experienced people in the world of crowdfunding. They can often give you an indication of how your campaign looks and where it may need improvement. This doesn't just apply to the actual listing but all aspects of the campaign. A mentor is a sounding board that can pass on wisdom without costing you money. Seek one out. Most of them would be honoured that you sought them out as a mentor.

Validation is an ongoing process The nature of a mentor is that they assist you over a period of time and you have an opportunity to implement the wisdom your relationship with them delivers.

## Validation through Product Prototyping

Sometimes you may be unsure of what colours will be most popular with your customers or which models or variations. You have the opportunity of testing product prototypes with your crowdfunding campaign. Each colour can be a separate reward so you will know by the number of people that ordered which are the colours or models you should produce. This gives you raw feedback direct from the market about what you should produce.

You can structure your rewards specifically around this objective without feeling you are manipulating your future contributors. See them as part of your campaign as it is because of their passion you are being funded. They want to have their say and assist you in delivering the best outcomes you possibly can

## Validation by Crowdfunding Consultants

If you have the budget a crowdfunding consultant can be a good addition to your crowdfunding campaign. Make sure they have worked with the platform you are using and more importantly have a list of campaigns that have given their client a successful outcome. Be wary of any consultant that hasn't actually been responsible for the running of a campaign. A good consultant can validate your campaign and improve it with you where necessary.

## Validate with a Pre-Launch Landing Page

One good way to get some good validation of the potential of your campaign is to set up a page before you launch your product and send it to your friends, fans, family and followers and get feedback on what you are attempting to do. Get feedback on the products, the rewards, the pitch you are using, the video, the story and anything else you consider is critical to success. Often you can be blind as to how your campaign will actually be perceived because you are so wrapped up in how wonderful it is. A few honest independent words can go a long way towards building a successful campaign for you.

## Apply to an Accelerator Program

While an accelerator program isn't normally associated directly with a crowdfunding campaign it can be very useful if you have lots of products to produce and fulfil as the result of a successful campaign. There are many campaigns that have done everything right up until it is time to produce the product and ship it. Witness the Coolest Cooler. The campaign owner did a great job with the marketing but didn't think through the fact that once they got the orders they were running a business and that products actually had to be produced and delivered within the budget of the funds received. This is easily done, as the excitement of the campaign is very seductive. If you work through the whole process as part of an accelerator prior to launching your product then the Accelerator team will help you structure the delivery and fulfilment side of things when your fundraise is successful.

## Validate with Crowdfunding Events

Crowdfunding events happen somewhere in the world nearly every single day. Often they are free or are at a minimal cost. If you make it to one of these events, you can wander around and ask people's advice about your proposed crowdfunding campaign. This may give you the type of validation you need to move the campaign in a successful direction.

## Validation Assistants

People all around the world are part of the Crowdfunding ecosystem and some of them have created tools that you can use to validate your campaign. Some are free and some cost but most of them deliver far more value than the small dent they may make in your pocket. Especially if they assist you in avoiding failure. Here are come on-line tools that may assist you.

- Prepare a "Business Model Canvas" to test your idea as an opportunity. The good thing about this is that you can test the ability of your opportunity as a real business. Will you be able to deliver on the promises you have made?
- Create life-like drawings of your product with "KeyShot". These can be very useful in gathering and testing interest for a product before you actually lay down the dollars for a prototype.
- Building a 360-degree image of your product with "ARQSPIN" will give your prospective validators a better understanding of your product.
- Research other campaigns and PR releases at "crowdsourcing.org" and "Crowdfundbeat". Other people's experiences of success and failure can be measured against your potential. There are hundreds of examples in these two publications.
- Leverage your PR with Joy Schoffler at LeveragePR or Robert Hoskins. These professionals will be invaluable in shaping the larger campaigns. However just a run through their websites gives you valuable insight into what to do and what not to do.
- Attend an event like "Crowdsourcing Week" where you can ask people about your product, if they think it is a good idea and if they could help you to make it successful.
- Design your campaign with "Crowdfund Mafia". One of the best ways to ensure success is to have a campaign mapped out, step by step. If you miss an important step it can mean a complete failure. Crowdfund Mafia has learnt the steps and processes that are essential to succeed.
- Get help from "Crowdfund Productions". They have participated in over 100 campaigns and will be able to give you valuable insight.
- Get advice from Kickstarter, Indiegogo and "Harness The Crowd". These three destinations are more general information but each have nuggets of gold in preparing and testing your campaign before it goes live.
- Work out if your budget will stretch to deliver the campaign you deserve. It is easy to be seduced into the idea that putting your campaign on say Kickstarter costs nothing. There are lots of small costs along the way you need to allow for to increase the chances of your campaign being successful.
- News releases can test how many people respond to your pitch. A great headline or pitch that is responded to from a news article is a very cheap way to validate the potential success of your product. Many journalists have the ability to take a complex story and boil it down to a compelling headline. Doing a few news releases will assist you to get gauge people's interest in your product.
- You can test mailings and email to validate your launch at Aweber and MailChimp. Both these options offer easy ways to get a professional mail-out together and also keep track of who opened your mailing and what they actually clicked on.
- Gather a test crowd to validate your offering with "Prefundia" the world's "coming soon" platform. Maybe before you go to a full-on crowdfunding campaign you run a "coming soon" campaign to validate true interest in your idea.
- Build, publish and test landing pages without needing to have computer skills. Often this is called A/B testing. You use this so that you can validate which campaign approach may be best. If you are doubtful about which approach you are going to use then why not test it and see which one could possibly work best.

- Nothing beats direct feedback from potential customers. If you are using your own site then you may want to get validation feedback from your visitors with chat software from "Olark" or "Zopim". Both these systems have a small chat window that appears in the bottom right of the web page you are profiling your opportunity on. In that chat window it is a good idea to prepare a question that will give you the exact feedback you are looking for. For example, "Is this the type of product you would recommend to a friend?"
- The world of data is growing. Big data is a hot term. You too can get great benefit and validation form your own data. All you need to do use the analytics services provided by "Krowdster", "Crazy Egg" and "Kicktraq".
- There are a lot of people who have spent thousands of hours gathering the best information worldwide about a running a successful crowdfunding campaign. One of the best guides is the "Kickstarter's Creator Handbook". This has been prepared by Kickstarter from their experience with thousands of campaigns.
- With the help of people like "Jamplify Contests" you can post a contest to validate your campaign and get people to contribute. This stops them from just clicking and going somewhere else. Human beings like contests and they like the prizes even more. You can gather valuable validation information from contests,
- Events used to be difficult to manage. You used to have to build a web site to capture event bookings but not anymore. Now you can manage pre-launch and launch events with Eventbrite. If you want to get real one to one feedback about your future Crowdfunding endeavour then a pre-launch event is a good strategy. At this event you can have prototypes of your product or service or perhaps a slide deck profiling your campaign and rewards. With this strategy attendees will give you direct valuable feedback before you risk it in the public domain. You may even get some pre-launch orders as a bonus!

## Validate by Surveying Your Market

One way to validate your offering is to test your product, service, niche selection or expected customer base by preparing a simple survey driven questionnaire. You need special tools to do surveys and fortunately there some good ones about. The ones most people use for this purpose are simple form based on-line survey tools like Survey Monkey and Google Forms. They make constructing a survey easy. Even if you want to have multiple choice questions? Start by asking key questions to validate your audience. You need to be asking the right people about your product.

Ask questions that give you direct feedback about whether or not the potential customers actually like your product or service idea. It is important to also find out what they don't like! One of the best questions to use, that is validated by significant practical results, is "Would you recommend this "Reward" to your friends"?

Where the rubber meets the road is whether they want to buy the product or not. Make sure you ask direct questions that gauge peoples actual purchase interest. You may even be brave enough to offer them the ability to buy the product as well. Many people are afraid to say directly they don't like the product or don't think it is a good idea. Most survey services like Survey Monkey or Google Forms have the ability for people to score answers on a scale of one to ten or similar. Why not do this on the question suggested above: "On a scale of one to ten how likely would you be to recommend this to a friend"?

With a strong focus on validation the wonderful benefits of crowdfunding can be yours.

There are many more benefits to running your own crowdfunding raise than just getting the money. There are also lots of ways to raise the odds of your funding drive succeeding by validating your campaign before it goes live.

Start by looking objectively at your pitch or story. Is it convincing, compelling and credible? If you had a barbecue, and invited a few friends along, would they say wow, yes please keep me informed about it. Then check your team. Are they balanced, passionate, capable and likeable. Once you have these two pillars under control then it is on to building your supporters and validating your offering in the process.

## REFERENCES

Niederer, P. (n.d.). *Simple Equity Crowdfunding Fundability Evaluation Tool Revisited*. Retrieved from http://www.paulniederer.com/2015/06/a-simple-equity-crowdfunding-fundability-evaluation-tool-revisited/

Sandlund, J. (2013). *The Rise of Meaningful Investing*. Retrieved from http://www.thecrowdcafe.com/rise-of-meaningful-investing/

# Chapter 11
# Crowdfunding as a Marketing Tool

**Fabio Allegreni**
*Crowd Advisors, Italy*

## ABSTRACT

*Most people, at a first sight, perceive Crowdfunding as a specific and exciting process to raise money. Of course, this is correct, but it is not the whole truth. A successful crowdfunding campaign must be addressed to a community of people, rather than to an indiscriminate crowd, which is also typical for any marketing strategy. Leveraging on this key element, crowdfunding can amplify almost any kind of communication activity, provided it is able to meet the needs of a community and to provide the benefits that the community expects. Therefore, if well exploited, crowdfunding is a powerful tool in the hands of marketers, whose potential goes far beyond the evidence of pre-selling a new product, to reach new marketing horizons such as branding, business development, CRM, credit scoring and even HR management.*

## INTRODUCTION

When Coolest Cooler's campaign on Kickstarter closed in August 2014, it ended up being the most successful Kickstarter campaign ever, raising $13 million from more than 62,000 backers. The target was originally set to $50,000, but, in only two days, the campaign hit $500k. Was such a huge success only a crowdfunding campaign? Clearly not. It was definitely a marketing campaign. Money strictly needed for putting the product into production was only a fraction of the amount raised. Besides funds, the main achievements of the campaign were both a large amount of pre-orders and a worldwide awareness. Are not they two typical goals of a marketing plan?

The takeaway of Coolest Cooler story is that crowdfunding is not only a matter of getting funded, but, also, enterprises can exploit it as a powerful tool to implement a marketing strategy. The ways to achieve this objective may assume different shapes, formats, and even processes.

Moreover, the use of crowdfunding as a marketing tool is not limited to enterprises raising funds, but can be adopted also by those eventually investing. As always, planning and executing a marketing

DOI: 10.4018/978-1-5225-0568-6.ch011

strategy cannot base on general rules, valid for every instance, but should be adapted to each single situation considering the opportunity, the environment and the strategic goals.

This chapter's purpose, far from aiming to present entrepreneurs and managers a set of rules following which their problems will solve magically, is to give them a structured idea and a few hints about how crowdfunding can widen their available weapons to reach their target.

## BACKGROUND

Let us start by defining what crowdfunding and marketing are. Other chapters of this book deeply treat the Crowdfunding concept, so here we only remind its main elements. The two key elements are included in its etymology: "crowd", a mass of people, and "funding", collecting money.

However, "a mass of people" is not enough to describe the success of the crowdfunding phenomenon. The crowd - each single individual composing the crowd - must be engaged. Engagement is as more effective as more a campaign is able to catalyze people around a common need or feeling. That means a crowdfunding campaign must leverage on a community, which might be smaller or larger depending both on the subject of the campaign and on the campaign's creator's ability to communicate it. In other words: crowdfunding is all about identifying a group of people by its needs, reaching them through an effective communication and, finally, providing them with what they want (a product, a service, a return on investment).

Let us focus now on the definition of marketing.

UK based The Chartered Institute of Marketing (CIM, n.d.) defines marketing as

*... the management process responsible for identifying, anticipating and satisfying customer requirements profitably.*

The American Marketing Association states (AMA, 2013):

*Marketing is the activity, set of institutions, and processes for creating, communicating, delivering, and exchanging offerings that have value for customers, clients, partners, and society.*

Marketing author and academic Philip Kotler (Kotler, 1972) defines marketing as:

*Satisfying needs and wants through an exchange process.*

These three respected definitions have one theme in common: meeting the needs and providing benefits.

It clearly appears, matching this latter finding with what we reported above about crowdfunding, that we are talking about the same thing: both definitions go in the same direction by stating that marketing as well as crowdfunding are tools to meet the needs of people and to provide them the benefits they expect.

The peculiarity of crowdfunding is that on top of satisfying needs and providing tangible benefits, it may reward people with intangible benefits too. Pledging money to help an entrepreneur dream come true or investing in a high potential startup to help its growth may eventually give a backer the opportunity to feel as an active part of the society's development; this may look rather more exciting than putting money on a listed though anonymous big corporation.

On the other hand, from the enterprise perspective, the process of aggregating a community in the name of both tangible and intangible needs and of its consequent benefits may be much easier and more rewarding. Easier because a communication campaign can leverage on a wider range of sentiments, more rewarding because, once engaged, people feel a stronger connection with the product and the enterprise, thus becoming a more loyal customer and a potential brand advocate.

According to a recent survey of Kickstarter projects by Wharton's Ethan Mollick and the University of North Carolina's Venkat Kuppuswamy (Mollick & Kuppuswamy, 2014), crowdfunding's ostensibly ancillary benefits are actually the primary reasons why successful startups use it. Author Kuppuswamy says: *Crowdfunding is not just a means for immediate funds, rather, it is a way to validate ideas, test markets, launch brands, find customers and impress investors.*

However, complications arise if we consider that there are different types of crowdfunding - i.e. reward base, equity based or social lending – and, additionally, that enterprises have different sizes and compete in many different industries. Marketing needs may then be quite different from case to case and, consequently, the opportunity of exploiting crowdfunding as a marketing tool may assume different purposes as well as distinct implementation modalities.

A structured approach to highlight the potential alternatives should be based on some classification. We suggest a first differentiation between marketing opportunities for Enterprises raising funds and those of Enterprises investing funds. This first distinction may help to detect the marketing opportunities of smaller and bigger enterprises. A further distinction within these two macros-categories will take into account the typical features of each type of crowdfunding, with a focus on Reward based and Equity based.

## MARKETING OPPORTUNITIES FOR ENTERPRISES RAISING FUNDS

### Reward Crowdfunding

Reward crowdfunding is usually the vehicle new entrepreneurs approach to raise the funds they need to bring an idea to reality. Usually, it works that somebody has an idea for a product or a video game, a book or an application. He/she uploads and describes it on a platform - say Kickstarter or Indiegogo – showing it off to the world and thus aiming to collect enough money to start the production. In order to stimulate donations, the entrepreneur offers some kinds of rewards whose value relates to the amount the donor is available to give. The most relevant is always the product itself, which will be produced and then delivered to donors if the campaign reaches its funding target.

Startups and entrepreneur are more and more using Reward Crowdfunding campaigns for the purpose to pre-sell a product, and this attitude, by now well-known and accepted by backers, opens a first set of great marketing opportunities. Companies are more and more applying crowdfunding not just to generate revenues, but more importantly, as a go-to-market strategy (Grell, 2015). Targeting a community of individuals with a strong sense of belonging to that group, with the aim of proposing them something exclusive, may be a good lever to get valuable information, which can be later used to consolidate the launch of a new product.

A company in the process of deciding the production of a new product, its development or its launch may consider a reward crowdfunding campaign as a component of the marketing mix, by setting clear and measurable objectives.

A general warning on employing reward crowdfunding for these purposes is that people expect to gain something concrete as a reward. Therefore, while a physical product is a perfect subject for such a campaign, the same does not apply to a service, such as an eCommerce site or a sharing platform, which is an intangible good and therefore hard to convert into a reward.

Market test. Likewise, a new entrepreneur or a start-up, a consolidated company may have a plan to launch a new product. It could be the new release of an existing one, or a brand-new or, else, an additional innovative item within a product line. The more innovative the product, the higher the risk is to invest in it.

Before launching an innovative product, marketing department usually tests it by submitting it to a focus group and/or commissioning expensive market researches. However, only larger corporations can afford these techniques. Small enterprises, often do not have enough resources to develop a product and, simultaneously, assess the market demand. Coincidently, those companies are often the most innovative, operating with an "out-of-the-box" way of thinking.

Launching a reward crowdfunding campaign may help to solve the problem. The main assumption is the availability of a group of people (a community), already known by the company and reachable directly through a mailing list or social networks. The larger and tighter the community, the higher the probability of succeeding. The most loyal customers, the company might have already raised and classified, for instance, establish a great community. A crowdfunding campaign in this case submits the new product to the most loyal customers asking them to pre-order the new product. If the campaign reaches the minimum pre-sale target, it means customers not only have demonstrated their appreciation, but, also, their willing to buy. An indirect effect is that the new product fans will tell their friends, thus enlarging the potential number of pre-buyers.

GE opened FirstBuild, a subsidiary of GE Appliances focused on developing the next generation of home appliances, which has launched 8 new products through various channels including Indiegogo (Dawson, 2015). Their primary goal has been to get products they believed in to market as fast as possible. Reducing the time to market means reducing investment for any one product while engaging in an exchange of value with consumers. FirstBuild launched Paragon Induction Cooktop crowdfunding campaign on Indiegogo early in 2015. The original concept for Paragon showed up as an idea posted on the FirstBuild website by Brad Walchak, a FirstBuild community member. First, they launched a design challenge to their community about 3 months prior to the Indiegogo launch date. What was that time used for? Product design, lead generation, and community engagement. However, despite the confidence they had built in the product through direct interaction with consumers, only a real-time test in the marketplace would validate the product. The crowdfunding campaign ended up in April 2015: the product acquired thousands of customers and more than $300k in sales, making GE confident in their value proposition and pushing them to move forward with production.

A great advantage of doing a market test with a crowdfunding campaign is that costs are mostly paid back: if the campaign is successful, it implies that people have pre-ordered the product and have paid for it. Margins on pre-orders can be attributed to cover the costs of the campaign.

In assessing the campaign, a company must clearly define its strategic objectives. With reference to testing a new product, they can be, e.g.:

- **Testing Market Acceptance:** Trying to reach a well-defined target in terms of the number of products sold gives a strong feedback on product's acceptability. Feedbacks are not only relative to how much people like the product, but, also, to how much they are willing to pay for it.

- **Testing the Features of a New Product:** Campaign may offer prototypes with different features as alternative rewards. The most pre-ordered prototype will be most probably the one the market will reward.
- **Reach a Minimum Volume for Industrialization:** Engaging engineers to project the product industrialization or committing with a supplier to build a mold is remarkable investment and can be a high risk for a small company. If, on succeeding with a crowdfunding campaign, a sufficiently high number of products are pre-sold, the risks dramatically decrease.

## Product Launch

A company, a small one, but a larger one as well, may have a new product ready to launch on its usual market. In this case, unlike the previous one we highlighted, the company has already invested on prototyping and industrialization, and it is planning a marketing campaign to launch it.

Marvell Technology Group, a semiconductor producer that took in $3.4 billion of revenue in 2013, on March 2014 used Indiegogo for a small fundraising drive. The company's objective: gauge demand for Kinoma Create, a software and hardware tool kit that includes a Marvell chipset to help build app-controlled gadgets. The campaign raised 5 times its original target form 552 backers.

A crowdfunding campaign can be a valuable additional tool in the marketing mix. A traditional communication campaign uses media to inform about the product, to assess its brand and to highlight its features. Usually, call-to-action is implicit, if traditional media channels employed ("if you like it, go out there and look for it"), or a little more explicit if digital media are chosen ("click to learn more or to find an outlet"). Only if the company sells the product online, the-call-to-action bring directly to a sale.

A crowdfunding campaign included in the marketing mix for the launch of a product adds the possibility to both communicate and sell. The warning here is that a reason for backers must be established. Why someone should order a product on a crowdfunding platform that is going to be available anywhere else? For instance, the crowdfunding campaign could be a sort of pre-launch where people have the chance to order the new product in advance with or without a price benefit.

In this case, the enterprise will concentrate part of its communication effort on driving the target audience to order the product, thus getting a measurable result. In addition, the following traditional communication campaign can leverage on the crowdfunding success with a new exciting story to help the main one (the product's features) standing out.

## Entering a New Market

A product may be new for a specific geographic market where the enterprise has strategically planned to expand its business. A crowdfunding campaign can be set up on a platform of the destination country, with the same goals highlighted above: testing a market or launching a product.

Additionally, the company may want to achieve a few more kinds of goal.

- **Get Distribution:** The reputation gained during a crowdfunding campaign and at its successful end, can be exploited to address the search of distribution partners in a new country. Not only brand awareness to consumers, but also to distributors is a goal the campaign can achieve. The lever is the evidence, provided by people pre-ordering, that the product can have space in the market and that, therefore, it can be a good business to distribute it over the country.

- **Reach New Customers:** Even if an enterprise is already operating in a new country market, a crowdfunding campaign may offer the chance to expand the market reach if the need is to improve revenues. A company can achieve this objective by leveraging on existing customers. In this case, a campaign should offer either a brand new product or the new release of an existing one. Or else, it can promote an additional item within an existing line of products.
- **Set a Track Record for Further Investors:** A successful crowdfunding campaign, no matter how high the funding target has been, proves that the company is able to attract customers. This is not only a direct benefit, but also an indirect one, in terms of appeal for future investors, who become aware of the enterprise's ability to raise funds. In fact, after the launch of a product in a new market, the company might need money to invest in further development and, therefore, might need to raise more consistent funds from Angels, VCs, Banks or throughout an equity crowdfunding campaign. A successful crowdfunding campaign is then a "jewel of the crown" the enterprise can add to its score, presenting its business plan for a financing round. Before its 2014 first successful campaign on Kickstarter, Pebble Watch could not raise funds from any VCs nor angel investors. After hitting $10 million from the crowd, Charles River Ventures invested $15 million (Pebble Techonolgy, 2013).

## Customer Relationship Management

Whichever is the main objective of a campaign, among those listed above, an additional benefit is commonly achieved: contacts. People may interact with the campaign in many different ways, from subscribing to following it, from pledging funds to leaving comments or asking questions.

Those interactions are ways for engaging people and can give back extremely valuable data and information, provided that, before the campaign launch, the company has planned to measure and track them. Depending on their origination, data can be useful in several ways, e.g.:

- **Lead Generation:** Of course, people pre-ordering the product are definitely buyers. Yet, all others who have been interacting with the campaign are potential buyers. Even if they did not place an order, they demonstrated anyhow some interest in the product by asking questions, following the campaign to get updates (if contemplated by the platform), leaving comments, sharing the campaign on social networks. Tracking them allows to build up a database of leads the company may contact afterwards with the objective to convert them into buyers.
- **Loyalty:** A crowdfunding campaign cannot run without starting from a solid network upfront to contact and engage. Company's customers, if any, may also compose the network. In this case, a crowdfunding communication campaign targeted on them, may consider leveraging on their sense of belonging to a community of "brand lovers" and, for instance, presenting the new product as an opportunity specially thought for them. Whether they will order the product or not, their binding with the brand will most probably increase.
- **Customer Profiling:** Another way to exploit the data gathered during a crowdfunding campaign is to use them as additional attributes to better profile customers (or leads). It is easy to figure out "buyer" or "not-buyer" as an additional attribute. Less easy, and trickier, is to associate with the data stored in the company's CRM any behavioral information related to the interactive activities tracked during the crowdfunding campaign. These include sharing the campaign, social networks

used, negative or positive opinions, number of interactions, etc. The more interesting side of it is that those kinds of information are dynamic and not static, meaning that they may change over time. For this reason, they must be added to CRM database only if the company is provided with an adequate technology and a CRM strategy.

Dodge did a great job with its DodgeDartRegisty.com campaign (Swart, 2014). Upon registration, customers could have parts of the car funded by backers. The campaign went viral, engaging more than 1 million social media impressions. Even if less than 40 cars were purchased directly, the company mined data from tens of thousands of potential new customers and saw sales more than double the following quarter.

## Equity Crowdfunding

Equity crowdfunding, more than reward, is definitely a way to raise funds, to the extent that the entrepreneur is willing to give away part of his equity in exchange of money for growth.

However, even a crowdfunding campaign may have marketing purposes in addition to funding. Like reward crowdfunding, an equity crowdfunding round starts from involving an initial community. So, once again like reward crowdfunding, it is a matter of engaging people and converting them. In this case into investors.

The amount of money involved in an Equity Crowdfunding campaign is considerably higher than that of a reward one and, hence, the relationship built with people must create a tighter link. People investing are careful about where they are putting their money and want to know details about the company. Therefore, an Equity Crowdfunding campaign is much more complex, but right because of that, it builds such a strong sharing of interests between the company and its investors, that they become the company's first brand advocates. This concept is the gist of Equity Crowdfunding as a marketing tool and the basis upon which enterprises can build other kinds of marketing activities.

## Setting a Benchmark for Further Investment Rounds

Companies are using the equity crowdfunding platforms as a way to market their offering in hopes of attracting further investments, beyond those collected directly. If the campaign succeeds, further to raising funds two more side benefits are accomplished:

1. The company obtains a certified value (a number of people invested their money in front of an equity valuation).
2. The business model is perceived as good enough to attract investors.

After the crowdfunding financing round, if the company, during the following months, is able to achieve the results it promised, a further and higher round can be set up and, most probably, it would be successful, as the business proved to be financed once and to be credible.

Furthermore, the valuation of the company at this stage can easily be higher than the one obtained in the first round, allowing the company to raise funds at a lower cost.

Someone went even beyond. Burgundy Winemaker Domaine Chanzy, a French winemaker, launched an Equity Crowdfunding campaign on the UK leading platforms Seedrs, as an alternative way to raise money for its IPO on the London Stock Exchange's (LSE) Alternative Investment Market (AIM) (Seedrs, 2015). Domaine Chanzy apparently believed that blending an equity crowdfunding round with a public listing would have expanded its investor base and attracted advocates for the brand.

## Entering a New Market

Launching a crowdfunding campaign in a new country can be a strategic move to enter a new market. The enterprise may have successfully marketed its products or services elsewhere and believes they are innovative enough even for the new destination country.

The main benefit of launching through an equity crowdfunding platform is the communication synergy: promoting the fundraising, communication activities promote the product/service as well.

An additional benefit is that campaign communication costs can be paid back partially or entirely by the new capital raised once the round reaches its target.

Finally, the "long tail" of new investors (people investing individually small amount of money) are the best witnesses of the product/service offered by the company they invested in. They are willing to have a return on their investment and will push their friends to buy the "their" company's products to let it be successful.

## Increase Creditworthiness

Financing a business with equity crowdfunding accelerates the process of raising money in front of equity rather than against debt. Equity crowdfunding is therefore an especially useful funding tool for innovative SMEs, as it can improve their creditworthiness and make them attractive to bank lenders by reducing their debt-to-equity ratio. An AFME (Association of Financial Markets in Europe) report states that in Europe *lack of equity is a key bottleneck to the provision of further overall SME funding, the US has a much more developed equity network for SMEs*[1] (AFME, Boston Consulting Group, 2015).

Crowdfunded companies, then, may leverage on their funding success to more easily access debt funds.

## Rewarding Current Customers and Business Partners

We stress the concept of how important leverage on an existing community is to run a successful crowdfunding campaign. An interesting way of doing it is to involve customers or business partners (e.g. distributors, b2b customers, suppliers, etc.) allowing them to become stockholders. It works particularly with large networks, where members are small enough and feel somehow in tune with the company.

Not only a campaign involving customers and business partners has a higher chance to succeed, but also it is a news itself and may have a better impact on media. Furthermore, partners and customers subscribing a funding call definitely certify that the company's business is attractive to the extent that even customers and suppliers are happy to invest in it. This looks as a great achievement for both the campaign and sales.

## Finding New Business Partners

Pizza Rossa, after a £440 thousand successful campaign on Crowdcube, a leading UK equity crowdfunding platform, immediately received requests for opening franchises from all over the world (Crowdcube, 2013).

When a company is new to the market, it is often very hard and time consuming to find effective business partners, e.g. distributors, re-sellers, suppliers. A side effect of an equity crowdfunding campaign is to help the entrepreneur or management in the search process. The communication campaign supporting the funding round necessary hits the company's industry. If the round meets the target, enterprises operating in that industry may then be more likely to get in contact and start a deal. This may be a side benefit or can be set as one of the objectives the company wants to reach through the crowdfunding campaign. In the latter case, the company must identify in advance measurable KPIs and must plan a specific communication strategy.

Sugru campaign on the Crowdcube platform broke two world records (Crowdfunding Buzz, 2015): it featured the first ever £1m crowdfunding single investment, and boasted the widest reach, with investors from 68 countries. One of the founder's objectives was to raise the profile of the company, allowing it to more easily find industrial partners for entering new industries and distributors worldwide.

# MARKETING OPPORTUNITIES FOR ENTERPRISES INVESTING FUNDS

Medium and large enterprises invest huge amounts of money in marketing activities and are constantly seeking new ways to stand out among competitors. Typically, they look for initiatives that maximize the exposure to their target and, at the same time, that minimize costs.

Reward and Equity Crowdfunding represent an alternative opportunity that may be extremely effective in both gaining visibility and containing costs.

## Reward Crowdfunding

Those running reward crowdfunding campaigns are usually individuals or small businesses. They plan their fund raising to finance the launch of a product, as we have already seen, but also for charity or civic initiatives. This opens up a bunch of opportunities that enterprises looking for alternative marketing tools may effectively exploit in accordance with their objectives.

## Brand Equity

Brand equity refers to the value of a brand. Brands are one of the most valuable assets a company has, as brand equity is one of the factors which can increase the financial value of a brand to the brand owner, although not the only one. Several elements can be included in the valuation of brand equity: changing market share, profit margins, consumer recognition of logos and other visual elements, brand language associations made by consumers, consumers' perceptions of quality and other relevant brand values. Therefore, brand equity is the value perceived by customers on top of a product or service. It can significantly affect the price and it derives from branding investments. A strategic objective, for any kind of enterprise, consists indeed in increasing its brand awareness, which implies enhancing both the visibility and the appreciation of its products or services. Marketing actions to achieve this strategic goal include traditional and digital advertising, sponsorship, promotion, public relations, social networking. All these kinds of communication activity have several common denominators: a profiled target audience, a benchmark (the product or service itself, a testimonial, a context etc.), and one or more communication channels.

A reward crowdfunding initiative presents all of these items. It has a target audience itself, which is built upon both the individuals investing in the project and the potential investors. It has a benchmark to compare to, which might be the context, e.g. the development of new brilliant tech ideas, the restoration of an important monument, or a social aid initiative through a charity. Finally, it has its own communication channel - the crowdfunding platform – although it also extends to several indirect channels, i.e. those, digital or traditional, used by the crowdfunding campaign promoter.

How can an enterprise leverage on these kinds of campaign? Here below we suggest a few examples, though, as always, the scope may be wider and depends only on the marketing team fantasy.

- **Promoting Directly One or More Campaign:** A company may launch its own reward crowdfunding campaign, not much for launching a product, as we have seen in the previous section of this chapter, but for a pure brand equity purpose. The opportunity may arise in relation to supporting a cause or an event where the company may want to be accredited as the main promoter. This way, the campaign target audience and, more generally, whoever is interested will associate the brand with a praiseworthy activity, as it touches deeply his sensitivity.
- **Co-Investing:** This approach is more complex, though more intriguing and full of interesting implications. The hinge of the approach is the opportunity for a company to finance one or more high-impact initiatives without allocating all the money needed but only a part of it. The target audience will perceive the company to be the promoter of the initiative, or its main supporter, but the Company is investing only a fraction of the entire budget. The role of the enterprise, here, is double. On one side it acts as the lead investor (or donor), letting other individuals or enterprises to cover the remaining budget. On the other hand, it also acts as an "advocate" and a driver for the fundraising campaign, leveraging on its reputation to involve other investors/donors. Depending on the positioning the enterprise wants to give its brand, it can explore several options, for instance:
    - **Startups Co-Financing:** A corporation may agree with a reward crowdfunding platform to launch a call for choosing a few start-up projects to be co-financed whether their crowdfunding campaigns reach a percentage of the funding target, (e.g. 50%); the Company will participate by covering the remaining 50%. The benefit here is the possibility to allocate a relatively low budget (the sum of the maximum amount of money allocated for each project), getting back a higher value in terms of spontaneous media coverage. This is what several Italian large corporations – Fastweb (Telco), Unipol (Insurance) and Poste (Italian Public Mail) – have agreed with Eppela, one of the biggest Italian reward crowdfunding platforms. Each of them allocated a budget form €100,000 to €150,000 to be spread among those of the previously selected projects that would have raised at least 50% of their target. On average, projects targets are lower than €20,000 allowing the sponsoring corporation to finance a relatively high number of new initiatives in a 4 to 6 month period.
    - **Co-Sponsoring a Monumental Restoration or a Social Activity:** Instead of paying the entire bill (e.g. for a monument restoration), whose importance for citizens could be either at a national or at local level, a Company might help promoters to set up a crowdfunding

campaign. The corporation will then help them to both finance a part of the initiative and find other co-sponsors as well as to spread the message among people. The company might also help in other different ways. e.g.: a retailer can help by exploiting the proximity of its outlets to consumers to communicate and remind them the initiative. A significant example is The Giant Flag: a legacy project promoted by the South African branch of FCB, a multinational advertising agency. It includes the building of a flag made up of millions of colored desert plants, and a four-megawatt solar field, on a tract of land that was previously completely barren. Thanks to its size equal to 66 soccer fields, it will be effectively viewable from space. It will cost about $20 million, which FCB will fund both with corporate efforts and with a Crowdfunding campaign that will let citizens participate directly in this giant project.

- **Supplying Rewards:** In a reward crowdfunding campaign the promoter must plan to give back rewards to donors, proportionally to the entity of money they donate, and, hence, to establish a scale of rewards. As mentioned before, the main reward, in case of a campaign set to launch a product, is the product itself. However, that is not always the case. Indeed, the campaign might have other subjects than the launch of a product or it might need to allocate rewards for the donation of smaller amounts of money. In both cases, the promoter is forced either to invent some intangible rewards or to spend money for buying gadgets. By the way, it would be a great benefit for a campaign promoter to promise and to deliver attractive rewards which the platform might make available for free. A Company producing goods or selling services may consider promoting them by collaborating with one or more platforms and letting them offer to a selected number of initiatives a choice of its products (or coupon to get its services) as rewards. Of course, the selection criteria should be established taking into account the mutual compatibility between the product/service offered and the subject of the initiative. In any case, the donors must perceive a strong link between the two. Domino's Pizza created a competition in which their start-up businesses could win Domino's Pizza gift cards. Entrepreneurial winners could even reuse the vouchers as an incentive for smaller investors backing their projects through crowdfunding (Moules, 2013).

## Research and Development

Among the classical marketing 4 Ps suggested by McCarthy (McCarthy, 1960), besides "*promotion*", which we considered above speaking about Brand Equity, "*product*" is another important item. A marketer considers a product in terms of its ability to satisfy a specific client's need. However, that need is not static, as it changes over time depending on the phase of the product/service in its life cycle. Therefore, among marketers' tasks, managing the product mix is a relevant one. Research is one of the main tools a company has to enhance its existing products or to create new. However, investing in research and development may be very expensive and risky for an enterprise, especially if the product or service is not part of its core line.

Crowdfunding can offer an opportunity to enhance existing products or services or to develop new ones without the need of allocating an outstanding budget. Indeed, it allows a bottom-up approach rather than a top-down. The disadvantage is a lower accuracy compared to that achievable with a dedicated research. That is why this solution may be a more suitable choice for the "*long tail*" products rather than the core ones.

A first draft distinction may consider a "market" approach and an "internal" approach.

- **Market Approach:** By market approach we mean the involvement of people outside the company, who can be consumers as well as professionals. Lego launched *Lego Ideas*, a platform where anyone can propose an idea for a new Lego set. If the idea gathers at least 10.000 supporters, it enters a special phase called the Lego Review. A review board of set designers and marketing representatives evaluates projects according to Lego's review criteria and pick those to become new Lego Ideas sets. Projects selected in the review go into production and receive a royalty on sales. Supporters can now own the set they helped make happen. This is not properly a crowdfunding process, but more a crowdsourcing one. However, the principle is similar as it leverages on the ability of the brand to attract fans who are rewarded if their idea proves to be successful. Similarly, other companies may set up platforms, or partner with existing ones, to let professionals or fans propose their project (e.g. a lamp, a game, a piece of furniture, etc.…) and pre-sell it. As long as the product has reached a pre-defined amount of pre-ordered pieces, the company will industrialize it and put it into production. An additional benefit of this approach is its ability to improve existing customers' loyalty.
- **Internal Approach:** Sony set up a business program encouraging employees to develop the most innovative ideas they can think of in an effort to find new hits. Then the company launched a crowdfunding and e-commerce platform called *First Flight* for products coming out of that project. The site aims then to take a project right from the idea stage, to fund it and finally to commerce it within the same platform. Unlike the "market approach", the internal one focuses only partially on customers (they are involved to validate the projects through the crowdfunding process offered by the platform), while it definitely points on the skills and passion of employees for raising new ideas. In such a large corporation, in fact, many good ideas may be kept hidden, simply because the procedure to raise them may be excessively complicated or it does not take into account several classes of employees.

The main common feature between the two approaches is the opportunity, given to either customers or employees, to let them raise their passion for a brand or a product. Skills and intuition beneath this passion can be disruptive, being detached from the rigid practices that a company usually establishes for its creative process.

## Equity Crowdfunding

We are used to consider Equity Crowdfunding with the eyes of a Startup or an SME seeking funds for its growth. We have previously highlighted how these subjects may take advantage of it in terms of marketing benefits. Indeed, a corporation may even look at Equity Crowdfunding as a marketing tool by assuming the role of an investor. Investing in a growing company may give not only an interesting financial return in the medium/long term, but also it can give significant results as a marketing investment. Of course, the hints we are highlighting below implies marketing objectives a company can achieve throughout other means than Equity Crowdfunding. Actually, a corporation may achieve the same strategic objectives by scouting the market directly to find the best opportunities to invest in. Nonetheless, the main advantage of resorting to Equity Crowdfunding stands in the opportunity to check the right horse (or horses!) to bet on among a wider list of initiatives, at considerably lower cost.

## Research and Development

Google has acquired more than 180 Start-ups and, most probably, it owns a stake in many more. Recently the search engine giant has found a venture capital vehicle. Their objective is clearly to speed up the technological research process in order to expand their business faster than it is reasonably feasible with their internal R&D department.

However, not all enterprises have such a financial power as Google to let them scout a large number of new technologies and eventually invest in them. Here is where equity crowdfunding platform can help.

The market is still at its early stage phase. Nevertheless, an increasing number of Startups and SMEs all over the world has been exploiting this new funding channel. Moreover, the industries involved are not limited to high-tech industry, going from biotech to food and drink production and distribution, to manufacturing, to business services.

A corporation has then the opportunity to choose among a large number of initiatives and, if the case, to allocate funds to invest in them. In addition, unlike running a VC vehicle, investment management in this case may be quite simple, as it does not necessarily require a financial return, namely an exit strategy, nor it needs substantial equity stakes to subscribe. The advantage, then, consists in the opportunity for the corporation to monitor closely the activity of the companies participated, to better learn how the team operates and to acknowledge the potential of their business solutions. It will come a time when the corporation may evaluate to either buy the entire company or at least acquire a majority equity stake or else to establish a stronger industrial or commercial collaboration (see below).

## Industrial and Commercial Partnership

Another purpose of investing a minority stake in one or more equity crowdfunding campaigns is to establish a commercial or industrial partnership with promising new businesses. The reason for doing that may be for either an industrial or a commercial purpose.

An industrial partnership, empowered by owning an equity stake, may rise as an opportunity when a Start-up or a SME launches a new product on the market. The corporation may evaluate the product is complementary to its offer or even disruptive compared to it. Else, the new product (or service) can be a semi-finished component able to improve the performances of the current line of products manufactured, or, conversely, an opportunity to expand their demand.

A commercial partnership may be an opportunity to enter a new geographic market if the Start-up is running a crowdfunding campaign and is operating in a market the corporation does not cover yet. Or else it allows expanding a product or service offer in order to better meet the customers' needs, thus helping to differentiate from competitors. Holding even a minority stake gives the corporation a better chance to affect the commercial policy of the participated Start-up and to evaluate the reliability of both its products and management.

## Brand Equity

Sometimes, investing in growing businesses and helping young entrepreneurs to make their dreams come true may be an additional motivation for large companies to fund enterprises by equity crowdfunding. Besides the primary objectives we hinted before – R&D and industrial and commercial partnerships – a corporation may want to achieve a secondary one: strengthening its brand image by appearing as a financial supporter of new businesses. Of course, in order to achieve such an objective, though being a side one, a company should plan a solid strategy. Investing spot in a few initiatives does not work. For this to impact on the public opinion and thus to affect its brand equity a corporation must plan to become a serial investor.

It implies planning an investment budget throughout more than one year and an investment policy in terms of setting clear criteria to both select the investment targets and acquire equity stakes. For instance, a plan should define whether investing in businesses at the early or growing stage of their life cycle, or else it should set a maximum equity stake to acquire.

Nowadays, this kind of marketing opportunity is still limited being the Equity Crowdfunding market at its infant stage worldwide and, therefore, it does not present yet a significantly large choice of initiatives for each type of industry. Even though, Equity Crowdfunding is growing double digit year over year such that a mid or large corporation should start considering this approach as a winning one in the near future and getting ready to fully exploit it.

For the Dutch electronics group Philips, the opportunity has been as much about sharing knowledge as funds.

Its Innovation Fellows Competition offered up to $60,000 to companies creating technological innovation. However, they provided not only funds but also mentoring support to help bring the winning idea to life.

This approach is based on the concept of "open innovation", the belief that big ideas can come from anywhere.

## CONCLUSION

Crowdfunding is certainly a way for raising funds, but it is even a powerful marketing tool. In fact, if marketing is a way to meet the needs and provide benefits to people, crowdfunding does the same. Therefore, depending on the way a company approaches and exploits it, crowdfunding can give a wide choice of alternative options for achieving the typical marketing objectives particularly related to two out of the four marketing Ps: *product* and *promotion*.

First, we highlighted that enterprises raising funds through a crowdfunding campaign have the opportunity to accomplish additional strategic objectives that go beyond fundraising.

This is especially true for reward crowdfunding campaigns, where marketing goals can have an even higher priority than fund raising. For instance, testing a new product by letting backers pre-order it, before putting it into production, is a great way to evaluate its market approval. Also, a reward crowdfunding campaign can be exploited for entering a new market, launching a new product or even strengthening the relationship with customers.

Equity crowdfunding campaigns may pursue a side marketing goal too. It may concern the financial strategy by setting the base for further future funding rounds or increase credit worthiness. However, it also deals with the company's products or services strategy. Actually, a company may look for entering a new market, or finding commercial or industrial partners.

Nevertheless, marketing opportunities are not limited to enterprises using crowdfunding to raise funds, but even for those that invest funds. In the latter case, the opportunity is more suitable for mid and large corporations that hold a relevant marketing budget and that may look at crowdfunding as a more efficient or additional way to implement their marketing strategies. Similarly to what we reported for enterprises raising money, for those investing funds both reward and equity crowdfunding are suitable for pursuing marketing objectives.

Investing in reward crowdfunding campaigns can be a powerful tool for increasing a corporate brand equity. This is especially true when a corporation is a pro-active investor, meaning that, for instance, it supports a program for financing selected initiatives as long as they reach a share of their funding target. People will associate the company's brand to its commitment in supporting the development and growth of new enterprises. Another way a corporation can leverage on reward crowdfunding investing is to scout new technologies or products, by stimulating either external or internal initiatives.

Investing in Start-ups or SMEs by buying their shares offered through an equity crowdfunding campaign enhances the acknowledgement of these objectives. Instead of concentrating the whole R&D funds on expensive internal teams, a corporation may also acquire a stock of one or more promising business, thus controlling more closely their ability of developing new solutions. In addition, interesting Start-ups or SMEs can become commercial or financial partners whenever their offer is complementary to the investor's one.

*Table 1. Summary of marketing objectives that crowdfunding can help to achieve*

| Raising/Investing Funds | Type of Crowdfunding | Marketing Objectives |
|---|---|---|
| Enterprises raising funds | Reward Crowdfunding | Market test:<br>• Testing market acceptance<br>• Testing the features of a new product<br>• • Reach a minimum volume for industrialization |
| | | Product launch |
| | | Entering a new market:<br>• Get distribution<br>• Reach new customers<br>• • Set a track record for investors |
| | | Customer relationship management:<br>• Lead generation<br>• Loyalty<br>• • Customer profiling |
| | Equity Crowdfunding | Setting a benchmark for further investment rounds |
| | | Entering a new market |
| | | Increase creditworthiness |
| | | Rewarding current customers and business partners |
| | | Finding new business partners |
| Enterprises investing funds | Reward Crowdfunding | Brand Equity improvement:<br>• Promoting directly one or more campaign<br>• Co-investing<br>• Startups co-financing<br>• Co-sponsoring a monument restoration<br>• • Supplying rewards |
| | | Research & Development:<br>• Market approach<br>• • Internal approach |
| | Equity Crowdfunding | Research & Development |
| | | Industrial and commercial partnership |
| | | Brand Equity improvement |

The advantage of implementing such strategies by scouting opportunities on equity crowdfunding platforms rather than directly, is a matter of time and money: not all companies can afford an expensive due diligence on a large number of small companies, which, conversely, equity crowdfunding platforms facilitate.

The opportunity crowdfunding offers as a marketing tool depends of course on a company's marketing strategy, but they also rely on the ability of management or entrepreneurs to catch them by taking marketing decisions with an open mind.

Crowdfunding is still at its early stage and, especially equity crowdfunding. It has not expressed its full potential yet. Nevertheless, it is fast growing in the US as well as in Europe. This will certainly give rise to new ways of exploiting it as an even more powerful marketing tool than we hinted. As more enterprises and initiatives will list on crowdfunding platforms and the more both individuals and companies will get used in investing in growing businesses, the higher will be the scope and thus the higher will be the potential of using crowdfunding for pursuing marketing objectives.

So far, the opportunities we highlighted in this chapter are just hints to let decision makers find some ideas, which they can base on to set off their creativity. Table 1 summarizes our hints and their classification.

## REFERENCES


AFME. (2015). *Bridging the growth gap*. Boston Consulting Group.

AMA. (n.d.). *Definition of Marketing*. Retrieved from: https://www.ama.org/AboutAMA/Pages/Definition-of-Marketing.aspx

CIM. (n.d.). *What is Marketing?* Retrieved from: http://www.cim.co.uk/more/getin2marketing/what-is-marketing/

Crowdcube. (2014). *Pizza Rossa campaign*. Retrieved from https://www.crowdcube.com/investment/pizza-rossa-ltd-12705

Crowdfunding Buzz. (2015). *Pmi in Uk Raccoglie più di £2,8 Milioni con Equity Crowdfunding di cui 1 Milione da un solo Investitore*. Retrieved from http://www.crowdfundingbuzz.it/pmi-in-uk-raccoglie-piu-di-28-milioni-con-equity-crowdfunding-di-cui-1-milione-da-un-solo-investitore-e-record/

Dawson, T. (2015). *Want To Prove Your Product? Validate On Indiegogo!* Retrieved from Indiegogo Blog: https://go.indiegogo.com/blog/2015/07/validate-your-product-on-indiegogo.html

Grell. (2015). *The Kingii Crowd-First Tactics & Go-to-Market Strategy with Crowdfunding*. Retrieved form Crowdfund Insider: http://www.crowdfundinsider.com/2015/08/72662-the-kingii-crowd-first-tactics-go-to-market-strategy-with-crowdfunding/

Indiegogo. (2014). *Kinoma Campaign*. Retrieved from https://www.indiegogo.com/projects/kinoma-create#/

Kickstarter. (2014). *Coolest Cooler Campaign*. Retrieved from: https://www.kickstarter.com/projects/ryangrepper/coolest-cooler-21st-century-cooler-thats-actually/description

Kotler, P. (2012). *Marketing Management*. Pearson Education.

Lego. (n.d.). *Lego Ideas*. Retrieved from: https://ideas.lego.com/

Market Wired. (2013). *Pebble Announces $15M in Funding and the Next Stage of Its Open Platform*. Retrieved from http://www.marketwired.com/press-release/pebble-announces-15m-in-funding-and-the-next-stage-of-its-open-platform-1792035.htm

McCarthy, J. E. (1964). *Basic Marketing. A Managerial Approach*. Homewood, IL: Irwin.

Mollick, E. R., & Kuppuswamy, V. (2014). *After the Campaign: Outcomes of Crowdfunding*. UNC Kenan-Flagler Research Paper No. 2376997. Retrieved from: http://papers.ssrn.com/sol3/papers.cfm?abstract_id=2376997

Moules, J. (2013). *Big companies show they are in with the in crowdfunding*. Retrieved form Financial Times: http://www.ft.com/intl/cms/s/0/1234e5f6-30e5-11e3-b478-00144feab7de.html#axzz3jHbGvQfG

Seedrs. (2015). *Domain Chanzy campaign*. Retrieved from https://learn.seedrs.com/seedrs-welcomes-domaine-chanzy-as-first-ipo-campaign/

Sony. (n.d.). *First Flight platform*. Retrieved from: https://first-flight.sony.com/

South Africa, F. C. B. (n.d.). *Giant Flag Project*. Retrieved from: https://www.giantflag.co.za/

Swart, R. (2014). *Why Crowdfunding Is a Must-Have Marketing Tool for Brands*. Retrieved from Marketing Profs: http://www.marketingprofs.com/opinions/2014/26060/why-crowdfunding-is-a-must-have-marketing-tool-for-brands#ixzz3jA6ZbNcv

## ADDITIONAL READING

Brown, D. J. (2001). *Heinemann Business Studies for AS Level*. Oxford: Heineman.

Farris, P. W., Bendle, N. T., Pfeifer, P. E., & Reibstein, D. J. (2010). *Marketing Metrics: The Definitive Guide to Measuring Marketing Performance*. Upper Saddle River, New Jersey: Pearson Education, Inc.

# Chapter 12
# Legal Aspects and Regulation in Crowdfunding:
## Comparisons across Countries

**Alessandro Maria Lerro**
*Associazione Italiana Equity Crowdfunding, Italy*

## ABSTRACT

*Crowdfunding started with the donation-based and reward-based models, but soon its disruptive capacity to innovate the traditional approach to finance made it evolve into an alternative finance instrument, suitable to fund new business. Peer to peer lending and equity crowdfunding are investment instrument that can truly sustain the growth of businesses and generate profits. As any investment instrument, lending and equity crowdfunding are heavy regulated in order to protect investors and maintain a proper trust in the system. This chapter will go through the American and European main regulatory principles, offering a picture of the overall approach to this innovative and disruptive financing tool.*

## REGULATION

The crowd economy comprises a number of different forces, trends, and dynamics originating from the crowd. Many of them are connected with money; this is why a large part of the crowd economy is identified as crowdfunding. Indeed, crowdfunding is a very broad concept that includes several very different operations, each having its own logic and metrics.

## "NON-INVESTMENT" CROWDFUNDING

### Donation

Thanks to the explosion of social networks and the wide use of the web, fundraising easily moved from offline events and efforts to online crowdfunding platforms.

DOI: 10.4018/978-1-5225-0568-6.ch012

Donation crowdfunding was the first crowdfunding model based on the web, and it is being used not only by charities and nonprofits to raise funds, but also by individuals to ask for help for their personal projects (business, entertainment, study, or any other purpose without limitation).

In donation crowdfunding, the crowd donates money to support a project with no expectation of financial return. In some cases, the crowd is offered some kind of rewards or incentives tied to their level of contribution, which have no significant intrinsic monetary value. In most jurisdictions, these perks do not influence the model, which remains a donation, and there is only a moral obligation to provide the promised reward.

However, in some civil law jurisdictions (originating with the "Napoleonic Code"), the eventual breach of promise might have consequences, like the obligation to give back the grant. In fact, a contractual obligation binds the donor and the project owner, simply because of the donation.

## Reward

Donation-based crowdfunding quickly evolved to a model in which the value of the reward is comparable or substantially equivalent to the value of the donation: the project owner offers to produce and deliver an object to the donor, provided that a minimum number of orders are placed and therefore break-even is reached.

This model perfectly answers the need of financial coverage for those who create innovative products outside a well-established company: they can test the market without committing the upfront marketing, production, and distribution costs required to fully commercialize a new product.

Moreover, the customer validation arising from a successful pre-sale campaign grants project owners a valuable tool when negotiating with investors. Several studies have shown that a successful reward-based crowdfunding campaign influences the proclivity of professional investors to sustain the project with an equity investment. A well-known example is Pebble, a smart watch that interacts with every kind of smartphone. Pebble's business model didn't attract investors, so its founders launched a reward-based campaign on Kickstarter. The campaign was a real success—it's still in the top 10 overall—and Pebble pre-sold enough smart watches to raise $10 million. This scenario totally changed the approach of venture capital firms, and after a few months, Charles River Venture invested $15 million in equity. One year later, a second crowdfunding round brought in $20 million more. The Pebble case demonstrates the strength of the interaction between crowd validation and venture capital analysis and strategy.

In most jurisdictions, when the perk offered has a value comparable or close to the value of the financial contribution, reward-based crowdfunding is nothing but a sale or, better, a kind of e-commerce. As a consequence, consumer protection rules are applicable to each subscription, as well as regulations concerning guarantees, delivery, and breach of contract. From a tax point of view, some governments have made it clear that this is not a donation and therefore VAT and similar local taxes are applicable to these kinds of sale, and the European Commission is considering the possibility of uniform regulation.

## Royalty

Using a different approach, a project owner might offer supporters a part of the profits that the project is going to generate. Without sharing in the ownership of the project, a funder could still share in part of the expected financial returns. This third and most complicated type of "non-investment" crowdfunding is the royalty-based model: funders are offered a return of a financial nature, calculated as a share of the revenues or profits generated, which will be payable under certain conditions and for a certain period of time. This profit-sharing formula can be used for a stand-alone project or for a specific business or product of a company (e.g., a software application) without involving any other business of the company. Risks are transferred to the funders, without designating an investment scheme, because the founder gets nothing as an immediate consideration—not even the promise to get the money back (otherwise, the "investment" would simply be a loan).

Funders fully take part in the losses, up to the value of their own contribution, which is non-refundable and can be completely lost.

Royalty-based crowdfunding is purely financial; campaign owners seek funds for their projects through the web. However, in some jurisdictions this practice is not totally new, and is already regulated as a specific kind of financing that might be either financial or in-kind (i.e., work for equity).

In order to offer uniform rules to all the funders, the project owner (or the crowdfunding platform) must propose a uniform agreement. Otherwise compulsory local rules might prevail and find discrimination among the funders as well as create unexpected situations for the project owner. For example, the project owner might open the business to new membership, accepting a large number of potential new supporters, without any kind of consent from the supporters already invested.

Funders do not fully take part in the earnings either: the project owner can propose a certain amount of shared revenues and the relative duration, fixed in a limited period of time (e.g., five years from the investment), regardless of the ratio of contributions to the value of the business.

Indeed, this is not an investment and funders don't buy shares in the business as if they were investors and shareholders. Moreover, the project owner can offer funders some rights: audit, access to documents, or even involvement in the conduct of the business.

In some cases, royalty-based crowdfunding has been used to circumvent prohibitions on solicitation of investment, rather than for strategic reasons. Otherwise, royalty-based crowdfunding is a viable solution for companies that do not meet the requirements to access equity crowdfunding or who do not wish to open their equity to third parties, and it allows companies to encourage their backers' participation in a useful and, possibly, an advisory role.

What is most complicated with royalty-based crowdfunding is the dissolution of the relationship (on both sides) upon the conclusion or sale of the project. In most jurisdictions, contract freedom allows the stipulation of an agreement which perfectly suits the purposes of all parties, but finding royalty-based crowdfunding that sets clear rules for these cases is not common.

Tax regulation may also be critical: some jurisdictions do not allow deducting royalties from the revenues to pay the funders, requiring the business to calculate profits, pay taxes and distribute royalties out of the net profits. This kind of tax regulation makes royalty-based crowdfunding rather expensive to the entrepreneur, because royalties paid would be lower than those resulting from a normal ratio of the share capital.

Royalty-based crowdfunding is useful for collecting money for a single project keeping the crowd out of the entire business of the proposer; e.g., the publication of a book, a movie or music. Sandawe allows the public to view a few draft pages of comic books to decide whether and what to finance. The Belgian company, which in July 2015 claims to have raised more than €1,607,000 for 56 projects, proposes an editorial plan and a budget (usually €35,000–55,000 per album), and the project starts only when the amount is collected. Funders do not become shareholders or get voting rights; they are remunerated with their share of 60% of net profits until full coverage of their initial investment, and with a share of 40% of net profits thereafter, in proportion to the funding.

Sandawe has a large community of active members (15,500 in August 2015, with a 100% annual increase rate): all comic fans. They are highly motivated to support projects for the sake of participating in their implementation, and at the same time are a relevant market for the distribution of the same products they have helped to publish.

A similar model of profit sharing is the German website SellaBand, one of the oldest European crowdfunding platforms, through which artists can post a series of music tracks: the public can listen and choose to invest in some of them, for a share price of €10. The money raised stays in a temporary escrow account until it reaches the artist's target amount, whereupon it is used for the production of a CD (or returned if the goal isn't reached). Some artists offer supporters, termed "believers," a share of as much as 10% of the CD's revenues.

The need to find alternative solutions to the subscription of risk capital has inspired some entrepreneurs to find solutions more closely compared to the model of profit sharing, by moving different levers. In this regard, Kappel (Kappel, 2009, p. 375) classifies "ex-ante" crowdfunding, which finances a goal yet to be achieved, in two variations: the investment model, which is classic equity crowdfunding, even when it results in a partnership or profit sharing; and the betting model.

To explain the betting model, Kappel presents the example of the British company Slicethepie, which is considered a licensed bookmaker according to English gambling laws. In addition to a classic model of donation focused on music production, the company offers a trading platform of "contracts": for each music album funded through the platform, the company issues up to 15,000 contracts: that can be purchased for £1 each. The contracts can be traded for two years; afterwards they are called up and cleared at a price calculated on the basis of sales of the album at a rate of £1 for every 10,000 digital copies sold. For example, if a purchase was made at £1.8, the break-even for the supporter is 18,000 album sold. Artists are free to sell their rights to a record company, provided that it awards the holder of any contract a 50% premium on the specific purchase price of each. The model is completed with a royalty of £2.50 to Slicethepie for each album sold, over two years.

One important side effect of this model is that the owners of "contracts" are encouraged to advertise the album and increase sales, if they want to get a higher terminal value.

This is an interesting solution, especially for its strategic approach: Slicethepie wanted to involve a segment of the public attracted to trading and betting, offering a chance to play, with modest amounts of money, in a particularly stimulating business. What probably draws the crowdfunder to Slicethepie is the thrill of playing the particular game that the company invented, more than the desire to invest in a company or the prospect of a margin business.

Risks arising from contract-holders' relying solely on their hearts might be huge, and governments can't help but consider them. Indeed, while some commentators in the past have referred to the U.K. Gambling Act 2005 as a possible legislative solution for certain types of crowdfunding, the betting model in the U.S. has been prohibited since 2006, under the Unlawful Internet Gambling Enforcement Act (UIGEA), which prohibits the transfer of funds to online betting.

## A Community Business

In any kind of reward-based crowdfunding (donation, reward, or royalty), the community is absolutely crucial to the success of the initiative and helps determine the form taken by the request for funding, which otherwise would become entirely similar to what one might expect at a bank.

In an interesting essay, based on preorder or profit-sharing paradigms, Belleflamme, Lambert and Schwienbacher (Belleflamme, Lambert, & Schwienbacher, 2012) developed a mathematical proof that explains that entrepreneurs prefer preorders for projects that require little capital, but profit sharing for projects that require a major investment.

In fact, by virtue of preorders, the ability of the entrepreneur to manage price discrimination between lenders and customers is influenced by the amount of capital needed to cover fixed costs. Beyond a certain threshold, the distortion of the price becomes excessive and profitability is reduced. Instead, if the capital requirement is high, following the profit-sharing mechanism, the entrepreneur is able to expand participation without compromising too much on profitability.

The same authors, however, point out that the mathematical analysis of profitability cannot take into account the objective side of crowdfunding that nevertheless is critical to the success of the project: namely the creation of a community, a chance to test the market, and opportunities for strategic marketing.

## INVESTMENT CROWDFUNDING

When crowd-funders are offered securities (shares, bonds, or debt instruments) as a consideration for their financial contribution, the crowdfunding model becomes substantially different, mostly in terms of legal consequences. This model involves the sale of a securities instrument (stock) in a company through a regulated, online platform.

Rather than simply providing financial support for the project, funders will expect some kind of financial gain from the transaction, so their solicitation must be balanced by a number of assurances required by the law, as for any kind of investment.

In principle, possible sources of funding for a company are two: capital (equity) and debt.

While equity grants some rights to control the business (first of all, the right to vote), it is the riskiest form of financing from the investor's perspective. On the other hand, loans don't bear the same risks, are liquid (i.e., repayable after a term) and are usually negotiated against guarantees.

### Equity

The categories of potential investors in equity are as follows:

1. **The Entrepreneur:** Who generally invests his assets or gets personal loans to fund his company;
2. **Family and Friends:** People close to the entrepreneur who invest in the business, generally in small amounts;
3. **Business "Angels":** Individuals (or funds, or companies) who want to invest money in small projects, often to diversify their portfolios;
4. **Venture Capitalists:** Professional investors who collect money from sophisticated but non-specialized investors, and invest in innovative projects and for a set time, with the aim of realizing a significant capital gain on liquidation of the investment;
5. **Strategic Investors:** Other companies that may have a tied interest in industrial, commercial or strategic initiative to be undertaken (e.g., suppliers);
6. **Crowd or Retail Investors:** Undifferentiated public that invests in companies and buys securities.

In the mainstream of business funding, it is the entrepreneur who initially funds the business, eventually with the help of family and friends. Very early-stage businesses usually have their "bootstrapping" period: the term is hyperbole derived from the fanciful "pulling yourself up by your bootstraps," adopted in the 19th-century U.S.; it refers to tactics of self-financing—using the cash flow or other stratagems—rather than capital or loans.

The second stage has typically been that of angel investors and venture capitalists. But most recently the investment time for such professional investors has shifted forward, and is being replaced by crowdfunding; reward-based first, and equity crowdfunding as soon as a certain maturity is reached.

Venture capital is provided by investment funds managed by advisors, specialized in specific areas (information technology, health, biotechnology, etc.). They have a very professional approach to risk and usually follow specific preset guidelines for the time of entry in the investment; the available amounts are generally enough to advance the target company, after several rounds, to maturity, or until a stock exchange or trade sale exit, and their structure is tailored to follow the investment.

Projects of even greater size are often backed by private equity funds, which generally prefer to concentrate on a few deals involving larger amounts, invest in advanced stages of a business, or facilitate acquisitions or mergers, adding a significant value to the target company. They only invest in validated businesses that have demonstrated the ability to generate significant revenues through a concrete history.

Anyway, the access to finance in the last ten years has become increasingly difficult. Startups prefer the use of venture capital, angels or strategic investors since subjects specializing in this form of investment are generally able to contribute more than just financial support, in terms of management, network, media production and distribution. Indeed, professional investment is usually referred to as clever money, compared to the "stupid money" coming from the bank system, which typically has no nonfinancial support. The effectiveness of this strategic choice is demonstrated by Keushnigg (Keushnigg, 2004, p. 239), who explains how innovative companies supported by venture capital funds have the highest growth rates.

The European Commission communication of 7 December 2011 entitled "An action plan to improve access to finance for SMEs" states that "the difficulties of access to finance is one of the major obstacles to growth for SMEs," as also evidenced by the ratio of the ECB, "Survey on the Access to Finance of SMEs," in February 2010. In fact, the EU funding with venture capital has great potential, but so far has hardly been tapped for the development of SMEs (small and medium enterprises). Despite the huge size of the asset management industry in the EU, there is no existing legislation specifically designed to channel investment in equity capital to SMEs.

The EU Regulation no. 345/2013 of 17 April 2013, about European funds for venture capital, states that "... providing finance and advice to these undertakings, venture capital funds stimulate economic growth, contribute to the creation of jobs and the mobilization of capital, encourage the creation and development of innovative businesses, increase their investment in research and development and foster entrepreneurship, innovation and competitiveness."

As explained by Bhidé (Bhidé, 1992, p. 109) in 1992 in *Harvard Business Review,* the institutional investor thinks in terms of default amounts to invest. Since 2002, venture capital funds have gradually increased these levels, since many small investments are more expensive in terms of monitoring, compared to fewer larger investments. Being listed on a regulated market is only possible in connection with significant investment thresholds, also keeping in mind the complexity and cost of the IPO (Initial Public Offering).

This distinction involves a practical consequence extremely important for startups: the equity market is effectively closed to projects that are too small, with a cash need insufficiently attractive to institutional investors.

Not least, it is necessary to consider the influence of macroeconomic conditions (market trends, trends in interest rates, etc.) on investment decisions, often more effective for small and medium-sized enterprises; in 1990 the credit crunch in the U.S. had reduced loan disbursements by 38%, from 144 billion to 88 billion dollars, as noted by Berger and Udell (Berger & Udell, 1998, p. 613).

## Debt

As an alternative, other organizations prefer to take out loans rather than dilute ownership by taking on investors. The players on the loan side of investment are:

1. **Banks:** Usually they don't lend money to startups without revenues, and in any case, they ask for guarantees;
2. **Leasing Companies:** They finance tangible assets (computers, technical equipment, etc.) and the business premises;
3. **Government Agencies:** In some countries they offer loans, usually at reduced rates, for particular classes of projects;
4. **Customers or Suppliers:** Entrepreneurs might succeed in getting financed through delayed payments accepted by suppliers or advance payments from customers.

Debt might seem more expensive than equity because of the interests to be paid, but indeed the problem is not the cost of finance, rather how to access it. The choice to get a loan seldom depends on the interest rate or the consideration required by the lender; it is more often determined by lack of alternatives or by a strategic choice of the entrepreneur. In some cases, indeed, he is not yet able to justify a significant business evaluation in order to get a fair value against equity shares of his business; therefore, he prefers to take on debt until a significant value may be demonstrated and capital is open to third parties.

Regardless credit crunch, borrowing money is increasingly difficult for start-ups, because while SMEs at least have a history and a turnover, a startup's history is usually not compatible with the requirements of financial institutions, being too short and limited to the building of a business yet to be demonstrated as successful; moreover, it is usually unable to offer assets as collateral.

This difficulty to access the loan market has led to the growth of a new kind of crowdfunding: in debt-based crowdfunding, individuals lend money to either other individuals (P2P) or to businesses (P2B). Debt platforms screen the applicants, underwrite the loans, assign interest rates (sometimes based on an auction of interest rates among willing lenders), and then create a market where investors are able to screen available loans, and supply parts of the total loan amount to the projects or businesses. P2P platforms might fund everything from debt consolidation to home improvement, to leases, to real estate, and even college tuition. Indeed, in some countries their business is limited to private borrowers; in some others, businesses are allowed to borrow; in others there are restrictions on the lending side. In still other countries, debt-based crowdfunding is heavily regulated through bank licenses, which substantially inhibits the expansion of the market.

Lending-based crowdfunding platforms are growing rapidly across the world and have amassed billions of dollars in transactions. For example, in 2013, Lending Club originated over $1 billion (USD) in unsecured loans from consumers, and now Prosper is following the same growth curve. Normally, unsecured loans come at a higher default rate, but experience has shown that the default rates on these loans are not greater than traditional loans, and they have a higher yield. The successful IPOs of LendingClub and OnDeck further demonstrate the acceptance of these financing platforms.

According to recent research, this could become a $1 trillion global market within 10 years—an enormous figure that would still represent only one percent of the $100 trillion global debt market.

The screening and underwriting process is faster than traditional loan underwriting, and the web-based technology is a great facilitator. This may be particularly beneficial to companies that otherwise would have to turn to expensive receivables discounting for the fastest approval when seeking capital. In many countries, where SMEs need to finance their working capital on credit in order to remain in business, and a late payment is too often a reason for bankruptcy, this could be the difference between life and death.

Lending platforms are leveraging all available information on potential borrowers in order to assess their creditworthiness, and are investing in new algorithms in order to gain the ability to select relevant information from big data. An Indian platform is focusing on credit scoring, analyzing candidate borrowers' behavior on the social networks, while African micro-credit platforms are starting to rely on a candidate borrower's online micro-payments history to assess credibility. Indeed, evaluation and risk assessment are of prime importance when providing unsecured loans.

Most recently, large corporations also started investing in lending platforms, by purchasing loan portfolios or by taking portions of a large number of loans, given that the growing success of the score system that these platforms are developing is often more successful than that of banks. This is why the availability of investors now exceeds the current demand for loans. Such an opportunity should lead even the most restrictive governments to be open to lending-based crowdfunding, and accepting investors from abroad.

A variety of lending-based crowdfunding is microcredit: started in Bangladesh by Grameen Bank, it is a way to kick-start microbusinesses in underdeveloped countries, through very small loans, that nevertheless make the difference between opening a business or starving—between life and death (United Nations Development Program, Evaluation Office, Essentials Microfinance, 1999). Thousands of women and men in Asia and Africa have successfully started businesses and changed their lives with as little as $700–800, and also contributed to the development of their own countries. It is a very risky business, given the lack of guarantees and the frequent absence of promising conditions; therefore, interest rates must be high enough to remunerate the risk. Nevertheless, the success of microcredit demonstrates that borrowers easily succeed in paying back loan and interest, and are quite eager to do that, provided that the overall amount is not huge.

One problem preventing the wide development of microcredit is that some governments ban its average interest rates (20–30%) as usurious, thereby prohibiting microcredit by definition. Such bans stand regardless of context, or terms and conditions of a deal, so that no discrimination is possible and no access to this precious instrument is allowed.

## The Regulatory Framework

Given the public solicitation operating through the web and its investment nature, equity crowdfunding is regulated within the framework of financial regulation and securities; on the other hand, lending-based crowdfunding is either subject to the same regulation as bank and credit institutions or non regulated at all.

The main topics for lending crowdfunding in those countries where it is regulated are:

- License to operate a platform;
- Who can lend;
- Who can borrow.

For equity crowdfunding, instead, there are more critical topics, including:

- License to operate a platform;
- Who can invest in equity;
- Who can receive an investment;
- Maximum limits per investor;
- Maximum limits per deal;
- Solicitation rules.

Hereinafter we will go through some different approaches to equity crowdfunding deals in order to have a broad view of different approaches and regulations.

## Equity Crowdfunding: The U.S. Legal Framework

Despite the title of the debate born from the pages of *The Wall Street Journal* ("Should Equity-Based Crowd Funding Be Legal?" by Lavinsky and Torrens) which offered opposing viewpoints of this form of investment, equity crowdfunding has never been illegal in the U.S., although it has been rather inaccessible in terms of cost and bureaucracy.

The dichotomy that has animated the American debate reflects opposing economic theories, interventionist and liberal, based on the "invisible hand" thesis of Adam Smith. One theory argues that it is a viable form of financing for companies that do not qualify for access to venture capital, mostly because of their limited size; the other states that if the target companies are not good investments for venture capitalists and angels, there are probably good reasons not to invest, and these reasons may be too subtle for retail investors to detect.

In the American system we must start from the concept of "securities" and their strict regulation. ("The term 'security' means any note, stock, treasury stock, security future, bond, debenture, evidence of indebtedness, certificate of interest or participation in any profit-sharing agreement, collateral-trust certificate, preorganization certificate or subscription, transferable share, investment contract, voting-trust certificate…" –Securities Act, 1933). Considering financial instruments stocks, bonds, investment contracts and other forms of participation, an investment contract means any collection of money that might explicitly or implicitly provide investors with a potential gain.

An investment contract is subject to the securities regulation depending on the Howey test; i.e., the verification of the conditions set by the Supreme Court in the Howey case: considering as an investment contract for the purposes of the Securities Act an agreement by which a person invests money in a company with other people or is solicited to expect profits because of a promoter's or a third party's advice. This definition covers investments involving real estate. The declared purpose of the Securities Act was to promote transparency in order to gain strong consumer protection: Justice Louis D. Brandeis is often quoted as saying: "sunlight is the best disinfectant; electric light is the most efficient policeman."

The U.S. law is divided into two regulations: the 1933 Disclosure Requirements and Registration Act and the 1934 Periodic Reporting Requirements Act. The result is a complex system based on three fundamental requirements:

1. Registration of the securities whose sale is induced;
2. Full documentary disclosure for the investor through a prospectus that contains detailed information about the entrepreneurial activity and verified accounts;
3. The obligation to produce periodic update reports.

A third act in 1934, the Broker-Dealer Registration Requirements, extended the registration requirements to intermediaries.

One of the main rules in the American system (both state and federal) is the difference between accredited and non-accredited investors. Accredited investors are defined in reference to the size of their assets, which must be larger than $1 million (excluding the value of their house) and current revenues exceeding $200,000 ($300,000 where cumulated with a spouse). In total, accredited investors account for about 3% of the American population, and their investments in non-listed private companies is about $30 billion a year, distributed among approximately 600,000 accredited investors.

In the U.S., as in many jurisdictions, what is most problematic and expensive is the prospectus requirement; i.e., the information to be provided to the public when soliciting an investment. Nevertheless, a number of exemptions occur, provided by U.S. legislation and the SEC to exclude or limit the rules for very small companies, or financial instruments having a small value or dedicated exclusively to sophisticated investors. Such exemption schemes, mainly built around the Regulation A and Regulation D, never exclude information and reporting duties.

When a U.S. website (BuyABeerCompany.com) tried to propose $300 million online funding for the purchase of Pabst Brewing Company, without registering the operation nor complying with the Securities Act, the SEC stopped the offer. The promoters immediately answered that they would collect investment intentions before actually getting investments. Since the money had not yet been collected, the case ended with an agreement based on the spontaneous withdrawal of the promoters from the initiative (SEC, Adm. Proc. File No. 3-14415). The promoters, whose communication of the offer was amplified via Facebook and Twitter, have reported investment intentions for over $200 million, from more than five million individuals.

The primary legislation regarding equity crowdfunding is the Jumpstart Our Business Startups Act (JOBS Act, April 2012), but it is not sufficient by itself to regulate equity crowdfunding: the SEC had been delegated to issue a special implementing regulation which will govern the process in detail and will be complemented by an appropriate regulation of the FINRA. Unfortunately, the deadline set by the JOBS Act had long expired by the time the SEC finally released a 500-page draft regulation and later, on October 30th, 2015, the implementation of the bill.

In the wide debate about the SEC delay, many questions were posed and no answers were provided. There was apparently some concern about an eventual mass diversion of investments from the regulated stock exchanges to the "real" economy of restaurants, shops and small businesses.

Meanwhile, while waiting for the SEC regulation, several states (15 at the moment) have issued their own regulation and exemptions for offering limited crowdfunding to their residents. This is certainly a great opportunity, given the large investment capacity of some states; nevertheless, the scalability offered by the web would better be exploited through a national unified framework, and even better at an international level.

While non-accredited investors individually may make only small investments, collectively they may create a huge boost to the capital markets.

Besides that, other instruments are provided by specific regulations and allow restricted forms of crowdfunding, limited to accredited investors. This might not be considered "true" crowdfunding, but the dimension of U.S. accredited investors' available funds is so huge that it is a very good first step (reportedly 40% of the $55 billion annually invested in venture capital worldwide).

## The JOBS Act

The legislative initiative at the federal level was taken by President Obama in April 2012 with the JOBS Act, with the stated aim of reviving the economy then in a worldwide downturn. Not surprisingly, Obama's policy and the support granted to the economy are generating important results in terms of growth and employment, in contrast to other industrialized countries (except Japan).

The JOBS Act contains three different pieces of legislation, tailored for accredited, non-accredited or both kinds of investors.

Title II of the JOBS Act, concerning accredited investors, has been implemented by the SEC, and dramatically contributed to the $1.1 billion equity raised worldwide in 2014 on-line. It allows accredited investors unfettered access to operate on online platforms.

Title III, in contrast, allows retail investors to participate in equity crowdfunding. Given that the U.S. has the broadest national economy, analysts are eager to check out how retail investors will react to this huge democratization of their access to finance.

Title IV, which became effective on 19 June 2015, permits investments by both accredited and non-accredited investors, albeit with some rather expensive strings attached for small businesses (hereinafter referred to also as Reg. A+).

The main principles of Title III of the JOBS Act are the following:

1. Crowdfunding platforms are subject to a registration requirement;
2. The amount retail investors can invest in a company is capped at $2,000 or a specified percentage of the company's annual net income or net wealth (5% for incomes up to $100,000, and 10% for higher incomes), whichever is greater;
3. Any company proposing an investment must provide information to potential investors and the SEC; such information includes:
    a. The objectives of the company;
    b. The capital that is being collected;
    c. The maximum term of validity of the offer;
    d. The price of the securities offered;

  e. News about the company and management (board of directors and managers);
  f. News related to major shareholders and the capital structure;
  g. Business plans and financial situation;
4. Special attention is to be paid to the risks, to be duly highlighted both to the potential investors and to the SEC;
5. The SEC may require written additional information;
6. If the offer of securities exceeds $500,000, audited financial statements are required; between $100,000 and $500,000, accounts might be verified by an independent expert in accordance with rules established by the SEC; up to $100,000 it is sufficient to publish tax returns and financial statements of the previous year certified by the CEO.

After a successful funding, information on the business and operations ought to be updated annually, together with the financial statements.

The JOBS Act has chosen to limit the collection of equity through crowdfunding, to a million-dollar maximum for each twelve-month period.

It's true that this amount is high when compared to the current values of crowdfunding, but the most common crowdfunding models in 2012 (donation and reward) are very different from equity. For example, the average successful project on Kickstarter is $8,390. In contrast, the world average for equity-based crowdfunding, according to data from Massolution report CF2013 in 2012/2013 was $190,000, which almost doubled in 2014.

The ability to raise equity online under federal guidelines is limited to companies that qualify as Emerging Growth Companies (EGC): this type of company is not identified in relation to its business activity, but the volume of its annual sales, which must not exceed the astronomical sum of $1 billion in the previous fiscal year.

ECG status is retained for five years after the public offering, unless the company issues non-convertible debt securities exceeding $1 billion in three years or does not become a "large accelerated filer" (Rule12b-2). These values are frankly very far from the best ambitions of a startup and appear to contradict a roof of $1 million a year, and were established solely to avoid limitations to a truly effective use of crowdfunding. The logic of the U.S. crowdfunding legislation, intended primarily for small businesses with deep local development, is intended to encourage a strong link between investors and communities. Indeed, the small and very small companies have a greater capacity to create jobs, and help achieve the main objective of the JOBS Act.

In this regard, it is interesting that Amy Cortese coined the word "Locavesting" just to explain the phenomenon of the investment located in a community and the benefit it brings to the development of the community's social components. According to this analysis, while a dollar invested in Apple will contribute to the development of a better iPad, the same dollar invested in an agricultural activity can stimulate the local economy, helping induced activities that remain in the territory.

From the perspective of small business, the million-dollar roof is sufficient (at least for a year), but most importantly, it enables the proponent to focus on concrete and effective financial needs, avoiding the distraction of projects that are too complex or ambitious. The repeatability of annual crowdfunding allows the business to postpone raising further capital in the immediate aftermath of the first year's successes. Also in the American logic, for companies with greater financial needs, given that the U.S. venture capital and angel investing are extremely popular and mature, other means of financing might be preferable. Not to be overlooked, finally, a generalist small business is less likely to fail than a high-tech startup.

This means that from the point of view of the American legislator, in order to create more jobs, and to limit risks for retail investors, it is useful to allow investments in "old economy" pure commercial companies, like a restaurant or a shoe manufacturer. It may not be the new Google, but it will most likely distribute dividends to investors and develop the local economy, advancing the goals of the U.S. government.

Each investor has a right to withdraw during the first 21 days after the initial crowdfunding offering to the public, regardless of the moment when the investor first learned of the possible deal. The reason for this rule lies in the great trust the American legislator puts into the network and in the ability of the people to expose fraud attempts: the thinking is that in a period of three weeks, the people will be able to discover and disclose any attempts at fraud. So, until the 21$^{st}$ day everyone can withdraw (even those who accepted the offer on the first day), but once the twenty-first day after the start of the campaign is over, no withdrawal right is allowed.

While equity crowdfunding access limits are modest, in order to allow the maximum access to capitalization, the JOBS Act sets significant requirements for the life of the company after the capitalization, and the sale of subscribed shares is restricted for one year. While it is true that the effectiveness of an emerging company may only be measured over time, this restriction seems to have no other logic than preventing these securities from being fuel for speculation and immediately traded.

A significant barrier to entry is the nationality of the target company, which must be exclusively American. This is a protectionist measure clearly designed to grow only American companies, as well as increase the gross domestic product and employment in the United States, in the spirit of the JOBS Act. The American system, subject to federal laws, is not therefore a candidate as an international marketplace for small and medium-sized enterprises, but is only a facilitator for its domestic market.

## Intermediaries in the JOBS Act

The typical crowd investment designed by the JOBS Act, both for accredited and non accredited investors, is through an authorized intermediary (broker-dealer) or through a crowdfunding platform registered with the SEC and at FINRA (Financial Industry Regulatory Authority). This excludes what happened spontaneously in different parts of the world at the dawn of crowdfunding, i.e., self-made websites to raise equity. The broker-dealer is responsible for the screening of the proposals, the subscription process and the delivery of the equity to the proposer once the financing goal is reached, according to the "all or nothing" formula. (The "take it all" formula, widely adopted in reward-based crowdfunding, is not allowed by the JOBS Act.)

The main charge the intermediary is entrusted with is the step-by-step compliance verification and the proper and deep disclosure of the offer and its supporting documents.

The broker-dealer is subject to a series of requirements set by the SEC at the regulatory level, and then:

1. Must inform investors of the risks;
2. Provides investors with educational tools;
3. Checks that investors examine the information provided by the offering party;
4. Checks that investors answer some questions and confirm that they properly understand the risks.

The only type of advertising allowed for crowdfunding is that directing the potential investor to the web platform (or the registered intermediary), and SEC/FINRA specifically regulate the kinds of possible advertising.

In addition, the remuneration of the intermediary (and thus of the platform) must be transparent for the investor.

## Crowdfunding Platforms

Equity crowdfunding platforms operating pursuant to Title II of the JOBS Act are not reserved only for broker-dealers or their subsidiaries.

The U.S. legislation directed the SEC to amend its rules to allow general solicitation and advertising for securities offerings being sold to accredited investors, stating that a platform offering securities under Rule 506, and permitting general solicitation and advertising by the issuers of such securities on its platform, does not need to register as a broker-dealer.

Such exemption is subject to the following requirements:

1. That the platform does not receive compensation in connection with the purchase or sale of any security;
2. That it does not hold customer funds or securities in connection with the purchase or sale of such security;
3. That it is not subject to statutory disqualification.

Otherwise, the intermediary must be registered as a broker-dealer and the FINRA regulatory system applies, along with all consequences in terms of expensive and time-consuming reviews, analyses, due diligence, and compliance.

Funders Club and Angel List are well-known and very effective equity crowdfunding platforms operating under Title II, though not registered as broker-dealers: in 2013, they received "no action" letters from the SEC, where the SEC Division of Trading & Markets staff agreed not to recommend enforcement actions. Since exempt platforms did not need to register as broker-dealers, under specific circumstances, these no action letters established a precedent.

Of course, Funders Club and Angel List offered securities to accredited investors only, did not earn transaction-based compensation (but rather "carried interest," as they were operating as investment advisors), did not handle customer securities or funds, and were not subject to statutory disqualification. If instead, for example, they were receiving commissions on securities transactions, compensation calculated on the success or size of the offering, or holding customer funds or securities, then they would have needed to register as broker-dealers.

Nevertheless, these platforms remain subject to the anti-fraud provisions of the federal securities laws, identification and anti-money-laundering regulations. Identification is particularly important in order to properly check the accredited investor profile.

The current legislation implicitly defines a primary difference between funding portals and funding platforms.

Funding portals are heavily regulated by the JOBS Act as miniature broker-dealers, and the October 2013 FINRA Regulatory Notice 13-34 set forth the rules for their FINRA membership, supervision,

AML and reporting rules. But until the SEC approves crowdfunding federally, there are no funding portals to be supervised by FINRA.

Instead, the American equity crowdfunding market is currently managed by funding platforms, websites where investments are advertised to accredited investors only. Such platforms are mentioned just once in the Title II exemption from broker-dealer registration and the SEC itself confirmed that they don't need to register as broker-dealers, by the aforesaid no action letters to Funders Club and Angel List.

Equity crowdfunding, therefore, is currently managed through three different business models:

1. Broker-dealers, who recommend investments and earn commissions;
2. Investment advisors;
3. Listing platforms, that simply list offerings without any recommendation, earn listing fees, and charge further fees for services or tools.

## JOBS Act, Title IV, and Regulation A+

Under the Securities Act of 1933, when a company offers or sells securities to potential investors, it must either register the offer and sale or rely on an exemption from registration. A longstanding exemption from registration is Regulation A: it allows unregistered public offerings of up to $5 million of securities in any 12-month period. In recent years, Regulation A offerings have been relatively rare (it is generally considered expensive and complicated) and the new regulation, that Title IV of the JOBS Act required the SEC to modify, adds a new set of rules commonly referred to as Regulation A+.

The SEC finally issued Regulation A+ on 26 March 2015, to take effect on 19 June 2015, providing for two tiers of offerings, plus some limits on secondary sales by affiliates:

**Tier 1:** Securities offerings of up to $20 million in a 12-month period, with not more than $6 million in offers by selling security-holders that are affiliates of the issuer.

**Tier 2:** Securities offerings of up to $50 million in a 12-month period, with not more than $15 million in offers by selling security-holders that are affiliates of the issuer.

For offerings of up to $20 million, the issuer could elect whether to proceed under Tier 1 or Tier 2. Both tiers would be subject to basic requirements as to issuer eligibility, disclosure, and other matters, drawn from the current provisions of Regulation A. Both tiers would also permit companies to submit draft offering statements for non-public review by Commission staff before filing, permit the continued use of solicitation materials after filing the offering statement, require the electronic filing of offering materials and otherwise align Regulation A with current practice for registered offerings.

In addition to these basic requirements, companies conducting Tier 2 offerings would be subject to other requirements, including:

- To provide audited financial statements.
- To file annual, semiannual and current event reports.

Moreover, a limitation has been set on the amount of securities non-accredited investors can purchase in a Tier 2 offering: no more than 10% of the greater of the investor's annual income or net worth.

The exemption is limited to companies organized in and with their principal place of business in the United States or Canada. The exemption would not be available to companies that:

- Are already SEC reporting companies and certain investment companies;
- Have no specific business plan or purpose or have indicated their business plan is to engage in a merger or acquisition with an unidentified company;
- Are seeking to offer and sell asset-backed securities or fractional undivided interests in oil, gas or other mineral rights;
- Have been subject to any order of the Commission under Exchange Act Section 12(j) entered within the past five years;
- Have not filed ongoing reports required by the rules during the preceding two years;
- Are disqualified under the "bad actor" disqualification rules.

Regulation A+ is essentially a good way to launch a mini-IPO to both accredited and non-accredited investors; it is considered expensive (commentators quote about $100,000), it takes a long time (about 6–10 months to get the paperwork ready), and it gives significant exposure to the target company, even if it fails (although a "testing the water" period is allowed). But the large amount that can be raised is a good prize for the attempt.

Years ago, fast-growing companies went public much earlier in their growth cycle (compare Apple's IPO in 1980, which raised about $100 million, to Alibaba's recent record-breaking IPO of $25 billion). In this new scenario, a first step before the stock exchange might be useful.

It might be also useful to compare the benefits of Regulation A+ to those of Regulation D (specifically Rule 506 of Regulation D), which is to date the most prevalent approach private companies use to raise capital: Rule 506 allows private companies to raise an unlimited amount of capital from accredited investors only (this is the main difference); it does not require federal and state review and approval of offering materials; moreover, it allows companies to generally advertise their offering and solicit investors.

The primary advantage of Regulation A+, then, is the ability to solicit the retail investor, rather than relying, as the argument goes, on the more limited universe of accredited investors. The consideration required for gaining access to the general public is mostly time and money.

Filing an offering circular to the SEC for a Regulation A+ offering is nearly as complex as filing a prospectus for an initial public offering. The SEC estimates about 10 months to approve the average Regulation A filing, with only a 23% acceptance rate for the filings it receives.

## Equity Crowdfunding: The European Framework

The European Commission is currently studying equity crowdfunding with great interest, given the continued strong commitment to the development of the internal market.

Back in 2008, the Commission issued a "Think Small First" communication (Communication no. 394/2008, 25.6.2008) which called for the establishment of a regulatory framework for small business, called the "Small Business Act," based on 10 guiding principles; the sixth principle highlighted the difficulties of SMEs in raising capital and in obtaining financing through normal banking channels, and supported the opportunity to evaluate possible private placement regimes to facilitate cross-border investment and reinforce the European market for risk capital.

More recently, with the "Entrepreneurship Action Plan 2020" communication (Communication no. 795/2012 of 01/09/2013), the European Commission signaled the importance of supporting SMEs, stating that each year they create more than four million new jobs in Europe. The Commission focused on the crucial nature of the initial stages of their life cycle, showing that about 50% of SMEs fail during the initial five years; a major cause of failure being the absence of an appropriate financial ecosystem that was favorable to growth. Consequently, the Commission highlighted its initiatives in the Programme for the Competitiveness of Enterprises and SMEs (COSME) and Horizon 2020 to provide financial support to startups and strengthen venture capital, angel investment and incubators.

This context also fits Regulation no. 345/2013/EU of 17.4.2013 relating to European funds for venture capital.

Finally, with the "State of the Innovation Union 2012 – accelerating change" communication (Communication no. 149/2013 of 03/21/2013), the European Commission noted that in Europe there is no shortage of innovative ideas to be converted in successful business models, but the main obstacle is access to finance, further aggravated by the current economic downturn: the collection of venture capital funds fell by 45% and the investment activities of business angels is currently about one-fifth that of the United States.

In this regard, the European Commission had some recommendations drawn up by the Expert Group on cross-border matching of innovative firms with suitable investors; among the recommendations is to "promote joint financing (crowdfunding)." The Commission also took some important steps, including one in 2014 when it created the European Crowdfunding Stakeholders Forum, a group of experts coming from both the private market and national financial authorities, meant to help the European Commission in understanding and regulating this instrument.

Before issuing uniform provisions for the internal market, however, one must first establish whether there are crowdfunding profiles already regulated at the EU level.

More recently the European Commission published the communication "Realizing the potential of the crowdfunding" (COM 2014/172, 27 March 2014), which was followed by the establishment of study groups among operators and experts in the field and the publication of a "Guide for SMEs. Crowdfunding. What is it?"

On 18 December 2014, ESMA (European Securities and Markets Authority) published an opinion to the competent National Authorities and Advice addressed to the European institutions in the field of investment-based crowdfunding, with the aim of achieving greater convergence in the EU regulatory and supervisory arena.

Finally, the Green Paper on the Capital Market Union focuses on crowdfunding as a tool that should make a decisive contribution to the financing of the European economy, even beyond national borders, asking whether there are obstacles to the development of adequately regulated platforms, even across borders. ESMA, in its response to the consultation, identified crowdfunding among the possible areas of regulation in the field of financial markets in the coming years.

In principle, there are no regulations prohibiting the models based on rewards or donation, and e-commerce rules apply to preorders. Lending-based crowdfunding is usually part of banking legislation, albeit with the necessary adjustments.

Instead, equity crowdfunding regulation is far more complex. At the community level, the Prospectus Directive (Directive 2003/71/EC as amended by Directive 2010/73/EC) regulates public offerings of risk capital larger than €5 million. Offers below €100,000 are exempted, but there is no agreement at the community level between these ranges, and each member state retains its own rules (and the right to change them).

An eventual directive on crowdfunding could harmonize operations up to €5 million, contain a specific exemption from the prospectus requirement, and delineate proper forms of consumers protection.

In terms of communication, ESMA issued a "Notice to Investors" which invites the public to:

- Be aware of the risks inherent in investing online;
- Check whether the proposing company is unauthorized (obviously depending upon the applicable laws);
- Be wary of the promises of high returns;
- Pay attention to software that generates automated orders.

However, the document is aimed at investments online via electronic platforms, regardless of the object of the investment, and not crowdfunding specifically. In fact, numerous operators recently started offering online trading platforms involving financial products freely available on the market. These operators often provide training with "wallets" or economic promotions to attract the public.

In addition, reference should be made to Directive 2002/65/EC of 23.9.2002 concerning long-distance marketing of consumer financial services, which plays a significant role in the harmonization of national laws, and has taken a large part in the drafting of national consumer codes.

In principle, where specific regulation doesn't yet exist, member states apply their internal finance regulations. This includes the application of Article 3 of Market in Financial Instruments Directive (MiFID) (n. 2004/39/EC) and Articles 1.2 and 1.3 of the Prospectus Directive, as also recently represented by ESMA during a workshop on crowdfunding organized in Brussels by the European Commission on 3 June 2013.

The main feature of MiFID is that it creates a system based on authorized and vigilant broker-dealers, subject to very strict regulation, and that it requires an authorized broker-dealer to carefully assess the investor's proclivity to make risky investments and the adequacy and appropriateness of the investment with respect to the investor's profile.

Article 3 of the MiFID Directive contains an exemption where the offer is run jointly with a broker-dealer, and the platform doesn't hold the investors' money. In 2016, the MiFID II Directive will become effective, and the member states will have to adapt their own legislation to the new rules. Although worded differently, Article 3 of MiFID II will continue to be the basis of equity crowdfunding exemptions in the European Union. Some European countries have already issued their own regulation based on the exemption provided by the revised Article 3.

The problem with this regulation is that nationally regulated equity crowdfunding platforms may not be "passported" throughout the Union, as have other financial institutions and intermediaries, as their operation is allowed only in the Member State where they got a license. This limitation strongly affects the scalability of equity crowdfunding, which indeed is one of the main features of this disruptive financial instrument. This is why the industry is calling for a specific European uniform equity crowdfunding exemption, enabling equity offers throughout the European Union.

Since crowdfunding platforms are usually IT companies, they rarely meet the requirement to apply for a broker-dealer or banking license, in order to be compliant with the MiFID Directive. This is why some European Countries (such as Italy) have chosen to require a platform to operate jointly with a broker-dealer:

1. The platform selects and publishes offers;
2. The broker-dealer receives and executes the subscription order under the MiFID requirements, holding the money until the offer is completely subscribed.

The exemption might help in the sense that national governments may set a limit below which the investment may be done without the broker-dealer. Italy has set this limit at €500 per deal and €1,000 annually (per individual), but after 18 months it was discovered that this amount is very far from the average transaction, which is above €10,000. This might lead to an increase of the exempted investment, in order to facilitate greater access to equity crowdfunding.

The Italian exemption of 26 June 2013 was the first in the world to be issued, even before the SEC regulation for accredited investors under Title II of the JOBS Act, but it followed an upside-down approach that didn't properly meet the needs of the market, perhaps due to the legislators' misunderstanding of how the market might have worked. A revision is scheduled for fall 2015.

Almost a year after Italy, the English FCA (Financial Conduct Authority) became the second national authority to issue equity crowdfunding regulations (effective 1 April 2014), but the FCA used a bottom-up process which started by observing the market and approving some provisional licenses to the first players who approached.

The U.K. is a vast financial market, and in fact the volume raised in 2014 exceeded €111 million. This large equity crowdfunding was also possible thanks to a regulation that left the investment process entirely to the crowdfunding platform, without involving broker-dealers or other intermediaries. Any MiFID-related verification and information is the exclusive responsibility of the platform, as are the identification and anti-laundering verifications, and they are properly run online.

There is no size or industry limit, provided that the target company is U.K.-based. Moreover, the English system allows the collection of the expression of interest (guaranteed by credit card), that becomes a true investment only once the target investment is reached.

The French exemption came with a regulation on 1 October 2014, also through a bottom-up process: platforms operate as CIP (Conseiller en Investissements Participatifs), registered with and under the inspection of AMF (the French financial authority); they don't need the help of a broker-dealer because they are registered intermediaries themselves, subject to the online verification duties specified by the MiFID regulation. There is no exempted amount, but the MiFID procedure is easier than in Italy, as it is run completely online and with a single goal: focused and specialized solely on equity crowdfunding.

The French exemption is working very well, as the market is growing 146% per year, and in 2014, €25 million was invested, with an average deal of €376,733. Nevertheless, there are restrictions concerning the size of the deal (€1,000,000) and the nationality of the target companies.

The German market clearly appreciated their national regulatory approach, as the market gained 50% in just the first quarter of 2015, when €12.4 million was collected, versus €7.9 million for the entire 2014 year.

The German regulation issued in April 2015 raised the limit for the prospectus from €1 million to €2.5 million, and set a personal investment limit for retail investors at €10,000, provided that they were found to be eligible for risky investments under the MiFID profiling rules. A huge effort has been made to move the entire investment process online, and to reduce the advertising limitations formerly applicable to any financial offering.

Spain was the last to adopt its national exemption and used a bottom-up approach, clearly resulting from two years' observation of equity crowdfunding in the world markets. The Spanish regulation imposed minimum capital requirements (€60,000) and professional liability insurance (€400,000) on platforms, to be increased proportionally to the investments managed.

From the American system, Spain imported the accredited investors profile: they have no investment limits, provided that certain revenue or estate criteria are met (€50,000 annual revenue for individuals). Non-accredited investors, on the other hand, have limits of €3,000 per investment and €10,000 per year.

The maximum offering is €2 million, unless the deal is reserved for accredited investors only, in which case the limit rises to €5 million.

## REFERENCES

Belleflamme, P., Lambert, T., & Schwienbacher, A. (2012). *Crowdfunding: Tapping the Right Crowd.* CORE Discussion Paper No. 2011/32.

Berger, A. N., & Udell, G. F. (1998). The Economics of Small Business Finance. The Roles of Private Equity and Debt Markets in the Financial Growth Cycle. *Journal of Banking & Finance*, *22*, 613. doi:10.1016/S0378-4266(98)00038-7

Bhidé, A. (1992). Bootstrap Finance: The Art of Start-Ups. *Harvard Business Review*, *70*(6), 109. PMID:10122688

Kappel, T. (2009). Ex Ante Crowdfunding and the Recording Industry: A Model for the U.S.?. *Loyola of Los Angeles Entertainment Law Review*, 375.

Keushnigg, C. (2004). Venture Capital Backed Growth. *Journal of Economic Growth*, *9*(2), 239–261. doi:10.1023/B:JOEG.0000031428.35711.fc

# Chapter 13
# How to Launch a Successful Crowdfunding Campaign

**Ronald Kleverlaan**
*CrowdfundingHub, The Netherlands*

## ABSTRACT

*This chapter provides insights into the important characteristics of a successful crowdfunding campaign. It starts with providing the fundamentals of crowdfunding and shows the reasons why it is important to find the right target groups and communities for a campaign. It describes in details the Return on Involvement model, what the funders want in return for their support. In the second part the additional results of crowdfunding are discussed, such as market research, marketing, brand loyalty and the possibility to use crowdfunding to raise additional funding. The last part of this chapter will focus on the background of setting up and running a successful campaign.*

## INTRODUCTION

Why are some campaigns very successful and do they raise millions of euros and have other campaigns huge problems to get their first investors? What are the secret ingredients of a successful campaign? In the chapter will we look into the do's and don'ts of launching a campaign. This will involve a good planning and preparation, finding the right community that will fund, creating a good pitch and finding the rewards that will convince people to support a campaigns.

## CROWDFUNDING = COMMUNITY FUNDING

The investors in a crowdfunding campaign are not an anonymous crowd of people that will automatically find a crowdfunding campaign and will support a campaign they find online and like. The majority of the funders are people that know the entrepreneur of project in advance, or get introduced at the campaign by someone who knows the. Therefore, it is important for a successful crowdfunding campaign to think about the communities that will be interested in the campaign and find out how to reach them.

DOI: 10.4018/978-1-5225-0568-6.ch013

## Find Your Communities

Every person and every organization has access to communities. Sometimes these are their own clients that are able to support the financing needs of the company, but it can also be business partners, current stakeholders of the organization, people from your own village or sport club up to friends & family and employees.

When the potential communities are identified, it is important to find the best way to reach out to them. Some communities are already grouped and can be easily reached through social media or on weekly/monthly regular meetings, but for other people it can be more difficult to contact them. For all of them it is important to identify some of the strongest connections in these networks. You can contact them in a later stage and try to convince them to help you promote your campaign in the community you are both part of.

## Return on Involvement (ROI)

The strength of crowdfunding is the trust people have in the entrepreneurs running the campaign. They will not just financially support a campaign, but want to get involved in the process of making the project or company a success. This involvement is very important for all people supporting a crowdfunding campaign. But what do they want in return? What is the Return on Involvement of their participation in a crowdfunding campaign.

While launching the crowdfunding campaign it is very important to focus on the potential donors/investors and their expectations. The investors are strongly connected and involved in the projects they support, but the return they are interested in is very different. Some of them are interested in a social return, some want to have a material return (products/services) and some are more interested to support crowdfunding campaigns to have a financial return.

For campaigners it is very important to know your community and your community members and know what kind of return they are interested in and define returns bases on these expectations.

### Social Return

People who are interested in a social return, only want to be recognized as donors and want to have a good feeling about the cause they supported. These kind of campaigns are normally launched by non-profit organizations that are able to focus on project based fundraising. These are typically the donation-based campaigns that are in line with the other type of fundraising activities of NGOs. This can be successful for some campaigns, but the average donation is rather small. In the fundraising world it is commonly know that providing tangible and unique reward to major donors, helps attracting higher amount of donations. This can also be applied to crowdfunding by offering small rewards in return.

When a project owner is offering small rewards, valued (much) lower than the price the crowdfunders are paying, this is called reward-based crowdfunding. These are campaigns where the project owner is offering for example a meet-and-greet with an artist or a VIP-dinner with the founders of a company or a public figure that will costs the project owner not much, but is worth a lot for the crowdfunders. For project owners it is important to look for people and companies in their network that are able to provide these products or services for them for free or for a large discount.

The main advantage of this strategy is that the average donation of a donation will be 3-4x higher than for average donations. In Europe the average donations are around 15-20 euro. In reward-based campaigns the average donation is 60-80 euro (see Figure 1).

### Project-Based Fundraising for Non-Profits

For a lot of non-profits this is very interesting, because it will give them the opportunity to upgrade their current donors. Typically NGOs use an one-to-many (mass) mailing strategy to reach a lot of small donors for donations without a reward and they have a one-to-one strategy with unique, tailored rewards to a attract a few major donors. There are a lot of small donors who are willing to support the organization with larger amounts of money if they are also able to receive some (small) rewards and are very interested to support specific projects of their choice. NGOs can use this, but one requirement would be that they are able to focus on project based fundraising.

The main difference for most NGOs will be that they have to move from fundraising for their brand, to fundraising for individual projects. This has a lot of impact on their organizations, because they have to work with earmark money that can only be used for the projects they are raising fund for and their organizations should be able to describe and provide the programs that are in need for funding in advance. This sounds logical for crowdfunding, but al lot of NGOs are not used working like this. They raise funds and based on the amount of money they raised, they will later decide which projects will get funded.

*Figure 1.*

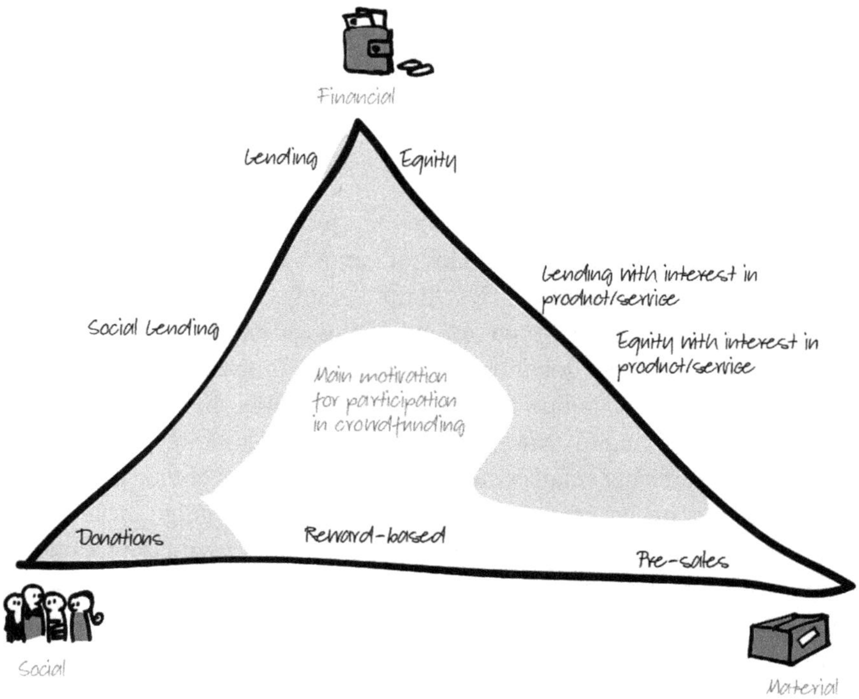

## Material Return

If a project owner is providing a reward that is valued at the same price of the amount of money that is provided, or is even worth more, we talk about pre-sales. Legally it is not really pre-dales, because in most cases the funder will provide money and finances the project that will give him ultimately (if everything goes as expected) a product or service in return. Therefore, it is not a a100% guarantee that the funder will gets its product, but because of reputation risks and possibilities for follow-up crowdfunding and continues support in his own communities, the entrepreneur will do everything what is possible to make sure he will deliver a good product or service in return.

This type of crowdfunding is mostly used by entrepreneurs, inventors and creatives who have an idea, and sometimes even a prototype, and need to have funding to create the first batch of their products. By pre-selling their products they get an immediate cash-flow which they can use to invest in creating these products.

An additional reason to use this type of crowdfunding is that it is a great way to do market research and find out of the entrepreneur is able to reach its potential market and if the potential clients are willing to pay for the product. This is the ultimate way of market research. It will not ask people about their intentions to buy a certain product, but let them vote with their wallet.

The risk in this type of crowdfunding is that the entrepreneur will need all the money to produce and distribute the rewards to his investors. After the products are produced and delivered he has made his first sales, but most likely will not have made a lot of profit. Therefore, this can be used very well for entrepreneurs to kickstart a business, but the company will need to get additional finance for further growth. This can also be done by crowdfunding or traditional finance. One of the additional benefits of having done a successful crowdfunding campaign is that attracting extra financing is easier when financers will see that you were able to attract a number of investors or customer in earlier campaigns.

## Financial Return

The last type of return that can be offered is a financial return. This is called Crowd Investing of Crowd Financing. When funders are interested to support the entrepreneur and his idea, but don't want to give him a donation, or don't need the product or service that will be made, a financial incentive can be used to attract them. There are a number of financial offerings that can be made, and most of them are only limited by the possibilities of the regulations in the country. Because providing a financial return will in fact mean that the entrepreneur will offer an investment opportunity, the local regulators will have specific rules applied on Crowd Investing.

In general, there are two types of Crowd Investing that can be used by entrepreneurs that want to offer a financial return to their investors: Loan-based crowdfunding and Equity-based crowdfunding.

### *Loan-Based Crowdfunding*

Offering Loan-based crowdfunding is the most popular way of Crowd Investing, especially in non-Anglo-Saxon countries, because this is most familiar with traditional bank loans. Instead of getting a loan from 1 bank, the entrepreneur will attract the loan from a group of funders. Normally a crowdfunding platform will act as a middle-man and collects the money, closes the deal and facilitates the interest payments to the funders. The exact terms of the loan (duration, interest, etc.) are unique per country and/or platform, but offer in most cases more flexibility than the standard loan models of banks.

### *Equity-Based Crowdfunding*

While offering a loan is the most popular option for traditional SMEs, for (fast-growing) startups or companies that need to have more equity capital, they can sell a part of the shares in their company to (small) investors. These can be business angels, but also individual investors who are interested to support the company and are willing to take the risk to invest in equity capital, but will also benefit from the capital gain if a company is successful.

There are typically two different ways the investment can be structured. The first one is that all investors will get direct shares in the company that is looking for an investment. The advantage for the funders is that they have a (small) say in the company and will have legal rights to attend shareholder meetings and vote. The disadvantage for the companies is that this is very difficult to manage and that all board decisions and shareholder meetings will be difficult to arrange. The second option is to use a Special Purpose Vehicle (SPV) that will represent all shareholders. This way all shareholders are grouped in one legal entity that has the total number of shares of all the investors. These investors will still have the benefits of capital gain and dividends, but will not have a direct vote individually. This structure is mostly interesting for small investors that are interested to participate in these companies, but are not interested in the (monthly) shareholder meeting and direct interaction and communication with the founders.

## Social Lending

A special way of providing the right rewards for potential investors is social lending. These are loans that are provided to the entrepreneur with a very low interest rate, sometimes even 0%. This is interesting for people who want to support an initiative or entrepreneur and don't want to get a high interest rate, but are not willing to give the money as a donation.

This can be used by causes that are lending money the entrepreneurs in developing countries, such as Kiva. People can lend small ($25) or larger amounts of money through Kiva to Micro Finance Institutes (MFIs) that will distribute it to local entrepreneurs who need to repay the loans. With an almost 99% repayment rate most funders will get their initial loan back, so they can lend it to another entrepreneur.

Another great example is using it to finance the installation for renewable energy for a local school or sports club. The people connected to the school of club will provide the money that is needed to finance the installation of the system. From that moment the school or club has a lower energy bill and is able to start repaying the funders without (or a very low) interest. After a number of years the organization will have structural lower cost and can invest that money in education and sport equipment.

## Lending/Equity with Material Return

Because investor feel really connected with a project, it can be a good idea to provide a product or service from the company they funded as part of the repayment. It can be a good idea to offer a product as interest payment. This way the funders feel even more connected with the company they supported, but it can also be beneficial for the funders and the entrepreneurs. The entrepreneur doesn't have to pay cash as interest payment but can provide the products or services it is producing already. Their cost prize is normally lower than the sales prize, so for them it is cheaper to offer that as a return, or they can be very generous and offer more than a funder normally would expect. This would be interesting for the funder.

An example would be an entrepreneur who starts his own local brewery. He will use crowdfunding to get a loan from people in his network. Instead of paying them 5% interest every year, he will send them a box of beer. The market value of this box is for example equal to a 7% interest payment, but the cost prize to produce it is lower than 5%. This will be beneficial for the entrepreneur and the crowdfunder and has the additional benefits that the crowdfunder will feel much more connected with the company and therefore will be interested to help promoting the company or even buy additional beers.

### Dunbar's Rule and Funding Target

Robert Dunbar suggested a maximum number of 150 people a person is able to create and maintain stable social connections (ranging from 100-250). Surprisingly enough this is also the average amount of people that join an average crowdfunding campaign. While preparing you crowdfunding campaign it is important to think about this to calculate the maximum amount of money you are able to attract. For example, at Kickstarter, one of the largest and most well-known reward-based crowdfunding platforms in the US, 85% of all the campaigns raise less than $20,000.

Every campaign is different and every project owner or group of project owners has different successes in their campaigns, but based on the average donation levels in your country, you can use Dunbar's Rule to make a calculation of the average amount of money you are able to attract and therefore to set the right goal in your campaign.

In Western European countries the average donation is for example 20 euro. Based on Dunbar's rule you will able to attract on average 3,000 euro per project owner. For a group of 3 people this can go up to 10,000 euro, although their will most likely be some overlap in these networks.

For reward-based crowdfunding this can be 3-4x more, because the average funding will be between 60-80 euro. This will result in 10-12,000 euro for an individual project owner up to 30,000 euro for a group of 3 project owners.

The highest funding goals can be reached by Crowd investing. On average funders will invest 500-1000 euro in a company, whereby it is normally much more difficult to attract investors in equity than in loans. An average campaign can, based on Dunbar's rule, raise 75,000-150,000 euro. A team of 3 people can raise on average 200,000-400,000 euro.

All of these examples are based on a good pitch, a good project and a team with a good reputation and no guarantee for a success. There are also always examples of projects and campaigns that are raising much higher amounts of funding, for example with reward-based campaigns that go viral, but the above calculations are based on averages.

## CROWDFUNDING IS MORE THAN MONEY

While preparing a crowdfunding campaign, it is important to think in advance about the additional benefits you can get out of it. A crowdfunding campaign is a marketing campaign that includes raising funds. The community that will support the campaign will be interested in the end result. Therefore, they are interested to support on more levels then just donating money during the campaign. The campaign is also a good test of the ability to reach the potential end users of the product and convince them to buy it.

## Market Research

Crowdfunding can be used as a great tool for market research. When an entrepreneur wants to launch a new product the traditional way is to invest in developing and building the product, start a marketing campaign and try selling it. This can take months or even years before you can really find out if people are interested and if you invested your time and money in the right product.

Another option would be to create a proto-type or describe your concept in a crowdfunding campaign and ask people at the start if they are interested in buying the product in advance already. This way you can find out if you can reach the potential customers and if they can be convinced in buying the product. If nobody will support the project, then nobody is interested or the entrepreneur is not able to find his potential market. In both cases it would be useful not to further invest in the project, or adjust it in advance. This way the risk for the entrepreneur is reduced.

By applying this concept to more products, the entrepreneur will in an early stage know about products that people are interested in and continue investing in these products.

## Marketing

Because a crowdfunding campaign is also a marketing campaign, it is important to look into the potential of using crowdfunding for marketing.

During the campaign there will be a lot of communication from the team to promote the crowdfunding campaign. Because the campaign has a clear start and end, all the marketing efforts will be bundled in a short period of time. The first supporters and the ambassadors of the campaign will also help in promoting it by spreading the word in their own network during the campaign. There will also be a possibility for free publicity of local (or even national) media willing to write about a successful campaign.

After the campaign is finished there is a strong network of ambassadors. They will be interested to help promoting the results of the campaign. If used wisely, these can be used as part of the marketing department of the organization.

## Brand Loyalty

People who feel more connected to a brand, are more willing to support the brand, also when things go less good. It is important that they have a strong connection and this can be realized by making a strong connection with the projects and keep them informed about the progress.

People who invested in a company through loans or equity feel themselves (partly) owner of the company. They will think twice before they will buy at a competitor.

## Access to Additional Finance

Because crowdfunding is a good market research tool, a successful crowdfunding campaign will make it much easier to attract additional finance from investors, grant from governments and loans from a bank.

When a company can show they have already raised funding from hundreds or thousands of people, it shows an investor that there is an interesting market potential. It will also show the investor that this entrepreneur is able to sell the products. One of the most important success factors for a new business.

For a lot of business angels, crowdfunding is therefore an important tool to find and pre-select the companies they are willing to invest in.

For a government providing grants to projects, it is useful to see if the general public is supporting them already and perhaps even they have funded a part of the project already. It can even be used as a pre-selecting tool in a competition for grants.

Depending on the type of financing raised, it can make it much easier to attract additional financing from a bank. Money invested in equity or junior loans make it easier for banks to lend additional money.

## STRATEGIC CHOICES OF A CAMPAIGN

### Crowdfunding Experts

Running a crowdfunding campaign for the first time is very time-consuming and it can be very useful to hire professional experts for support. These experts can help you preparing your campaign, but also advise on the legal and fiscal consequences of crowdfunding and help you to select the right platform for your project of start.

When choosing a crowdfunding expert for support, it is important to check in advance how much experience this person or organization has. Because this is a new industry and there are no certifications for professional or certified crowdfunding professionals yet, a lot of marketing and financial experts are promoting themselves as crowdfunding experts.

Based on the type of financing by crowdfunding an entrepreneur is looking for, a different type of advisor need to be selected. For a donation, reward or pre-sales campaign, it is important to find a campaigner who is able to setup a strong marketing and communication campaign. The background of these advisor are mostly in fundraising, marketing or community management. For investment campaigns, either loans or equity, such a campaign advisor is also necessary, but the focus will more be on the financial part of your proposal. You will need to have a financial advisor who can help you with the financial part of your business plan and can create a good overview of your cash-flow for the next years.

Separately a tax advisor need to be consulted in advance to know how to structure your campaign regarding VAT and afterwards for tax filings.

### DIY Crowdfunding or Using an Existing Platform

Most people will use a crowdfunding platform to publish and promote their crowdfunding project. This has a lot of advances and it makes your life much easier, because you don't have to build a website yourself and you know that the legal and fiscal elements are taken care of. A well-known crowdfunding platform can also create an additional level of trust to your potential donors/investors. The downside is that you have to pay some fees for putting your project on their platform and success fees when reaching your goal Depending on the size of your campaign and the type of crowdfunding you want to use this can go up to 5-10% of the amount of money you are raising.

If you are expecting to raise a lot of money it can be useful to build your own crowdfunding campaign website for it. This is normally interesting while raising 100.000 euro or more, but this depends on the country where you are based and what type of crowdfunding you are using. In general investment based crowdfunding (lending/equity) is much more expensive to setup, and sometimes illegal to do it yourself, then donation-based crowdfunding or reward-based crowdfunding. Other reasons to start your own crowdfunding campaign site are that you can brand your crowdfunding campaign site much more in your own look-and-feel and you have 100% control over the contact details of the people that donated to your campaign, because normally they will be stored in the database of the crowdfunding platform. Especially with companies that have a very strong community of people that know them, or if you want to use crowdfunding often for different projects, it can be very interesting to do it yourself.

## THE PLANNING OF A CAMPAIGN

The more time and energy spend on the preparation of a Crowdfunding campaign, the more successful it will be. A crowdfunding campaign is a marketing campaign, so a campaign need to be planned thoroughly and you need to know the target audience you want to reach to maximize the results. Because a good crowdfunding campaign costs a lot of time and resources it is important to know the additional benefits of a crowdfunding campaign to get the most out of it.

### Pre-Campaigning

Before starting a campaign, it is very useful for an entrepreneur to take a look at other campaigns to find out what other successful and less successful campaigns in their country have done before. It will give inspiration and insight for the campaign. If possible it would be very beneficial to meet with the founders of these campaigns to hear their side of the story. In most cases much more happened then can be seen from a campaign site. You can contact them directly or go to local crowdfunding events where entrepreneurs tell about their experiences in the campaign. In a lot of cities these events are organized by banks, accountancy firms or by independent organizations such as a Chamber of Commerce or CrowdfundingHub.

In this phase it is also useful to focus on the goal of your project and prepare your pitch. What is the main goal and why do people want to support you? What will be the outcome of the project for you and your investors? Talk with people about the plan and the campaign. Don't be afraid someone will steal your idea. Most people will offer you their help and these are already your first ambassadors for your campaign later.

You will also need a team of people who will support you. You shouldn't do crowdfunding alone. When people are very enthusiastic you can ask them to become part of your crowdfunding team. You will need to have ambassadors promoting your project, but also people helping you in your (social) marketing activities, organizing events, etc.

### Preparation of a Campaign

When the decision is made to start a campaign, the most important part of the campaign will start: the preparation phase. The biggest mistakes are made in this phase. Mostly because of underestimation of the process and not preparing for all the additional values of crowdfunding.

## Storytelling

An important part of Crowdfunding is to convince people in your (extended) network and outsiders who don't now the entrepreneur and the project before, to support the idea. Therefore, the entrepreneur should be able to explain clearly the reasons why this project should start, what will be done with the money that is being invested and what the investors will get for it in return. It is important to make the message very clear and powerful.

The best way to convince people is by storytelling. Build a story about why it is important for you, the community or society that this needs to be done and why you and your team would be the best to realize it, whether it is a new company, a small local community project or a new product you'd like to create. This message should be written down and recorded on video, so it can easily be distributed.

For recording a video, the passion of the entrepreneur is very important. People like to give to other people who are passionate about their projects. Make also sure the video is short (max 2-3 minutes) to keep the attention of the viewer and show you key message in the beginning of the video.

## Identify Your Target Group

The next step would be to identify your target groups. Who are the people in your own network or external network who would potential be interested to support the project? These can be different groups with other expectations and therefore it is important to adjust the marketing message for the different target groups.

Some groups of people are closely related to you and you have a strong personal relation with them (family, friends), some have more a business relationship with the entrepreneur (customers, business partners, LinkedIn contacts) and some have looser connection, because they are part of the same (local) communities or social media network.

For the campaign it is important to identify these potential networks in advance and make sure that the right networks will be approached on the right moment. It is important to first start with the people who know you personally and who you can approach with a personal appeal (1-to-1) and later use one-to-many communication for people who you don't have a very strong personal connection, but will still know you. In the last phase you will use mass communication to approach the large communities about your campaign. For details about this planning, see the "campaigning" paragraph below.

## Rewards

People have different motivations of supporting a crowdfunding campaign. As discussed earlier in the chapter on Return on Involvement there will be different expectations for the people what they will get in return. Especially for reward-based crowdfunding it is important to be creative in finding interesting rewards. Not only will this increase the number of people supporting the project, it will even motivate them to spend more on the campaign.

The first rewards can be found within the project itself. If the result of the campaign will be a product or service, then this can be given back as a reward. To make it special it is possible to create "first-editions" of the project or add special features to the product that will not be available for the retail edition. These can also be VIP services for cultural/creative projects.

The second type of rewards are rewards from the entrepreneur or the team. This starts with small personal thank you messages to a diner with the founders or unique services the entrepreneur can give from his/her own background. These can also be other products of the company, or activities like a product tour through a factory, research lab or company.

A third type of rewards can be found at companies closely related to the campaign. Companies and people who like to support the campaign, they can also donate some of their own products and services as a possible reward to offer in the crowdfunding campaign, instead of supporting the campaign with money.

### Planning Communication

Because a campaign is very intense, it is very important to prepare a lot of work and activities already in advance. You can plan a lot of (speaking) events, publications in advance and of course it is important that the flyers will are ready in time. Something most people don't think about is that it is also possible to prepare a lot of the social media communication and emails that will be send to people who are interested and as a thank you to people who invested in the campaign. When these texts are written a bit in advance, the entrepreneur will not only write nicer and better messages, he also will have more time to think about the possible follow up actions that can be done by the investors. A simple message on social media will be shared by the funder within his/her own network, if you make it really personal, perhaps even with an personal image or video. And an email with a simple description how your funder can help you to spread this message will stimulate him to forward your email to a number of friends and show them your campaign.

In the months before the campaign you will gather all the people in your community and find the best way to communicate with them. Try to reach out to them far in advance of your real campaign and inform them you are working on a new product/service and want them to get involved. Build up mailing lists, online (Facebook) groups or other platforms that give you the opportunity to easily interact with your community.

## Pre-Campaigning

A few weeks before your campaign starts it is important to test your final pitch. You should reach out to the inner circle in your community, the people that know you best, and test your final pitch and ask them directly if they will support your campaign. If they do, you ask them for how much they will support you and what rewards they would like. This way you can not only test your pitch and rewards, but this is also a good test about the possible funding goal you are able to reach.

This test will show the capabilities of the entrepreneur to reach out within his/her own network and the ability to convince people to support him. It will also show how much they are willing to invest. Because a large part of the funding and support of first-time campaigners will come from their own inner-crowd, this is a good prediction of the total amount of fund that can be raised. Before the start of the campaign, the entrepreneur should have at least 20-30% of the amount pledged by people from his own network. If this is (much) lower, the funding goal should be lowered, because it would otherwise be very difficult to reach that goal.

One of the additional requests for these investors is that they will fund the campaign at the start. This way the campaign will have a good start.

## Campaigning

It is very important for the first people to invest at the start of the campaign. When they do this, it will give a boost to the campaign and from the start it has a more positive vibe. This is very important, because most people are risk-averse and will not join a campaign or project when they are one of the first. They want to join when they see it is already a success and they want to become part of that success.

Therefore, it is also very important to build up the campaign in 3 phases. Because the strength of crowdfunding is community funding, you should also embrace the strength of the community to support the campaign. At the start of the campaign the communication should be focus on the Inner Crowd. These are the people that know the founder best and can be reached individually. After these people are connected it is important to reach out the rest of the community of the founder. This can be done my mass e-mailings, communication by social media or newsletters. Only in the last phase people will be reached by mass media, traditional newspapers and support of other networks.

### Inner Crowd

In the Inner Crowd are the people that can be reached personally. The more personal and intense these connections are, the more successful they are. The most successful communication methods here are physical meetings or phone calls. The conversion rate here is very high, because it is much harder to say no to a personal friend who is reaching out for support in this phase. Because this is a very time consuming effort, his should only be done by people that are well known, but because it is so successful, every campaigner should do this by at least 10 people to get the first funding.

The next step would be to approach people personally with an individual and tailored message, but contact them online. This can be done by a personal email or DM messages through social media. Spend enough time on every message to make it tailor-made for the recipient. The more personal approach, the higher the conversion rate is. A follow-up call is of course also very useful.

For all these steps it is important to keep an overview of the people you contacted, how much they committed and if they indeed contributed already in your campaign. A lot of people are really willing to support you, but because of a lot of different reasons and a busy life, they tend to forget it. Because the campaign is only running for a small amount of time, it is important to do follow-up calls or send them an email if you notice they didn't contribute yet.

Therefore, a daily check on the people that donated is very important. This will also give you the possibility to reach out to the people that contributed and thank them for their support. If you do this through social media, they will also reply on it and help you spread the message of your campaign to a larger audience.

### Own Community

When you've reached out in the first phase to the people you know best and contacted them personally, it is time to move to the next phase and use mass marketing to reach the people in your network that still know you, but the connection is not strong enough to send them a personal update (or your network is too big to reach out to all of them personally).

At this moment your campaign is running for a number of days and because of the support of the people in the pre-campaigning phase and the support from you inner crowd, you already have a good start in your campaign. When people that know you and hear about your campaign from your mass emailing or the communication you send to them through social media, they notice that you have a successful start of your campaign and therefore they are much more willing to join.

### External Communities

After you have reached out to the people you know, you can start to find people who don't know you yet, but would be potential funders of your campaign. The best way to do this is to identify the characteristics of the people that would be most interested to support you. In most cases these are the people in the networks and communities that are closely linked to your cause, such as sustainability or specific niche groups interested in a specific sport, activities, pets or hobby's.

Before starting the campaign, it is important to identified these networks already and contacted them about your plans. This way it is easier for you to contact them during your campaign and you can show them some real progress and success. They will be able to spread your message in their own (social media) network, but will also be interested to publish an interview in their internal newsletter of magazine.

At this moment you will also plan your speeches and presentations at external events. Contact far in advance the organizers of events you know take place during your campaign and are interested for your target audience. The conversion rate at these events is not very high, but in most cases it will generate a few more funders, but at least a lot of extra fans. These fans can be beneficial in a later stage of your company. In most cases it will also connect you to other people you never were in contact with before and who can help you to introduce you by other investors, potential clients, partners or employees.

This is also the right moment to publish your story in newspapers and other mass media. Most entrepreneurs send their press release to as many blogs, newspapers and tv stations as possible, but the success rate us normally very low. And even if it is successful, it will not help the campaign. The reason for this is that in general media is not interested to hear about your ideas. They want to publish successes. Therefore, it is better to contact them later, when you show some success and your reached your first milestone already. For your potential funders who have never heard about you before is it also important to show them how successful you are and that a lot of other people already funded you. People are not interested to be the first funder when nobody else is supporting you. But when they see that it is a success and already a few hundred people participate, they follow the rest much easier.

## End Sprint

The last days of your campaign will be very hectic, but if you use them wisely it can provide you with a lot of extra funding. A lot of people are lazy, also in supporting you. They really want to support the campaign, but they waited because they were busy doing other things. Now it will pay off if you kept records of the people that were interested. Send them an urgent last-minute email telling them that they only have x days left and tell them exactly when the campaign will close and thisis their last chance to join. Make it as easy as possible with a direct link to your payment page, perhaps even pre-selected the amount of money and/or reward they told you they were interested in.

This is also the moment to send a last reminder to your current funders about the end of the campaign and ask them to think about upgrading their funding and asking all of them to ask 1 or 2 friends to join the campaign.

This will normally give a boost in the last phase in your campaign that will help you to reach finally the target or will even give you a large boost for over-funding.

## Post-Campaigning

Directly after your campaign finished, send a thank you message to all your funders and congratulate them with the success you made together. Continue building the relationship with your (new) community. These people are very strong connecting to your goal and are willing to help you in the next phases also. Keep them close. Make sure you update them in the next months and years about the progress of your company, because they made it possible.

When you offered rewards, now it is time to start distributing them to your funders. Group your funders in categories, so you can update them on the individual rewards they will get and give them, if possible, a good estimation when they can expect their rewards. If you can send them already, do as quickly as possible. If they pre-financed some product or service that first need to be created, give you funders a clear planning when they can expect it. And keep your promise! If anything goes wrong and you expect a delay in delivering your reward, be open and transparent about it. The worse thing to do is not tell any of your funders about it. If you are transparent and open about it, they will understand in most cases.

## Next Steps

After finishing a campaign, this is not the end. This is the start of the preparation if the next campaign(s). Companies can use serial-crowdfunding and start new campaigns in different phases of the growth if their company. They can start with an equity campaign to raise some seed-funding, pre-sell their first products and services by reward-based crowdfunding and lend money through crowd lending when they are in need for some growth capital and don't want to give away additional equity.

Companies using this concept on a structural level will also transform into a CrowdBusiness business model, where they not only use their network to raise funding in one or more campaigns, but also involve their community in marketing activities, crowdsourcing for new innovations and will even move into co-operative models where their investors will become owner of the organizations.

# Chapter 14
# Real Estate Crowdfunding: 2015 and Beyond

**Steven A. Cinelli**
*REMARQ, USA*

## ABSTRACT

*Like a juggernaut, "crowdfunding" has hit the media, the financial markets and the common narrative by storm. Incipient in many ways, the intermediation of financing transactions online has become a billion-dollar industry. As technology has advanced, even recreated, industries, there seems none more primed than "finance", an inherently information business. By creating improved efficiency, both art and science of finance are enriched. A 21st century vestige of the 1980's syndication business, real estate seems to be enjoying the fruits of the crowd, with $1 billion of property financings conducted online in 2014, with an expected $2.5 billion this year. For sponsors and investors, there appears legitimacy to the online approach, underscored by the level of venture capital now finding home in this burgeoning sector. Yet, like its progenitor, might real estate crowdfinance find legislative, regulatory, and practical headwinds, stunting its progress? Still early, with business models, scalability and sustainability still suspect, the current momentum seems promising.*

## WELCOME THE CROWD

While coined as a term only a few years ago, "crowdfunding" certainly has taken the markets by storm. Now in its pubescent period, the investment and financing methodology of aggregating small amounts of capital through online interactive platforms now accounts for billions of dollars (or Euros or Yen) in capital formation and a breadth of new investment opportunities for global participants. Gaining visibility in non-securities based fundraising, *ala* perquisite platforms such as Kickstarter and IndieGoGo, the power of technology to efficiently expose private investment offerings is creating a changed landscape for existing (and largely analog) financiers and intermediaries. This is particularly the case within the real estate realm. 2014 was a breakout year for real estate crowdfunding, with over $1 billion being infused into property-related investments through a hundred or so platforms, with an expected 250% growth this current year (2015) (Massolution, 2015). Whether the growth of real estate crowdfunding is

DOI: 10.4018/978-1-5225-0568-6.ch014

a function of discerning and executing a new and improved way of intermediating capital providers with capital users, or rather taking advantage of an opportune moment as swelling regulatory and legislative oversight hampers the commercial banking mother lode, it does appear that the market, as of now, shall be characterized by exceptional progress. But we do remain judicious in our view of the quality and expanse of such advancement, and core to this view, is the sustainability of the underlying crowdfunding business model, particularly for real estate.

## IF THERE'S A HISTORY, THERE'S NO MYSTERY

As I commented in the 2015 Massolution Real Estate Crowdfunding Industry Report, what we are witnessing is largely a 21st century manifestation of the real estate syndication model heavily utilized during the 1980s, at least for commercial real estate endeavors. Working either as a principal or as an intermediary, groups would raise funds from "syndicates" of accredited investors through the private placement of limited partnership offerings, with proceeds being used to acquire, rehabilitate, construct, or reposition commercial or multi-family real property. In the effort to aggregate such funds, the "syndicator" or say, sponsor, would prepare a placement memorandum, sometimes consisting of hundreds of pages, which it would provide individual investors generally by snail mail, and then "dial for dollars". Participant investment sizes were set at a level to induce broad participation, but not so low to the extent of creating a "herding cats" phenomenon. The verbal exchange between fundraising sponsor and prospective investors tended to be laborious, as many of the questions or comments were redundant from investor to investor, regardless the size of a particular investment commitment, whether $25,000 or $250,000. Telling the same story time and again, was a characteristic of the syndication process. And then there was the monitoring of who received a book, what the nature of the conversation was and finally allocating time to complete the one-on-one sales process. Post fundraise, the communication between sponsor and investors was equally arduous, layering more paper, snail mail, and repetitive repartee.

Enter crowdfunding, or more explicitly, the analog nature of fundraising communication between sponsor and investors going digital. Rather than authoring a paperback tome, crowdfunding platforms expose dozens if not thousands of investors simultaneously to an investment proposition through engaging and informative websites, with customer relationship management (CRM) systems to support both outreach and ongoing investor management. Simply, this has become the process of telling the story singularly to a plurality, underscoring the efficiency of digital communications. Consider the cost and time savings of creating a single digital copy and allowing through invited online access compared to the mailing out 500 individual books, and then following up on each. And yes, pursuant to securities regulations, each of the old style books were to be manually numbered and controlled. Further imagine if such "book" required an amendment or introducing an update in the project. The efforts and costs to distribute what needed to be shared for informed investment decision making drastically have been reduced. This has wide ranging implications: in reducing the cost of intermediation, smaller transactions might be economically "syndicated", as one can only assess so much in intermediation fees so as not to severely impact investable funds, nor gouge the economics in the end for the investor.

Another aspect of online syndication is that a broader audience can be reached. There has been nomenclature applied here, such as the "democratization" of investing, which respectfully has a semblance of truth. Reducing the cost of intermediation, and specifically on a per investor basis, smaller minimum investment sizes can be handled, leading to larger syndicates, which allegedly can be managed efficiently.

A 1980s syndicator would be glassy eyed with a thousand participants in a million-dollar deal, and probably couldn't charge enough to make it worth his while, let alone spend time on property requirements.

Certainly one key difference from the syndicators of yesteryear is that the motivations on investors' part to invest has changed. Back in the 1970's and 1980's, individual marginal tax rates were elevated to 70% (Taxfoundation, 2015) for incomes over $200,000 and a key selling point of the syndications was that the periodic passive losses from owning real estate (largely through the depreciation expense) could be used to offset an investor's other ordinary income, thereby reducing taxes. Thus real estate syndications were real tax driven, as tax shields, and potentially in the end, if the property materialized a gain on sale, even better. Because of the huge demand for such tax shields, there was an onslaught of demand and syndicators continued to aggressively accumulate investor capital to put to work. What transpired though leading to a compression of this industry were: (i) the excessive level of investment led syndicators to invest in weaker properties and the actual losses ensued; and (ii) the federal government changed the tax laws, so that no longer could passive losses offset ordinary income. In effect, the syndication industry came to a halt.

A key difference between that period and now is the offerings made by the crowdfunding syndicates are investor motivation and the common usage of debt as a financing structure for the underlying properties rather an ownership equity position. Rather than owning the property as an equity holder, and advantaging the passive losses, the typical financing structure is a first lien secured loan, which would get paid back with interest upon the disposition or refinancing of the property. The interest rate on the loan, typically between 8% and 14%, addresses the appetite for yield that current investors maintain. We do recognize that many platforms offer equity type investments, but over 80% of 2014 RECF financings were embodied as secured promissory not

But back to the democratization spin: this is really not a new twist. In 1960, under an act of Congress, the Real Estate Investment Trust Act (Taxfoundation, 2015) was instituted, creating a mutual fund methodology for real estate investing. It was observed that real estate, particularly large commercial buildings and diversified portfolios of income-producing properties were the exclusive province of institutional investors, and legislators developed REITs as means for small investors to aggregate their funds to be deployed into larger real property. And to support the use of REITs as an investing aggregate, they were assigned a favorable tax treatment (namely the REIT itself was non-taxable subject to it distributing 90% of its income). The REIT methodology has proven to be advantageous to income-seeking investors, by offering portfolios of both real property ownership interests (equity REITs) and real property mortgage loans (mortgage REITs). The REIT industry is significant, with over $900 billion (REIT, 2014) of REITs outstanding, ninety percent of which are in equity form. In fact, taking the US lead, there are now REIT genres in over thirty countries.

## APPETITE FOR ALTERNATIVES, REAL ESTATE IN PARTICULAR

Over the last twenty years and in particular the seven years, spurred with the convergence of the financial crisis and a broader quantitative easing from the Federal Reserve and other central banks, there has been dramatic shift in investment portfolios to alternative assets, and in particular real estate. With interest rates resting at an all time low, many investors are not only seeking out higher returns yet are desperate to find them given their own requirements. Think about pensions funds, whether public or private, that must address actuarial payout requirements to their respective retirees. Close to half of public pensions in the US are underfunded, so not only are they challenged to meet quotidian payouts, but they have catching up to do for the accruals that continue. Bonds, or debt obligations of private and public sector entities, including the sovereigns, may be less risky than holding real estate, but investors have recognized that spreads between prime commercial real estate assets and high quality bonds have increased significantly over the last ten years, and their risk/return thesis has steamrolled the growing interest into enhanced real estate allocation. Currently we have seen interest rate differentials of close to 300 basis points (3.00%), between prime assets and high rated bonds, thus in a low interest environ, one can understand the move.

In general, alternative assets are increasingly becoming a staple in larger investment portfolios. For decades, traditional or core assets, such as public equities and bonds, represented the bulk of an investment portfolio, whether that is an individual investor or an institution overseeing billions of dollars of investable funds. But certainly over the last ten years, we have seen a demonstrable allocation away from core assets into many things alternative, such as private equity, venture capital, real estate, hedge funds, and even commodities. This has been a function of macroeconomic challenges and conditions, coupled with a growing comfort in what had been heretofore considered illiquid assets. And we are witnessing this practice across investor classes.

From a current macro view, public equities are priced up, elevating to price/earnings (19x-27x for S&P500 based on the analyst, The S&P500 PE Ratio, n.d.) and other performance ratios to levels not reached in since the tech bubble years in the early 2000's. With a historical ten-year average of a P/E of 15, certain analysts believe that the equity market is over priced by 30%. This could continue to propel upward though assuming a return of the retail (small) investor into the market, who has been sitting on the sidelines since the financial crisis of the late 2000s, and the elongation of quantitative easing by the Federal Reserve Bank, keeping interest rates low and liquidity available. But low interest rates have softened the appetite for fixed income investments, like bonds, as investors are seeking higher returns that what is readily available. Further, as analysts suggest equity prices are high, they also anticipate that market interest rates will gradually rise, negatively impacting bond prices. All said, investment capital must find a home and with the challenges of generating targeted and required returns, we are witnessing a marketed shift to alternatives.

But this is not defaulted assertion, one of chasing yield, for the reallocation to alts. Alternative assets provide investors new avenues to manage portfolio risk and return, correspondingly targeting certain outcomes, addressing exogenous factors such as interest rate movement and inflation. In fact, alternatives have become a strategic imperative in most larger portfolios. Many sizable investors whom once allocated a small portion (<5%) in search of pure "alpha" (above market returns), have now developed increased internal capability to pursue alternatives as a fundamental strategy. This is not manifested as a merely an increased allocation to managed venture or private equity funds, which still command expensive costs with management fees and carried interests. This is rather the desire to commit more funds, with better economics, including high levels of direct investing in these asset classes.

For institutional investors, alternatives will account for over 26% of their portfolios, according to a McKinsey & Company 2014 research report (McKinsey Company, 2014), and this escalates to about 30% for those investors with over $10 billion of assets under management. And in respect to such portfolios, nearly ten percent is dedicated to real estate.

## REAL ESTATE AS AN ASSET CLASS

In assessing asset classes, investors' focus is generally on return, but in the context of other characteristics and criteria, such as liquidity, principal risk, comparability to other classes, interest rate levels, and volatility to name a few. And the basic premise is that the higher return being sought, the higher the risk being taken. With respect to real estate, some of its attributes are as follows:

- Its a tangible asset, i.e., you can see it and touch it.
- It typically generates return in two ways - through current income and capital appreciation
- With the exception of investing through publicly traded REITs, it is generally considered illiquid, meaning it will take some time to monetize the asset into cash, and when it is monetized, there is a high friction cost, such has there are typically selling or brokerage commissions.
- It is generally viewed as a complementary addition to the portfolio as it has low correlation with the core asset groups, i.e., its value moves independent of stocks and bonds, and thus can diversify or reduce overall portfolio risk.
- It is considered a hedge against inflation, as its income may rise through rents as overall price levels increase.
- It is financed with a combination of equity and debt. Debt investors may enjoy a fixed return through interest and principal payments while mitigating their risk with a secured interest in the property. Equity investors capture a variable and hopefully higher overall return, through rents and property price appreciation, which is enhanced by using the "leverage" of the aforementioned debt.
- And lastly, real estate's value tends to be highly localized due to market conditions, credit availability, and the timeliness and breadth of information available on the property.

In terms of investing in real estate, there are various methods, such as investing in pools of properties with others or on an individual property level, and in the financing instrument itself, namely equity or debt. As discussed, there are REITs (Real Estate Investment Trust), which enable investors to pool resources together and purchase a larger property or its underlying debt.

Other mechanisms to invest in real estate are through smaller limited partnerships or other Special Purpose Vehicles (SPVs), which also enable individual investors to pool funds to acquire medium sized properties, generally outside the range of the single investor individually. Sponsored generally by a syndicator, this further allows for diversification as well as passive investing, as sponsors serve as managers, for which they are compensated. Crowdfunding makes use of LP and other SPV structures to hold investment interests on behalf of the small investors providing for some level of bankruptcy or liability remoteness. And as stated, with either REITs or LP structures, real estate crowdfunding is the newest way to aggregate funds, but typically using online communications to facilitate capital raising and ongoing reporting back to the investors.

An additional way to gain real estate exposure is by acquiring property on one's own, be it residential or commercial, which is then rented out to a tenant. The investor, now landlord, bears all the responsibility of being the sole owner, such as being obligated on the mortgage financing, property taxes and insurer, dealing with maintenance and repair issues, and generally managing the relationship with the tenant. Certainly, there is a tradeoff in such active investing with doing something more passively, but there are costs and return limitations in the case of aggregated investing.

## TYPES

Real estate is not a one size fits all proposition and the crowdfunding industry has been exposing most to its investor communities. There are many subclasses of real estate, each enjoying different attributes, risk elements, and return expectations. Further, depending on the subclass and the purpose of the underlying financing, which investment capital is deployed, time horizons may differ, as well as the nature of return being sought. As earlier described, real estate can offer two means of return – current income and capital appreciation, and based on the structure of the financing program and type of property, returns could emanate from either or both means.

Amongst the various subclasses are (Massolution, 2015):

- **Commercial and Industrial (C&I):** This subclass generally consists of office buildings, warehouse and distribution facilities, and retail space, which are leased out to private and public parties, thereby generating rental income. Such leases can be structured on short-term or long term basis, consist of flat payments or have periodic adjustments, and can even be tied to the operating performance of the lessee, in the case of specialty retailers with "percentage rent" clauses. In recent years, there has been considerable investment in redeveloping existing properties as many of the most active urban areas are gentrifying and repositioning themselves, creating incremental lease rates, and thus yield for investors. This subclass constitutes the largest proportion of the REIT industry.
- **Hospitality:** Think hotels and resorts: these properties generate income from providing lodging paid on a short-term basis, while offering amenities, such as food and beverage services, i.e., restaurants, golf, spa, social function services and business conference facilities. Generally suffering during recessionary times as both business and vacation travel becomes anemic, hospitality participants have enjoyed a substantive upsurge since 2011, with improving economic conditions buoying both business and consumer spending. During 2014, one RECF platform offered a partial equity interest in the Hard Rock Cafe Hotel, combining both economic incentives with patronage elements.
- **Multi-Family Residential:** This subclass refers to those properties deriving rental income for periodic residency, such as apartment buildings, but also cooperatives, duplexes and condominiums, which may be individually owned, and serve as rental property or primary residences. As purchase prices of single family residential have risen over the last three years, the demand for rentals has escalated significantly, which in turn has provided compelling income streams for multi-family investors.
- **Single-Family Residential (SFR):** This represents properties independently owned and detached from other properties, which serve as either primary (or secondary) residences or used for rental purposes. The value of US SFR properties approaches nearly $20 trillion, making it the second largest asset class in the US. Globally, residential real estate has been estimated at over $110 trillion, witnessing a considerable move upward over the last four years, bouncing off the temporal nadir of the crisis years. New means of SFR exposure have been practiced during such time as large real estate private equity firms and sponsors, such as Blackstone, Starwood Waypoint, Colony and others have accumulated large SFR portfolios or have purchased bulk packages, known as "tapes", of SFR properties owned by mortgage lenders, a result of foreclosing on delinquent home loans.

Such bulk purchases are generally done at deep discounts, with properties then becoming part of a growing rental home portfolio.

- **Infrastructure:** Generally, this subclass includes projects funded by either public sector entities or in collaboration with private sector participants and non-governmental organization (NGO). Included projects would be roads, bridges, storage facilities, educational facilities, student housing, agriculture, including vineyards and forest lands. In many cases, municipalities may take the lead by "floating" tax-exempt bonds, or some combination of NGOs and for-profit organization participate with complementary funds. We have categorized this as potential revenues may be derived from such projects to repay the capital provider, such as a bond issuer.
- **Community Projects:** A subset to "infrastructure", there is a growing public sentiment and participation in financing community projects for the purpose of improving the socio-economic condition, rehabilitating dilapidated structures or creating a structure (e.g. fountain) honoring a community leader or organization. These developments are generally not "revenue generating", so no economic return or reimbursement is being sought by capital providers.

## FINANCING AND INVESTMENT STRUCTURES

There are generally two approaches used in the structure of the applicable financing. The first approach is for capital providers to invest as equity participants, wherein they own a contractual percentage of the property being financed. Depending on the legal structure being used (to be discussed herein), such equity investors are "shareholders", much like stockholders are in an operating company, and whose economic return is largely based on the appreciation of such company. Certainly, such property "equity" holders may also participate in the current income that is generated from said property, such as corporate shareholders may enjoy periodic dividends. With the current interest rate environment, many real estate equity investors are primarily attracted to the stream of payments from lease income, and secondarily to the overall appreciation of the property, which would be realized upon its disposition. Net rental income as a percentage of the original investment principal may compare favorably with other fixed income alternatives, such as treasury bills and bonds, and corporate obligations.

The other approach is to serve as a lender on the project, i.e., provide capital in the form of secured debt. This has been the predominant form of finance for projects finance through crowdfunding platforms. From a risk and return perspective, investing in debt is generally lower risk, given that the property as collateral is pledged in support of the loan and such debt ranks senior to the equity in the case of liquidation. But with lower risk comes a lower anticipated return. Such return is generally limited to an interest rate, which in most crowdfunding cases, is fixed during the life of the investment. But given some apparent lack of credit availability from conventional sources, such as commercial banks, various platforms (like Patch of Land) provide high yield short term loans up to 12 to 18 months. Another area of debt financing which we will discuss is in the residential mortgage space, with platforms following the lead of non-bank mortgage originators offering attractive, price-driven, low friction mortgage loans to homebuyers and homeowners.

In either the debt or equity structures, sponsors or the platform will frequently set up an investment entity, a Special Purpose Vehicle (SPV) as earlier referenced, which serves as the aggregator of investor funds and acts as the nominal owner of or lender to the project. For legal, regulatory and management reasons, the SPV will be the single owner of the investment structure and be overseen by the sponsor, who will manage the project on an ongoing basis. The SPV structure has been used in large part to have a single "voice" or vote on select matters, and would represent as a single shareholder for corporate governance purposes. While there are many SPV structures, the typical form is a Limited Liability Company/ Corporation (LLC), which blends elements of a corporation and a partnership, providing creditor and liability protection for LLC members, and pass-through tax treatment so that the SPV is not taxed as the entity level, thereby avoiding two layers of taxation – at the corporate level, and when distributions are made, as the member or shareholder level.

Table 1 is a comparison of the two financing approaches (Massolution, 2015).

It is worth noting that with the onslaught of crowdfunding, approximately 80% of the platforms and 90% of the funds raised to date have been in a loan or debt structure. On one hand, firms like Patch of Land, have made a market in providing short-term, high-yield loans to sponsors and developers seeking to renovate and "flip" a property. In 2014, Patch completed approximately sixty-five transactions, with an average transaction size of $350,000 (Massolution, 2015). Since then, both its overall capacity and average transaction size have increased precipitously.

The largest of the RECF platforms in terms of capital deployed, LendInvest, has focused on providing mortgage loans to aspiring homebuyers and homeowners, predominantly in the greater London market. During 2014, it is estimated that over $200 million of mortgage loans were originated by LendInvest, with a global reach continuing.

*Table 1.*

| Structure | Debt | Equity |
|---|---|---|
| Role of Investor | Lender | Owner (Shareholder) |
| SPV Structure | Limited Liability Company, Limited Partnership, S-, C-, B- Corporation, Delaware Statutory Trust | Same |
| Fundamental Characteristic | Loan collateralized (secured) by real property and assignment of leases | Ownership Interest in the Property |
| Documentation | Promissory Note and Mortgage or Deed of Trust | Title of ownership on Grant Deed or Warranty Deed. Shareholder Agreement |
| Rights and Defaults | Creditor Rights – can pursue foreclosure in the case of default, such as non-payment of interest and/or principal when due | Shareholders Rights – Subject to underlying Investor Rights Agreement |
| Terms and Conditions | Fixed Maturity and Payment timing; Limitation on additional debt; balance sheet and operating covenants | Evergreen Maturity; voting rights; registration rights; and rights of first refusal on subsequent financings |
| Investor takes Property Price Risk | No. Debt is a fixed principal obligation | Yes. If the price rises, the investor enjoys gains. If price drops, investor incurs loss. |
| Economic Return | Periodic payments of interest on outstanding principal balance of loan, plus the return of principal. | Periodic payments of Income or dividends, plus participation in the price movement of the property, either up or down. |
| Liquidation Preference | Priority position over Equity. | Subordinate to Debt. |
| Transferability | Yes, subject to securities laws | Yes, subject to securities laws |
| Legal Recourse to Owner | Possibly, depending on legal structure of investment vehicle and terms of Debt by Lender | No. Generally limited to investment amount. Sometimes Equity Agreement includes capital calls, i.e.., request for additional equity investment. |

## PLATFORM APPROACHES

While we have identified during 2014 nearly 100 platforms engaging in some level of real estate finance intermediation, such platforms take different approaches to how they conduct their activities. This is really a matter of what they do, and for whom, as well as the types of economics they may derive.

We distinguish between "objective" and "subjective" platforms, or in more colloquial terms, "marketplace" facilitators and "direct issuers". In the former case, there are numerous platforms that have constructed their respective platforms to serve as a *utility* for sponsors and other types of fund raisers. They have built the infrastructure for sharing an investment proposition with their online network of investors, and handle most activities through the fundraising process. Simply, their sponsor client is "renting" the platform to enable the fundraise, wherein the platform is compensated in usage fees. From a business model standpoint, these platforms are driven by volume of transactions and increase fee generation.

In the "subjective" category, this is generally sponsor-based, i.e., the platform is directed as an extension of the sponsor, possibly even owned by the sponsor to efficiently raise funds. Relationships with the investor network likely pre-exist, with now such relationships being moved from an analog engagement to one digitally. This is very much the 21$^{st}$ century version of the older syndication model, where technology is used as the means of communication, but the economics of the syndicator remain the same – earning various layers of fees, coupled with a significant profit participation.

There is a third approach, considered a hybrid, which in most ways is a marketplace utility as described above. However, as an added feature to the service offering, certain platforms have arranged for interim financing, via lines of credit, where the platform will pre-fund the investment opportunity, assuring the sponsor has the capital to pursue his effort, and then take it upon themselves to remarket on a fractional basis the project to its investor base. Taking principal risk in terms of pre-funding, such platforms remain confident that their investors will afford the necessary take out. A number of the platforms that have raised venture and operating capital, have used their balance sheet liquidity to enable such warehouse financing to take place, rather than third party credit facilities. Of course for taking such risk, these hybrid platforms earn higher fees and may in fact participate in the overall performance of the financed project.

## THIS IS STILL A FINANCE BUSINESS

While we can wax on about the merits of the internet in capital formation, a most pressing palaver is one that is fundamental within the media industry: what is more crucial - content or distribution – to the success of a platform? Certainly we can default to affirming both, indicating that one needs to develop strong and absorbable content, while cultivating reach. But in the end, finance is an information business, and although the 'net enables investors to be engaged globally at tremendous pace, the quality of the content will determine winners from losers. Democratizing the investment process by introducing more levels of investors, one need take heed that while enabling investors to participate, are they being delivered appropriate information to make an informed investment decision, and further, do they have the acumen to comprehend the content and its caliber? And do the new world intermediaries, in their betrothal to online engagement, actually assess and, as many espouse, "curate" the transactions with the appropriate level of insight and discipline, or shall we see the exposure of deals adversely selected, as fundraising sponsors see this as a new, and maybe only, path to raising funds for their projects?

As a directed aside, there has been an active, though lengthy, exercise at the Securities and Exchange Commission to reassess the current definition of "accredited investor" (Accredited Investor, n.d.). As of now, nearly every RECF platform exclusively directs its opportunities to the accredited investor. Developed as part of the Securities Act of 1933, the SEC provided an exemption to public registration of securities if issuers were to sell to those investors of means, namely those that had both the financial capacity to lose the capital they invest, but also had the wherewithal to investigate the merits of the opportunity. While earlier amorphous as to what constituted "means", in 1982 in a partial rewrite of the '33 Act, Regulation D was established defining an 'accredited investor' in terms of specific income and net worth levels. Yet since 1982, the absolute levels of income ($200,000 for single natural persons; $300,000 of married couples) and net worth ($1 million) have not been adjusted, with the exception of the personal residence exclusion from net worth. Well, over thirty-three years, one does recognize the value of the dollar has been impacted by time and inflation. Inflation adjusted, income and net worth levels to remain on par with 1982 would see a "bump" in income levels to $600,000 and a net worth threshold of over $3 million. This would in effect "de-democratize" the private investing world, removing nearly 70% of the currently classified "accrediteds" from the market.

In current Commission discussions about the definition, a key proposal deals with the financial acumen of the investor to determine whether or not s/he is capable of assessing the risks, given professional status, academic training, or germane credentials, such as a CPA or CFA. While not promulgated, the concept of credentials and observed acuity to determined accredited status makes tremendous sense, and, as stated, the ability to understand and assess the risk was core to the original definition. It does not go unnoticed that most of the wealth of the US is based in privately held companies and the ownership represented in unregistered securities. The existing exclusion for "insiders" (i.e., directors and officers) is predicated on knowledge of the company and its prospects.

This diversion into how the SEC may reconstitute the definition of accredited investor is apposite to the role of content provided by the crowdfunding platforms. Depending on the platform, and how they consider their role as a procurer of content, it is implied that is a level of responsibility on these screens to investing to provide adequate and comprehensible content for decision review and making, particularly as they are being paid to facilitate

## CUSTOMER PROPOSITIONS

While we recognize that certain platforms are appendages of direct sponsors to supplement their direct fundraising activities, we will really address those platforms that are "fee for service" providers. Specifically, a sponsor of a real estate investment project is engaging a crowdfunding platform to assist in and manage the fund raising process, as would an investment bank or registered intermediary. To wit, the platform is compensated, largely on its success to raise the requisite funds through its online intercession with its installed investor community. Because of securities regulations, this is a 'closed system' where investors need to be registered within the platform and communication on any offering is done within a secured environment. In effect, while the platform gets compensated by the sponsor (user of capital), the sponsor engages the platform given its base of investors, prescreened with a hopeful appetite for the 'product' the platform avails. In essence, as with any intermediary, there are two customers - the buyer and the seller, in other words the sponsor and the investor.

## LET US REVIEW THE PROPOSITIONS FOR AND TO BOTH

Why would a sponsor engage a platform? Simply, to raise capital on its behalf. The use of intermediaries to drive the fundraising process is an old discipline, but one that typically has scale issues. The time and effort to develop content, package, distribute, engage and close investors is considerable, and sponsors may generally choose to allocate that responsibility to "hired help". Intermediaries, particularly investment banks, are engaged given a demonstrated ability to work with stables of investors, understanding their respective desires and inclinations for investment opportunities.

Sponsors, or users of capital, have a series of motivations for using a real estate crowdfunding (RECF) platform. Fundraising becomes much more efficient, as the financing process is the creation and delivery of information passed on to investors for review. By using an internet-based platform, the reach "horizontally" and "vertically" is tremendously enhanced. From a nomenclature basis, "horizontally" addresses the breadth of reach using online means; in other words, as simple and timely as communication can take place with an investor in San Francisco, the same communication concurrently can take place with investors in New York, Los Angeles, Chicago, even Rome and Hong Kong. Rather than using analog means, simultaneous distribution and reception of content through digital means has altered the landscape measurably. This is about telling a story "singularly to a plurality", as earlier stated.

Additionally, the use of technology allows the investment story to be shared in a more robust way, rather than merely in written format. The ability to depict the story "vertically" can enhance the investor engagement, deepen the knowledge of the project for the investor, which together can improve the likelihood that the investor may participate. Being able to use content beyond that which is "flat", such as script and photographs, sponsors and platforms can utilize videos, 3-dimensional depictions, interactive modeling, and provide online interviews with the principals: all of which creates more resonant story-telling.

Beyond the sharing of the investment to the network, technology also creates efficiency in ongoing investor management. Certainly, pitching the opportunity is key, but the administrative benefits through the fundraising process, say which may require updates, amendments or edits, can be performed with greater efficacy, and including post-transaction reporting. Do note that applied digital technology also addresses the ability to deal with more participants which may have smaller individual investments. Imagining an analog offering of $5 million, with a minimum investment of $5,000. Translating into up to 1,000 investors, this could be an absolute nightmare if dealing solely with analog (paper and phone) communications. This ability to deal with more fragmented networks enables broader reach to more levels of investors as compared to requiring larger minimum investment sizes.

And from the sponsor's standpoint, by having the ability to aggregate smaller sums from a larger contingent of investors, may in fact improve the overall economics of the deal for the sponsor. Many times, when a single large investor, *ala* an "anchor" investor, participates, such investor may desire to improve its own economics as it may view itself as critical to getting the deal done. This may translate into lower management or transaction fees to be earned by the sponsor, or a reduced profit participation. Either way, the "democratization" of investment can avail benefits to both the smaller investor and the sponsor.

Lastly, a growing motivation for sponsors to use such platforms is that financing may be committed up front. Whether a sponsor-driven platform or a hybrid, numerous platforms are taking "principal" risk, wherein they are prefunding the investment, intending to lay off the investment to their investors afterwards. For the sponsor, he receives his targeted funding without the risk of a crowd market's dismissal of the opportunity. Quite true that there is a cost for such "bridging", but for the sponsor, his returns are more clearly based on the funded project's performance, so eliminating the risk of the initial step (of funding), it appears to be a worthwhile tradeoff.

## AND FOR THE INVESTORS

We also much consider the motivations from the other side of ledger, namely that of the investors. There are a number of reasons for the growing participation in crowdfunded real estate opportunities. One observation is that investors can have access to incredible numbers of investment alternatives, delivered efficiently and comprehensively in a web environment. Rather than searching out investments through reams of paper prospectuses, an investor can review either on one RECF site or various sites a multitude of investments. Like angel investing and the syndication activities of years gone by, much private placement activity was highly localized, i.e., that investors with relationships with sponsors were generally geographically nearby, and thus the opportunities were also characterized by such local flavor. Now, with the expanse of online communication, investors can be exposed to sponsors and projects from hundreds of locations, which allow for broader opportunity and diversification.

Additionally, within the breadth, investors also gain access to varying structures and return elements. Different platforms may offer short-term debt while others, longer-term debt, or possibly a hybrid of debt and equity to enhance a prospective return. Also, as with syndications, investors can benefit from the active engagement of the sponsor, rather than pursuing real estate investing on their own, dealing with "toilets, tenants and taxes". Sponsors, and hopefully the platforms, conduct significant review of the project, its potential and its risks, disclosing such to enable informed decision making. For many investors, they desire to deploy capital on a passive participation basis, relying on those sponsors to deliver the returns being advocated.

But this is an investment process that is not without challenges for the investors, which must be recognized. In general terms, the investments being facilitated by the platforms are unregistered securities, and as such, have a high degree of illiquidity, in other words, they cannot be resold on a simple and efficient basis. While there are discussions by some platforms of developing secondary markets and trading, this shouldn't be expected in the near term, and thus investors need realize that an indefinite holding period is likely.

And as mentioned above with respect to an "anchor" investor possibly impacting the economics of the sponsor, the lack of one can prove deleteriously expensive for the participatory investor. Whether through egregious fee levels, profit participation, and administrative expenses, investors are subject to the financial whims of the sponsor, which according to certain accounts, sponsor and non-sponsor platforms may enjoy as much as 40% of an investment's economics, at times without any principal exposure. This can serve to dampen the normalized returns to the capital provider.

One other area that has yet been broadly tested has been the role of the platform in a transaction that may not perform as anticipated. While sponsor-directed platforms have an economic motivation to deal with challenges on behalf of their investors through profit participation, other "objective" platforms may find themselves in a conflicted position. For example, while the platform is raising capital for its paying client, the sponsor, it has developed and brought to the table its investor network to participate in the sponsor's project. What happens in the case difficulties ensue with the project, such as cost overruns, financing or marketing issues, or even suspect activities? What if the financing needs to be restructured or refinanced or compressed from a loss? Which side of the table does the platform sit on – that of the sponsor, who has compensated the platform or that of the investors, which need to feel that their interests are being addressed in a fiduciary capacity, in hopes of mitigating risk? This to date has not yet been tested, with some of the platforms indicating the "outsourcing" to others for risk-mitigation and restructuring assistance, we do believe the jury is out on how sideways deals may be moderated and remedied.

## SIZE OF THE MARKET

Over the last three years we have seen explosive growth in the real estate crowdfunding industry. Initially buoyed by a mere handful of platforms engaging in short to intermediate term commercial real estate loans or meager equity participation in relatively small rehabilitation or development projects, we have now witnessed a seismic shift in both number of platforms as well as types of financings being conducted. This phenomenon further has global breadth, as markets through Asia and Europe, Latin America and Canada, have developing platforms, particularly in the area of residential mortgage lending. In fact, one of the most notable crowdfunding projects was consummated in Colombia, where Prodigy Networks raised on behalf of the sponsor, approximately $172 million from 3,100 investors. A $240 million, 66 story, mixed-use residential/commercial building, BD Bacatá, is the tallest building in Colombia and certainly marks the power of the crowd.

The real estate crowd funding market continues to accelerate its growth and we anticipate this to continue as key segments, like originating single family mortgages, benefit from continued restrictive behaviors in the commercial banking market. From meager beginnings in 2012, where less than $20 million of financing was intermediated through online platforms, according to Massolution 2015 RECF Industry Report, we witnessed over $1 billion in 2014 volume with 2015 expected to advance by 2.5x that level, achieving about $2.6 billion across real estate asset classes and structures (Massolution, 2015). A snapshot of the market's growth is shown in Table 2 as we look at it from different perspectives.

### Type of Real Estate

Since the beginning, C&I real estate has dominated the RECF product suite, with the exception of the BD Bagata project in Colombia in 2013, a mixed use (retail/residential) property sponsored by Prodigy Networks. Multi-family and single family residential properties continued to find interest within the platforms either in terms of "fix-and-flip" short term financing, or more recently, the growing appetite for first mortgage loans requested by sponsors in longer term rehabilitation and repositioning projects or homeowner/homebuyers in permanently financing their homes. Certainly, LendInvest, to date the most prolific of the platforms in terms of volume, has arranged over $200 million of mortgage financing in the London market.

We do anticipate that the growth in the home mortgage space will continue at a solid pace and to include the RECF players, as alternatives to bank originated mortgage loans have been on the rise since the financial crisis of 2008-2010. Non-bank originators now account for over 40% of US mortgage loans, of which 90% are ultimately being purchased by the government sponsored enterprises, Fannie Mae and Freddie Mac (Light, 2014). Notable US student loan platform, Social Finance (SOFI), expanded its product offering in 2015 by introducing higher loan-to-value, interest only owner-occupied purchase money and refinance loans, and is estimated to close nearly $400 million during this calendar year, with

*Table 2. RECF growth (in millions)*

| | 2012 | 2013 | 2014 | 2015 | 2016 | 2020 |
|---|---|---|---|---|---|---|
| **Type of Real Estate** | | | | | | |
| C&I | $11.0 | $113.0 | $742.0 | $1,425.0 | $2,870.0 | $35,000.0 |
| Hospitality | $- | $10.2 | $27.0 | $55.0 | $150.0 | $5,800.0 |
| Multi-Family | $5.0 | $170.0 | $53.0 | $254.0 | $450.0 | $18,000.0 |
| Single Family | $2.0 | $77.0 | $175.0 | $825.0 | $2,350.0 | $65,000.0 |
| Other | $1.0 | $7.0 | $18.0 | $55.0 | $125.0 | $3,500.0 |
| **Geography** | | | | | | |
| North America | $12.0 | $184.4 | $530.0 | $1,320.0 | $2,940.0 | $47,000.0 |
| Europe | $6.0 | $32.0 | $426.9 | $1,070.0 | $1,945.0 | $26,500.0 |
| Asia | $0.5 | $2.8 | $21.0 | $150.0 | $850.0 | $43,500.0 |
| Latin America | $0.5 | $157.0 | $35.0 | $70.0 | $155.0 | $8,500.0 |
| Other (inc.Africa, Oceania) | $- | $1.0 | $2.0 | $5.0 | $55.0 | $1,800.0 |
| **Financing Structure** | | | | | | |
| Debt | $7.0 | $159.0 | $768.0 | $1,923.0 | $4,399.0 | $87,000.0 |
| Equity | $4.7 | $218.0 | $187.0 | $570.0 | $1,346.0 | $39,500.0 |
| Royalty | $0.3 | $1.7 | $16.0 | $45.0 | $75.0 | $250.0 |
| Hybrid | $7.0 | $17.0 | $42.0 | $77.0 | $125.0 | $550.0 |
| Total Financing | $19.0 | $377.2 | $1,014.9 | $2,615.0 | $5,945.0 | $127,300.0 |
| YOY Growth | | 1985% | 269% | 258% | 227% | |

institutional investors purchasing the growing pools. Through the European Union, Asia and elsewhere, mortgage originators are still largely the commercial banks, with covered bond programs offering the liquidity for refunding, with the commercial banks participating with many of the platforms as they become new source of originations.

Geographically, the US still dominates with over 90% of the North American volume. Of the 85 platforms identified by Massolution, 70 were domiciled in the US, with close to a dozen new platforms getting ready to launch in 2015. Europe, led by LendInvest, is witnessing additional activity across sub-classes of real estate, and the current year should see more than a doubling of total financing activity. Asia, while slower to develop in the real estate market, notes that dozens of platforms have launched or are soon to launch to hopefully follow the exceptional growth experienced in the peer-to-peer lending environment, which has built a $20 billion annual market with close to 1400 platforms in China alone.

As mentioned, the financing structure of most real estate crowdfunding projects takes the form of debt. Whether by ease of use or capturing the demand for higher yield, debt structures have been used across types of property and geographies. This is not uncommon across other financing segments in the crowd and P2P lending arena. P2P consumer lending, led by stalwarts LendingClub and Prosper, carved a niche for yield hungry investors, while SME (small to medium sized enterprises) platforms, or small business lending, has also attracted yield players as well as small banks, which are asset hungry, but can't originate on the same cost structure. In fact, OnDeck Capital, a leader in SME lending and a 2014 IPO, had disclosed that its average loan interest rate is over 51% per annum in its S-1 filing in (now since lowered to 28%). Why with interest rates as stratospheric as that are platform originations steamrollering financing across classes and geographies? Certainly investors are keen, but the overall availability of credit has led many borrowers to trade access to credit over cost of credit.

There is a confluence of factors that we believe account for the tremendous rise in marketplace lending, including of course within real estate. While we believe the marketplace lending is disruptive and providing a new avenue for capital to reach desirous capital users, we believe this is largely a response and opportunity afforded due to the financial crisis of 2008-2010 and the legislative and regulatory oversight placed upon the commercial banking sector. Whether from the Dodd-Frank Wall Street Reform and Consumer Protection Act of 2010, or the Third Basel Accord of similar year, more stringent capital levels, risk management and compliance requirements, stress testing guidelines, and market liquidity risk assessment have collectively contributed to how commercial banks conduct themselves and practice the art of banking and finance. In simple terms, the banking industry was subjected to a major scolding by the regulatory regime, which viewed the practices and results of the major banks threatening capital markets and amplifying systemic risk within general economies, leading eventually to a pull back on the industry's traditional business line, the extension of credit. From 2009 through 2013, lending fell off tremendously and then began to rebound, but this laid an opening for others to enter the field, drawn largely to areas where the banks had taken a significant hit, such as commercial real estate loans, consumer credit, and single family mortgage loans. This has become a bastion of the crowdfunding contingent.

With blistering penalties, added compliance costs and requisite reserves, coupled with an interest rate environment making it difficult to generate needed loan/deposit spreads, the lending activity of most banks, large and small, witnessed slower growth. Certainly, coming out of the recession period, centered with the financial crisis, there has been growth, but we see non-bank entrants playing increasingly larger roles. Many major banks generated the preponderance of their respective income from activities outside of conventional balance sheet lending, but whether due to the Volcker Rule or other disciplines, such activities have been required to be tempered if not eliminated. What's left is low margined business, that of credit extension, particularly if these lenders do lend and hold the loan. True, there is aggressive loan sales, particularly in the mortgage space (to the GSEs), but for markets that either securitizations or sales cannot take place, it seems like the industry would rather sit on the sidelines, holding (somewhat) gilt-edged sovereign securities, avoiding upped capital and reserve levels if they choose to reenter the lending streams as once they did. It is worth noting that over the last thirty years, the number of commercial banks have been reduced from north of 14,000 to about 5,500 currently (see Figure 1).

This recent retrenching of the banking industry during a period of external capital demand has given increased rise to what's become know as "shadow banking", i.e., the extension of credit outside the regulated commercial banking framework. According to recent reports, credit extension outside the banks now rivals loans on bank balance sheets in the US (59% for banks to 41% for non-banks, per Goldman Sachs) (Nash & Beardsley, 2015). And in fact, total liabilities in the "shadow" are estimated to exceed $120 trillion, including swaps and other derivatives.

But to regress, the changing and challenging banking environment has given rise to new players to address the funding needs of consumers, businesses, sponsors and investors. We do believe this to continue indefinitely and shall assume that in the near term, the buoyant competition between banks and non-banks will assimilate into various partnering efforts (see Table 3).

Table 3 represents the top producing RECF platforms according to the Massolution 2015 RECF Industry Report (2015). Some of the notable findings are:

*Figure 1. US number of commercial banks graph*

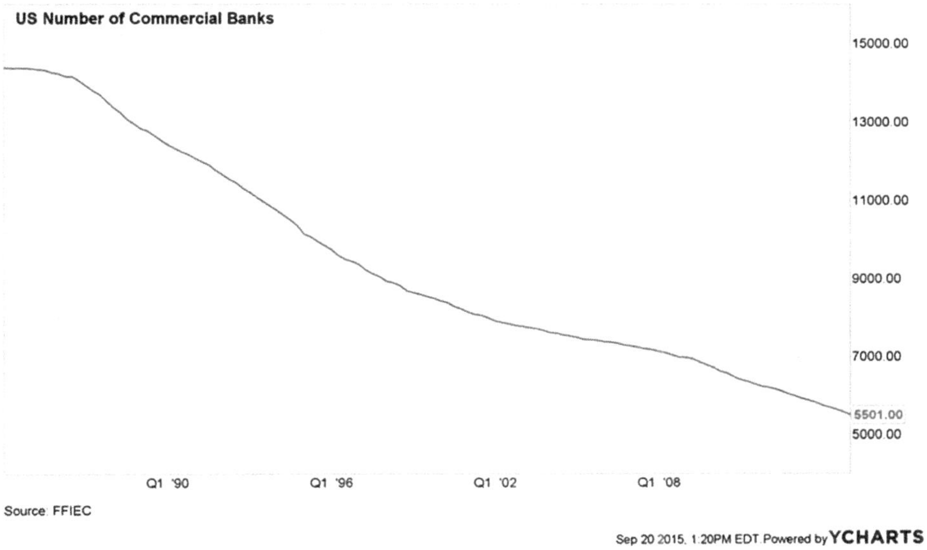


*Table 3. RECFP volume ranking 2014*

| | | Debt | Equity | Hybrid | Volume | Country |
|---|---|---|---|---|---|---|
| 1 | LendInvest | x | | | >$250m | UK |
| 2 | Prodigy Networks | x | x | | >$125m | US |
| 3 | WeAreCrowdfunding (Carlton) | | x | | >$50m | US |
| 4 | Groundbreaker | x | x | x | $25m-<$50m | US |
| 5 | RealCrowd | x | x | | $25m-<$50m | US |
| 6 | RealtyMogul | x | x | | $25m-<$50m | US |
| 7 | SOFI | x | | | $25m-<$50m | US |
| 8 | Money 360 | x | | | $25m-<$50m | US |
| 9 | CrowdMason | x | x | x | $25m-<$50m | US |
| 10 | Wealth Migrate | | x | x | $25m-<$50m | US |
| 11 | DiversyFund | x | x | x | $25m-<$50m | US |
| 12 | Asset Avenue | x | | | $15m-<$20m | US |
| 13 | Early Shares | x | x | | $15m-<$20m | US |
| 14 | Ifunding | x | x | | $15m-<$20m | US |
| 15 | Fundrise | x | | x | $15m-<$20m | US |
| 16 | Patch of Land | x | | | $10m-<$15m | US |
| 17 | CrowdBaron | x | x | | $10m-<$15m | HK |
| 18 | CoAssets | x | x | x | $5m-<$10m | SPORE |
| 19 | CrowdStreet | x | x | x | $5m-<$10m | US |
| 20 | Sindeo | x | | | $5m-<$10m | US |

- Most platforms indicate a capability to avail both debt and equity finance on behalf of the sponsoring issuer. Ultimately, we have seen that 90% of the financing that has been done by the platforms is in a debt-like structure. The most common is a short term first lien loan, generally being used to finance a rehabilitation project.
- There is a growing movement into the owner-occupied single family residential mortgage space, with SOFI and Sindeo, playing off the success of LendInvest in the UK.
- Volumes continue to increase, particularly for those that have received some form of venture funding. They have had the ability to leverage their balance sheets, and either commit capital to close deals up front, later to syndicate the investment to their investors.

There has been additional movement in the recent year with a number of platforms now funded, beginning to experience some level of scaling. It is still suspect as to the overall revenue generation and sustainability, but we are seeing progress. Some of the RECF platforms (Times Realty News, 2015) that are demonstrating recent growth and deserve some notable profile are:

- *Realty Mogul* is a well-funded RECF platform which recently completed a $35 million Series B financing round, moving its total funding to north of $45 million. It has apparently consummated $80 million of property financings, largely debt, though there is a mix of equity participations in larger deals which it provided a minority tranche of the total equity. The company reports that it has invested in 240 different properties, which have a total project cost of nearly $500 million.
- *Fundrise* garnered quite a bit of visibility when it raised a significant funding round ($40 million) early in its development, from investor social media company RenRen and a series of New York commercial real estate developers. Uniquely, Fundrise has not limited its activities to accredited investors like the rest. It was first (and only) RECF platform to conduct securities offerings under Regulation A, i.e., under $5 million of funds raised, and before the Regulation A+ limits were established. It is estimated that Fundrise has completed $60 million of capital raises.
- *Patch of Land* has carved itself a niche is high-yield short term debt investments. Reducing the size of the participations to $5,000, accredited investors can participate in these short term loans, which yield from single to mid-teen digit returns. To differentiate its efforts, it has been practicing bridging the financing, using its own balance sheet, to fund a loan, later syndicating it to its investor network. Having raised over $24 million in venture financing, it has the capability to warehouse many of the loans it originates. Additionally, it is one of the few firms that is investing as a principal in its transactions. As of this writing, Patch has funded nearly $27 million in over 100 projects.
- *RealtyShares* began its activities in 2013 and has experienced some decent growth by offering both debt and equity participations in commercial and residential projects. Noteworthy, it has even included a bitcoin payment option. Recently, the company raised a $10 million Series A round. Thus far, it is estimated that RealtyShares have raised approximately $10 million for projects this calendar year.
- *iFunding*, another player in the real estate space, further reduces the minimum investment to $1,000 for accredited investors, placing such money into a variety of projects, whether multi-family residential, office or warehouse. It continues to experience growth as it placed just over $15 million in invested capital last year, and has seemingly doubled the volume through 2015.

- *EarlyShares* is another emerging platform that actually began as a corporate finance platform, raising capital form early stage growth companies. Catching the wave of real estate investing, it began to focus on high-yield real estate investments, marketing average preferred returns of 7% and targeted overall internal rates of return of 18%.
- *Sharestates* is a New York based platform offering one-year senior secured debt investment opportunities, largely along the East Coast. The type of real estate focused on is generally multi-family, but the firm has moved into other subclasses. Of recent note, it inked a partnership with Ranger Capital, a mixed strategy asset manager, that has been increasingly committed to expanding its alternative asset portfolio. Under the partnership, Ranger is planning to invest up to $30 million in Sharestates deals.

## HOW DO THE RECF PLATFORMS MAKE MONEY?

Many platforms have different operating approaches and as such they derive revenues in different ways. It is almost tautological on how the platforms state the benefit to each of their respective constituencies. There have also been certain accounts that the fees charged along the way have been excessive, on a deal by deal basis, where as much as 30% of the economics of a deal may be subsumed by the platform, rather than enhancing investor returns. Table 4 is an overview of the variety of fees or methods of compensation derived by the platforms.

## HAVE THEY DELIVERED?

While we have discussed the growth in the market in terms of the number of operating platforms as well as growing volumes of transactions, can we state that the platforms are really delivering a "new world" of real estate financing and investing? Here are some takeaways from our research over 2014 (Massolution, 2015):

- Total project costs per deal ranges from $300,000 to $10,000,000.
- Funded participation by the crowd ranges from $150,000 to $4 million.

*Table 4.*

| Compensation to RECF Platforms | % of Industry Revs by Fee Type |
|---|---|
| Application/Advisory Fee | 3.0% |
| Construction/Development Fee | 1.0% |
| Platform Usage Fee | 20.0% |
| Transaction/Success Fee | 32.0% |
| Financing Fee | 10.0% |
| Management Fee | 18.0% |
| Administrative Fee | 8.0% |
| Carried Interest/Profit Participation | 8.0% |

- Average crowd funded deal size in 2014 was $800,000.
- "Crowd" is being redefined – there are more institutional investors.
- Over 500 campaigns were run in 2014.
- Platforms averaged between 4 and 25 campaigns, but this is changing with size of platform and type of financing offered. Certain platforms, such as LendInvest consummated nearly 350 campaigns in two years.
- Success rate of completed funding – 65%, with that number improving with more platforms moving towards prefunding deals.
- Average all-in yield was 11.2% - 7.8% current + 3.4% capital appreciation.
- And – investment results are promising, as most deals are short-term loan, with good performance, yet equity is a longer term hold, which is too early to tell how returns will be.

## DRIVERS TO FURTHER GROWTH

By all accounts, this is a growth space. While we recognized under 90 platforms in 2015 within the US, we now are identifying close to 150. Again, this doesn't address the dozens of international platforms through Asia and Europe. And many of the existing platforms are accelerating their respective growth, given both macro- and micro- conditions in the market. But growth, quantitatively and quantitatively, can be much different based on strategies and resource availability. With an apparent "wind at their backs", here are a number of drivers to continued growth:

- Relatively low interest rate environment leads to higher allocation to real estate investment allocation. The search for yield is being partially satisfied with offerings of high-yield short term secured loans or payouts on attractive lets.
- Many sponsors are developing their own platforms as the ability to connect, control, converse and even cajole investors through online vs. analog communication has proven beneficial. We are also seeing that a number of intermediary platforms are now deciding to take "principal" investment risk, and hopefully add to their company value through portfolio returns.
- Continued regulatory and capital pressure on commercial banks, a fundamental source of commercial real estate (CRE) lending, has created a bit of credit pull back, opening opportunities for crowdfunding and other shadow banking endeavors.
- Improving sophistication of platforms in terms of what they offer to sponsors and investors. Whether that is bridging the financing, or offering solid aftermarket reporting, those differentiators can accelerate preference to certain platforms.
- Capital availability is a huge stimulant to growth. And that is capital in two flavors. A number of platforms have raised internal growth capital from the venture community, which allows them to build capacity in deal review and processing, as well as using their balance sheets to securing bridge loans to pre-fund the crowd offerings. Further, institutional investors, similar to the P2P lending space, are increasingly participating in this space, with dedicated capital commitments to select platforms. In one respect, this negates the "crowdfunding" descript, as one or two larger players taking down an entire issue would not be characterized as a "crowd". So this reverts to the earlier observation of RECF is the digitized version of the 1980 syndication model.

- With the aforementioned capital, and pre-committed and sizable investor participation, various platforms are doing larger deals. Like the P2P industry, the processing costs of a transaction, independent of size is largely the same. With larger deals, where transaction fees based on size are paid, the economic model improves.
- We anticipate seeing more strategic alliances and alignment between these *avant garde* platforms and their banking brethren, particularly in areas that origination costs for the banks has become unworkable. This is particularly expected in the single family residential mortgage market.
- While the mantra thus far has been about "deal flow" and "upsizing the deal", we do believe that in the end, within the finance world, the primary proposition of value will arise from data production and monetization strategies. With the exception of a mere handful, data monetization has not registered as important. We believe this will change.
- And while real estate is contained within the illiquid alternative asset grouping, as transaction volumes in size of deal and number, we anticipate seeing a return of the securitization market, with private label securities, which may include pools of crowdfunded offerings. SoFi has successfully securitized nearly $1 billion of student loans, and is intent of doing the same with their direct mortgage business.

And in discussing opportunity, we cannot be naïve in assuming clear sailing for unbridled growth. True, there will be obstacles. Some of the concerns to growth emanate around risk management, which in terms may alter the success of any particular crowdfunded investment. Could it be poor underwriting of the risk up front, i.e., adverse selection of deals; or post-funding oversight, i.e., if a deal goes sideways, how is risk mitigated and managed? Here are some of the issues that we believe will arise, which could have a dampening effect on the growth:

- While real estate crowdfinance may facilitate $2.5 billion in fundings for 2015, it is less than 1% of overall real estate financings conducted in the US. With an estimated $1.5 trillion in single family mortgages and just over $1.2 trillion in CRE financing estimated in 2015, the role of the "crowd" is *de minimus*. Nonetheless, the marketplace finance platforms are operating in the domain of a highly regulated commercial banking industry, and as one pundit suggested "If it looks like a bank, it could get regulated like a bank". One of the provisions with the Dodd-Frank Wall Street Reform Act addressed substantive non-bank financial firms, and in its "too big to fail" concerns established oversight on selected "systemically important financial institutions" (SIFIs). Using largely balance sheet tests (e.g., asset size, leverage, debt levels), these SIFIs are subject to the compliance requirements of their commercial banking compatriots. Among the acknowledged SIFIs are Metropolitan Life, Prudential, AIG, and GE Capital (Wikipedia, n.d.). Will any of the marketplace platforms become the size to warrant SIFI oversight? Highly questionable. But the regulatory and legislative tails into those operating in the "shadows" is a possible concern, independent of size.
- Risk management, particularly for problematic transactions, will reveal a needed capability and sustainability of many platforms. As we saw in the 1980s, as capital was abundant, syndicators were deploying funds is less than admirable projects. There compensation structures were premised on getting the deal done, with front end fees, only to see the ultimate project waste away. Might we witness the same with the current portfolio of RECF platforms? One mitigating factor is that most transaction structures involve senior secured debt, as opposed to the earlier equity

participations. Certainly deploying creditor rights as opposed to being subjected to them offers a risk mitigant this time around.

- Further to the above point, we are currently seeing a weakening in credit performance in P2P and SME lending. Is this due to stretching one's underwriting protocols to "do the deal" and sustain growth, or are broader economic conditions weakening? Finance is as much science as it is art. We incurred the wrath of over-lenient financing from the banks leading to the financial crisis of 2008-2010. Might the quest for growth lead to underwriting leniency?
- Currently, the transaction sizes within RECF are relatively small. The average transaction size in 2014 was under $800,000 with the mean transaction size about half of that (Massolution, 2015). While we know there are platforms that have raised capital, are using their balance sheets, and have institutional commitments for what they offer, most are still undercapitalized and focused on very small transactions. This raises the question of business and revenue models, and their respective sustainability. If a platform successfully consummates a $500,000 senior loan, it may make in total fees up to $35-$40,000. With personnel review, infrastructure, technology and general overhead, dozens if not more transactions need be processed to achieve breakeven if not ongoing profitability. Note that such transactions are generally not review per a simple credit scoring module, as each property, sponsor, risk factors and other variables are different. So we ask, at what point does this become a scalable model? Intermediation in these CRE and other transactions can be laborious, even if you use online connectivity with investors. There is still front end analysis, structuring, and monitoring to be done. How much margin can one derive if one only generates $40,000 in gross revenues?
- And as a subset to the above, how should these platforms be valued by the investment community? Should they be valued off of transaction volume, revenues, earnings, or other self derived metrics, such as size of investor base, possibly ongoing (recurrent?) portfolio management fees? Late last year, as Lending Club and OnDeck Capital prepared to price their respective IPOs, this author questioned the market valuations place on them. In comparative terms, while in the consumer or small business lending businesses, they were being valued at revenue and asset multiples 10-20x other companies with larger dedicated franchises, such as Wells Fargo, Capital One, and JPM Chase. Yes, those are banks, but they have deployed significant technologies to conduct their business, and they are diversified. Should LC or ONDK enjoy what can only appear as "technology multiples" vs. financial institution multiples, even though the fundamental business is lending. Since the IPOs and the immediate aftermarket run up, both have settled at about half their IPO prices, and are doing whatever they can to perform and justify even the current values.

## CAPITAL MARKETS, ENTERPRISE VALUES AND EXIT STRATEGIES

As earlier references, capital access for the platforms has been a key driver to their respective growth. We have witnessed many raising venture capital even some strategic funds to progress. Further, firms like Realty Mogul and Sharestates have, like the top P2P lending platforms, received committed capital for usage within their markets to insure capacity to fund offerings.

## Dot.com Redux?

During the early 2000's, the investment markets got whisked up in the next generation of commerce, namely the *dot.com* phenom. It appeared that a new value proposition was placed on companies that decided to move their business intercession to the internet for reasons of more efficient and less costly distribution. In fact, many upstarts decided to go directly on line to sell goods and services bypassing conventional distribution and sales channels. In many ways, there seems to be similarity in the growth of online financing platforms. During the dot.com days, there were those that succeeded (and still are) while there were many crash-and-burns. Rather than espousing the digital connection as the driver, the winners knew their basic businesses: Ebay in auctions, PriceLine in travel, Amazon in logistics, and eTrade in brokerage. But then we saw others like Webvan.com, which should have learned better the economics of the grocery business or Pets.com, which seemingly misgauged just how big its market was and its pet owning customer behaviors.

We know that venture capital has found a nice home in crowdfunding, particularly in real estate. But we ask on what basis are models being defined and valued. Recall the dot.com valued "eyeballs" even it there was no way to monetize such. Is investor network size or sponsor relationship bases to value? We have alluded to what the business of these platforms is – namely efficient intermediation for which fees are paid for success. The economics are still suspect with small transaction sizes, level of review and transaction volume.

A question that is raised refers to the potential value creation of such platforms for their investors. Many which have raised operational funds, upwards of $50 million, and at valuations that place post-money values in excess of $130 million. While we know the space is claiming to be "disruptive" to other real estate financing methodologies, such as banks and direct funds, we question whether the business and revenue models support such lofty values, and if so, what supports this. Unlike the P2P space, which is characterized by replicable financings, where some level of credit scoring and actuarial analysis can be performed, thus availing scale, can the RECF platforms be operated with the same mechanisms? Each transaction is seemingly different, requiring certain "laborious" analysis, assuming that the platform is doing some level of review and diligence. So the question is how many transactions can a platform undertake in a given week, a month, a year? As reflected in the mortgage origination business, it used to be that a loan officer could process 8-10 transactions a day. Now, maybe two, with the continued adds of regulatory and compliance requirements. With a cost of labor, and the size of deals which drive revenues per deal, how scalable is the model?

But we ask how should these companies be valued, based on their outlook for revenues and profitability, or on the use of technology to intermediate financial transactions? To expedite the review and processing of transactions, facilitated by technology is a plus, but will the investment results prove out in the end, or will, as some speculate, that adverse selection will creep in. Major financial institutions, all of which deploy technology across their organizations. But yet, they are still valued as financial institutions. For them, technology is a way to reduce costs, thereby providing desired financial services more swiftly and hopefully profitably. But that is their score card - delivering a product or service on a profitable basis. Referring to the platforms, technology too is an enabler, but should not be viewed as the end-all. These companies need to be measured the same way. What do they sell? How do they generate revenues? And can they generate revenues, servicing their clientele, on a profitable basis? We believe the jury is still out.

Let's ask what does create value for the platforms, all of which will hopefully evolve into revenues and ultimate profitability. We see four main pockets of value, which we believe that platforms should position for:

1. **Differentiation:** While many platforms are offering the same "product" to their respective investor communities, such as short-term high yield notes, we ask what's the difference between one and the other, separating a platform from the pack? A good example of differentiation is SOFI, which began in refinancing student loans, but its recent move into mortgage lending is notable. Rather than offering what thousands of mortgage originators (online or offline) are pushing, it has structured a high loan-to-value (~90%) without the historically required mortgage insurance. Its underwriting process and focus is creating a unique niche, yes, differentiation, that continues to set it apart.
2. **Unique Customer Engagement:** Certainly nicely designed and user friendly websites are a given, but how and what the websites may do to enhance the lifetime value of the customer to the platform is a key. Besides generating fees on completed transactions, how frequently does the investor (and sponsor) return to the site and are there further ways to monetize that relationship. Might this be through "paid" analysis and research, or other products that could be cross-sold? How "sticky" and "monetizable" is the environment that the platforms engage their customer?
3. **Revenue Model and Scalability:** We have discussed this earlier. Do the platforms have a unit economic model that is profitable on a fully-loaded deal by deal basis, using applied technology? Or do they have to reach a certain level of volume, either in total dollar volume or number of transactions? Like the P2P industry, profitability is becoming visible largely with increased loan sizes, not sheer numbers of deals. What must be the volume, transaction size and fee level to have positive unit economics?
4. **Data and Analytics:** Finance is an information business first and foremost. The use of technology can move and refine information much more nimbly than humans, and with higher frequency transactions, ability to track performance, introduce indigenous and exogenous factors into the mix, new insights can be introduced. Improved risk management, pre- and post- deal, as well as improved pricing the risk can be the result. While some platforms understand this in basic terms, few have developed data strategies, including monetization of the fruits of their activities. We do believe this to be the ultimate Holy Grail of technology enabled intermediation.

Through various conferences and articles, platform principals have been queried as to where is the inherent value within their models. The responses are deemed suspect and in fact, we believe that in the end, are just basic business practices, coincident to all platforms, offering limited incremental value. We don't believe they necessarily "create value". Here are some of the typical responses:

1. **"Quality Deal Flow":** Many platforms espouse the nature of their deal flow, i.e., financial opportunities that they have been introduced to for consideration of funding. Building a successful portfolio of deals, proving execution history and ultimate investment performance, like a Goldman Sachs, builds reputation and more outside interest to "shop there". But stating that one sees $10 billion of transaction opportunities to yield $10 million being pursued, becomes a logistic issue as well as tough to place value on. Establishing relationships with highly successful sponsors that become "in house" exclusive clients, has some merit, but in the nascent stage of the industry, it would seem that few platforms have earned the desired reputational capital.

2. **Diligence:** Being exposed to new transactional opportunities is critical, but more so is the review and analytical process to determine whether an opportunity should be pursued and if so, how to structure it. A challenge is that large intermediaries have teams to assess and deliberate, and ultimately choosing to move forward on its selections, for which they are compensated upon completion. The fees paid to larger intermediaries tend to reflect the relative size of the deal, i.e., bigger fees for bigger deals, which allow for broad upfront effort. When it comes to smaller transactions, fees tend to reflect such size, so is the level of diligence more limited? Or given the structures of the offerings such that diligence can be compressed, e.g., with short term secured loans. And as technology is applied, can the diligence become enhanced with lesser levels of labor, and produce improved decision making?
3. **Investor Network Numbers:** A key benefit of using an intermediary is the breadth of its relationships to funding sources. Rather than a principal who is seeking funds directly "dialing for dollars", using an intermediary with established contacts in the investment community facilitates the fund raising process. Further, it is hoped that the likelihood of successful fundraising is improved, as intermediaries may have a unique sense of their investor sentiments and appetite for a particular issue. When comparing to an online platform, such investor relationships may not be as close knit and generally, the sheer numbers of online investor networks, while impressive, may not be characterized by the same strength of relationship. While the numbers may suggest the law of average in terms of more investors lead to a higher chance of success, one may ask is it better to have 10,000 small investors rather than 1,000 active larger investors to effectuate at deal? And are such networks proprietary to a platform, or do investors, seeking new places to deploy investable funds, sign up to many platforms? Numbers alone seem not to tell the story, rather, how active are the networks to get the job done.
4. **Visibility to Break Even Operations:** This is a little discussed topic, in part as the industry is in development and growth mode. As referenced, small transactions, still requiring a level of analysis and processing, tend to generate smaller fees. Has the economic model, on a deal by deal basis, been assessed and what might have to be achieved to attain breakeven operationally?

## PROGNOSTICATIONS

As an industry in development, with new entries and methodologies, what might we experience in the coming periods? While we have alluded to the significant growth over the last year and forecasting similar rates over the next few years, what may be possible futures for this space? Here are some of the observations that may impact, positively and negatively, for further expanse. To characterize the "what if's", allow us to deploy a couple metaphors, black swans and butterflies. This descriptive we used in the Massolution RECF Industry Report which was released in the Spring 2015 (Massolution, 2015).

A "Black Swan" hails from Nasim Taleb's manuscript on Black Swan theory, which addresses hard-to-predict impactful events that with hindsight could have been expected. And the "butterfly effect", articulated by Edward Lorenz, implies that small changes in conditions in one area can result in huge differences elsewhere.

Some of the black swans we may reflect upon are:

- The demise of the "2 and 20" partnership model. Much of the institutional capital invested in alternative assets are generally managed through investment funds, where the manager garners a management fee, e.g., an annual 2% fee based on funds being managed, and a profit participation (or carried interest), which generally tracks at twenty percent (20%) of the earnings after return of initial capital. Over time, based on the performance of the funds, this can be a very expensive proposition to institutions, and many have developed internal capabilities to invest directly, thereby bypassing the aforementioned economic costs. With broader access to opportunities availed online, with professional underwriting, disclosures and portfolio management, it is conceivable that institutional investors will see the marketplace platform approach to deploying capital directly as a viable option, rather than through funds.
- While "crowdfunding" has typically been identified as small sponsors, entrepreneurs and other fundraisers using platforms to aggregate disparate capital for diminutive ventures, we have seen large corporate entities jumping into the fray. Not necessarily motivated by capital formation, but rather striving for better connectivity from their end consumer or client. Companies like Nike, Warner Brothers, CocaCola and Proctor & Gamble, have used the power of the "crowd" to facilitate engagement benefiting product development and marketing. As with consumer products, will larger real estate sponsor groups tap into crowd methods, harvesting a larger portion of the market for their own fund raising. Very likely, as we have already experienced
- Unsavory litigation. While most of the crowd platforms are operating under the 'accredited investor' exemption, i.e., offering investments to those that can 'fend for themselves', it is anticipated that eventually some form of Title III of the JOBS Act or Regulation A+ with enable non-accredited investors to participate in private and/or illiquid offerings. As such, might there be an occasion that a small investor, whom loses money in a deal, conclude that disclosures or overall management of a deal were less than adequate, casting a legal pursuit. Have the platforms accounted for such contingency? Or rather, while espousing the broader "crowd", will the platforms just remain affixed under the Regulation D exemption?
- Many platforms have successfully raised capital as referenced, with lofty valuations. With the IPO aftermarket performance of Lending Club and OnDeckCapital, will this put a damper on fundraising and valuations of companies pursuing marketplace finance. Recent exception is SoFi which consummated a $1 billion private financing led by SoftBank, valuing the business well into "unicorn" status. We remain cautious of both the flows and valuation of capital into this space. In fact, we do anticipate that there will be more strategic investment directed here, led by larger financial institutions.
- In the wake of The Dodd-Frank Wall Street Reform Act, financial institutions, large and small, have been subject to increasing levels of regulatory scrutiny as a result of the financial crisis of 2008-2010. Even large non-commercial bank financial institutions, characterized as Systemically Important Financial Institutions (SIFIs), such as Metropolitan Life, Prudential and GE Capital, are now subject additional layers of reporting and review. We have seen recently that the largest players in the P2P lending industry are now in dialogue with regulators about better or best practices, as well as risk management. We assume that as the RECF platforms gain in size, and intruding on the historic landscape of the banks, there will be the prospect of regulatory review, beyond the SEC's oversight of securities issuance. How this effects the practices and viability of the platforms is subject to speculation.

And now, what about changes that might cause broad ripples, i.e., having a butterfly effect on the industry and its future. Among the "butterflies" we might just see are:

- Changing the definition of Accredited Investor. Having not adjusted the definition since 1982, the SEC has been reviewing what characteristics should define a "natural person" investor to participate in private (unregistered offerings). On one hand, by inflation adjusting the financial requirements (income and net worth) from the 1982 level to levels 2.5x higher would reduce the overall number of possible investors that could participate in such private offerings by almost 70%. In a constructive and historically sound manner, the SEC is reviewing other methods of "accreditation", such as professional and academic acumen as the tests for participation. We mention "historical" as the original exemption provision in the Securities Act of 1933 referred to not just the financial capacity of the investor, i.e., the ability to absorb losses, but also the ability to conduct an intelligent review of the investment opportunity. This "informed decision making" viewpoint is critical, particularly now in the "information age", and the ability to disseminate broadly requisite content to investors. This new approach may open up private markets, while underscoring the responsibility of platforms to articulate offering materials for investor consumption. Ultimately, this could enhance the overall numbers exponentially.
- While Title III of the JOBS Act has not been promulgated as of yet, there have been over twenty US states that have developed intrastate constructs for crowdfunding, thereby allowing unaccredited investors to participate, albeit limitedly, in private offerings. Will the states continue to take this up on their own in wake of the sluggishness at the federal level? We believe so.
- Crowdfunding is a subset of the growing "shadow banking" evolution, namely the extension of credit outside the regulated commercial banking market. We do see that both borrowers and lenders have a growing comfort in transacting outside the banking system. And with improving controls, risk management and accessibility, we see this to continue and accelerate.
- While access to investments through online mechanisms can avail investments to larger investor communities, this is not just about access but about opportunity. While the SEC's mandate has been proffered to mitigate risk through disclosure and reporting requirements in the '33 Act and '34 Exchange Act, crowdfunding should also be viewed as contributing to investment opportunity. Heretofore, venture and other alternative investments have been the province of the "upper class" which has the "qualifications" to participate in such higher-yielding investments, where the common investor could not gain access. In the narrative about income and wealth disparity at venues like the World Economic Forum in Davos, we need to find ways for broader communities to have access to compelling investments in order to build wealth, though the approach need be couched with certain protections and limitations. But of the three million businesses in the US, only 5,000 have public registered stock for investing and trading. Given that the preponderance of wealth has been created in privately held companies, might not a thoughtful approach to availing a more expansive miscellany of investments be constructive and economically elevating? We believe that the SEC, while in its quest to wrestle with risk, should also create frameworks for broader participation.
- Lastly, and in particular to the millennial generation, the crowdfunding phenomena has embellished philanthropic activities. This has been positively felt by small non-profit organizations, as well as community development real estate projects. We expect that this trend will continue at an accelerated pace.

## FINAL OUTLOOK

The real estate crowdfunding industry will continue with its explosive metric of growth. From new platforms, to expansions of product offerings and improved risk management protocols, and finally, growing investor comfort in the online investing practice: this bodes well for this burgeoning market both in the US and abroad.

We anticipate seeing a phenomenal growth in the residential mortgage market, which has already seen the impact of non-bank originators absorbing over 40% of the market's demand. New US entrants, such as SoFi and Sindeo, will further the competitive landscape in a trillion dollar setting. Already, LendInvest in the UK has demonstrated leadership.

There will be regulatory creep, as more activity is conducted through these platforms, with increasing scrutiny to protect both users and suppliers of capital. As one pundit exclaimed, "if it acts like a bank, it should be regulated like one"

2016 will mark a year of wins and losses. We will witness platforms gaining in size and capabilities, becoming formidable members of the financial markets. We will also see that the current frothiness of value, driven by amorphously described "disruption", see some dampening, and result in numerous platforms being shuttered or consolidating with others. That assumes that they recognize what the value they are creating really is.

Many platforms, and in particular their venture investors, may have had visions of public offerings during the coming 12-24 months at lofty valuations. Early litmus tests (LC and ONDK) and equity market conditions may cause rethinking in monetization strategies. Might we see two step partnerships with larger institutions, desirous of a "fintech" exposure begin to take investment positions, later fully absorbing these platforms: quite possible.

Overall, while playing in the "shadows", there will be much light shining from this industry over the coming years.

## REFERENCES

Accredited Investor. (n.d.). *Securities and Exchange Commission*. Retrieved 30 August 2015 from http://www.sec.gov/accreditedinvestor

Light, J. (2014, August 27). Nonbank Mortgage Originators Bounce Back. *Wall Street Journal*.

Massolution. (2015). *Real Estate Crowdfunding – Industry Report*. Author.

McKinsey Company. (2014). *The Trillion Dollar Convergence: Capturing the Next Wave of Growth in Alternative Investments*. New York: McKinsey Financial Services Practice.

Nash, R., & Beardsley, E. (2015). *The Future of Finance Part One, The Rise of the New Shadow Bank*. New York: Goldman Sachs.

REIT. (2014). *Market Caps*. Washington, DC: U.S. REIT Industry.

Taxfoundation. (2015). *U.S. Federal Tax Rates*. Retrieved from http://www.taxfoundation.org

The S&P500 PE Ratio. (n.d.). In BloombergBusiness. Retrieved 1 September 2015 from http://www.bloomberg.com

Times Realty News. (2015). *TRN Realty Crowdfunding Report 2015*. Author.

Wikipedia. (n.d.). *Systemically Important Financial Institutions*. Retrieved 12 September 2015 from http://en.wikipedia.org/wiki/SystemicallyImportantFinancialInstituions

# Chapter 15
# The Application of Crowdfunding to the Energy Sector

**Chiara Candelise**
*Bocconi University, Italy*

## ABSTRACT

*Crowdfunding in energy begins as a response to reduced investments (both governmental and private) into the transition to decarbonized energy systems and to the spread of innovative business models and approaches conductive of greater participation of citizens and communities in distributed renewable energy projects. This chapter presents results of a worldwide overview of the use of crowdfunding in the energy industry. Evidence gathered from available energy crowdfunding platforms highlight a very new, but quite dynamic sector. The crowdfunding tool has been applied in most of its forms, ranging from peer-to-business lending to pure donation, with strong environmental and social mission and the explicit aim of increasing participation of citizens in sustainable energy investment. Evidence also shows that, despite maintaining their environmental and clean energy focus, some energy platforms have begun to move from niche, grass root initiatives into larger projects and collaboration with energy private sector and institutional finance.*

## INTRODUCTION

This chapter provides an overview and analysis of the development of crowdfunding within the energy industry. It looks into the trends and dynamics of this emerging sector and presents results of a worldwide review of energy crowdfunding platforms. It aims at providing a better understanding of the application of this innovative financing tool to the energy industry and of its performance to date. It intends to provide a reference analysis to policy makers, regulators as well as finance, industry and general public stakeholders about the state of the art of this emerging sector.

DOI: 10.4018/978-1-5225-0568-6.ch015

## ENABLING FACTORS

One of the drivers behind the growth and spread of crowdfunding worldwide has been the reduced access to capital to citizens and firms for new investments due to the financial crisis and restricted access to credit since the second half of the 2000s. Indeed, crowdfunding platforms have begun to spread around the world in 2008, along with other emerging alternative finance instruments outside the traditional banking sector, including mini bonds, social impact bonds, microcredit, impact investing. A similar trend has affected the energy sector, with the financial crisis impacting on investments in energy projects and infrastructures. In particular, both governmental and private investments in sustainable transition have been decreasing (Geels, 2013), despite the growing need for financial resources to support clean energy projects in response to climate change reduction policies and strategies. Thus, as in other sectors, crowdfunding in energy could be seen as an answer to the need to mobilize financial resources, and support the transition toward cleaner energy systems.

Another important enabling driver to the use of crowdfunding in the energy industry is the strong transformation undergoing in the sector since the late 90s. The restructuring and liberalisation of the industry has allowed the entry of new stakeholders in the energy markets. Contextually, climate change mitigation strategies and the need to decarbonize energy systems by increasing the proportion of renewable distributed generation and reducing energy use through the implementation of energy efficiency measures has set the basis for a shift in the energy system paradigm, from a centralized "top-down" system to decentralized generation, a "bottom-up" system. An immediate difference between the two paradigms is the role of the energy end user, who in a decentralized generation system could become both investor and generator. The modularity of renewable energy projects has allowed smaller size investments (than centralized generation plants) and the entrance in the energy market of new generators and investors (including citizens, local authorities, small firms) previously set outside of the industry mainly dominated by large energy companies (Kempener et al., 2015). This trend has been particularly evident for small/medium scale renewables and energy efficiency measures which have generally higher transaction costs (and comparatively low equity return rates) which makes them less appealing for large companies. In addition, large companies often lack for experience with small scale projects and can have split incentives: decentralized generation and energy efficiency can have a negative competitive impact on their core business. These changing scenarios have thus left space for the entrance of new and smaller players in the energy industry.

The transition to clean energy systems is challenging the role of old generators and utilities' business models. In Europe the share of renewables with dispatch priority in electricity markets in increasing which, in conjunction with post crisis reduction in electricity demand and fossil fuel generation overcapacity, has led to a reduction in wholesale prices and in the profitability of utility owned conventional power plants. Utilities are losing money on their electricity generation core business and gradually reducing their role of investors in energy systems.

Such transformation of the energy industry is also seen in the financial market. Utilities and large energy companies were traditionally seen as low risk investments by institutional finance (large investors, pension funds) as they operated in a regulated and stable market. However, the changing paradigm in the energy sector has seen the entrance of new players in the energy finance market and the development of new financing instruments, such as green bonds issued by development banks, corporate investors and municipalities or yieldcos (publically traded companies formed to own power plants and pass revenues to investors as dividends) (Frankfurt School-UNEP & BNEF, 2015). Utilities weight in the finance market decreasing, top 20 energy utilities were worth in 20013 less than half of 2008 value (of around €1 trillion) and have been the worst-performing sector in the Morgan Stanley index of global share prices (Gray et al., 2015; The Economist, 2013).

Nonetheless, much of the transition to decarbonized energy systems has been led by major investors and large companies, but smaller players as well as citizens and local communities are increasingly playing an active role in delivering clean energy investments. Decentralised energy is now part of governments' and societies' 'localism' agenda, as citizens, communities, firms, local authorities and other public sector organizations can become energy producers and not only 'passive' users of a service delivered. As such, they could fully harness benefits and incomes originating from energy investments implemented in their premises and territory. Indeed, the last decade has seen the increasing development of community energy and shared ownership approaches for investments in the energy sector worldwide (ILO, 2013, van der Schoor and Scholtens, 2015, Yildiz, 2014). These approaches are new business models and structures conductive of greater participation of citizens and communities in energy projects, as such are a form of democratization of energy investments and policy. However, it has been pointed out how such community led scheme are often under-resourced (Seyfang et al., 2014) and lacking of sufficient access to finance to scale up activities (Phimister & Roberts, 2012). Crowdfunding, which aggregate financial resources, people and potential investors on specific energy projects have the potential to expand the applicability and the scope of such type of approaches. It can guarantee additional source of funding, where conventional financing channels do not always deliver, through participatory approaches and enabling the possibility of reaching a higher number of potential investors thanks to the communication power of internet and social networks.

Through crowdfunding platforms, a company, a local authority or a community developing a renewable generation plant or an energy efficiency measure can not only raise capital but also involve citizens and other stakeholders in the investment, allowing them to invest, become shareholders and benefit from the return on the investment itself (according to the type of crowdfunding model adopted - see Section 2.2). Crowdfunding platforms allow people to invest even small amount of money on energy projects and proponents of projects to get access to a large audience of potential investors. It thus provides the possibility to those interested in investing in clean energy, but without suitable conditions (e.g. lack of sufficient capital or of a suitable area to develop the project) to directly contribute to $CO_2$ emissions reduction and benefit from the returns on the investment. In some cases, also thanks to its communication and public engagement potential, could help in gathering financial resources to support investments less suited for traditional project financing (such as smaller size investments or projects aiming at delivering long-term solutions such as regeneration, social housing or generation in remote areas).

Indeed, crowdfunding is an interesting financing tool for the energy sector and there exist several active platforms worldwide 'vertically' focused on the sector, which are already proving to be a powerful method to raise capital in particular for renewable energy projects.

## A WORLDWIDE REVIEW OF THE SECTOR

This section presents results of a worldwide overview and analysis of crowdfunding development within the energy industry to date and of the initial results and trends coming out of this new and dynamic sector.

Energy crowdfunding platforms have been searched for and selected through analysis of online grey literature (McCann, 2013; Renewable Energy Crowdunding Conference, 2014; Solarplaza, 2014a, 2015, 2014b) and systematic web browsing running Boolean keyword searches for the terms 'energy', 'renewables', 'crowdfunding', 'community shares, 'equity', 'solar', 'wind', 'biomass', 'energy efficiency', 'shared ownership'. The platforms identified and included in the review have at least one project focused on energy and, if not active yet, express clear focus on the energy sector in their mission.

Such systematic review has allowed identifying forty two energy related platforms worldwide. Through close reading of all published projects on each active platform (i.e. the published material on each project web page) a data base of 389 energy crowdfunding campaigns has been created, accounting for several variables including: project name; location; technology type; opening and closing date of the offer; target and amount of money raised; type of investment offered; returns guaranteed. Data have been gathered up to October 2015.

## A New, but Dynamic Sector

The application of crowdfunding to the energy sector has been relatively recent, within the short lifetime of the crowdfunding sector itself which appeared as a wide phenomenon around 2008 when platforms such as Indigogo or Kickstarter (IndieGoGo, n.d.; Kickstarter, n.d.) have been launched. Table 1 lists the forty-two platforms searched for and identified worldwide. Apart from the pioneering experience of the German platform GreenVesting in the 2009, the first relevant platforms and campaigns have begun to appear in 2012 (the non-energy specific platforms such as Milaap and Microgenius despite being launched in 2010 and 2011 have both been publishing energy related projects only in 2015). Since then the number of initiatives and platforms have been increasing overtime. By July 2015 twenty-nine energy crowdfunding platforms were active with projects online and thirteen new platforms were in the pipeline.

*Table 1. Platforms overview: country, launch date, and technology focus*

| Platform | Country | Launch | Platform Focus | Technology Focus |
|---|---|---|---|---|
| GreenVesting | Germany | 2009 | Renewables | Solar |
| Milaap | India | 2010 | General | Solar |
| Microgenius | United Kingdom | 2011 | General | Mix |
| Lumo | France | 2012 | Renewables | Mix |
| Crowdener.gy | Germany | 2012 | Renewables | Solar |
| WindCentrale | Netherlands | 2012 | Renewables | Wind |
| Greencrowd | Netherlands | 2012 | Renewables | Mix |
| Abundance Generation | United Kindom | 2012 | Renewables | Mix |
| SunFunder | USA/Tanzania | 2012 | Renewables | Solar |
| Bettervest | Germany | 2013 | Renewables and energy efficiency | Mix (Mostly Energy Efficiency) |
| Econeers | Germany | 2013 | Renewables | Mix |
| GreenXmoney | Germany | 2013 | Renewables | Mix |
| We share Solar | Netherlands | 2013 | Renewables | Solar |

*continued on following page*

*Table 1. Continued*

| Platform | Country | Launch | Platform Focus | Technology Focus |
|---|---|---|---|---|
| Gen Community | United Kingdom | 2013 | Renewables | Mix |
| Trillion Fund | United Kingdom | 2013 | Renewables | Mix |
| Collective Sun | USA | 2013 | Renewables | Solar |
| Divvy | USA | 2013 | Renewables | Mix |
| Mosaic | USA | 2013 | Renewables | Solar |
| Village Power | USA | 2013 | Renewables | Mix |
| Enerfip | France | 2014 | Renewables | Mix |
| Lendosphere | France | 2014 | Renewables | Mix |
| LeihDeinerUmweltGeld | Germany | 2014 | Renewables | Mix |
| Duurzaaminvesteren | Netherlands | 2014 | Renewables | Mix |
| Veolis | Switzerland | 2014 | General | Mix |
| Solar Schools | United Kingdom | 2014 | Renewables | Solar |
| Clean Reach | USA | 2014 | Renewables | Ocean Energy |
| Re-Volv | USA | 2014 | Renewables | Solar |
| Coopernico | Portugal | 2015 | Renewables | Mix |
| GridShare | USA | 2015 | Renewables | Mix |
| Citizenergy | Europe | Non active | Renewables | Mix |
| Joukon Voima | Finland | Non active | Renewables | Mix |
| Crowd2win | France | Non active | Renewables | Mix |
| Benoolend | France | Non active | Renewables | Solar for rural electrification |
| Babyloan/Total | France | Non active | General | Rural electrification |
| GreenCrowding | Germany | Non active | Renewables | Mix |
| Ecomill | Italy | Non active | Renewables | Mix |
| Fundera | Italy | Non active | Renewables | Mix |
| Trine | Sweden | Non active | Renewables | Solar for rural electrification |
| Crowdgeneration | United Kingdom | Non active | Renewables | All |
| EnergyDonwell | United Kingdom | Non active | Renewables | Mix |
| Palmetto Direct | United Kingdom | Non active | Renewables | Mix |
| CrowdEnergy | USA | Non active | Renewables | Ocean Energy |

**Note:** 'Launch' is the online launch date of the first energy project on the platform, despite several platforms have been reported to be conceived couple of years before the date of launch of the first campaign. Unique exceptions in the list are Milaap and Microgenius General platforms: platform launch date is reported, but their first energy related campaigns have been launched in 2015.

Europe, in particular Germany, United Kingdom and the Netherlands have been the initiators of the sector in 2012, with USA following in 2013 and currently hosting the highest number of energy related platforms (Table 2).

Crowdfunding in the energy sector is currently associated with and serves only clean energy investments, with most of the platforms specifically focused on renewable energy (the few non-thematic platforms

*Table 2. Number of platforms per country*

| | USA | Germany | United Kingdom | Netherlands | France | Switzerland | Portugal | Italy | Finland | Sweden | India |
|---|---|---|---|---|---|---|---|---|---|---|---|
| Active platforms | 8 | 6 | 5 | 4 | 3 | 1 | 1 | - | - | - | 1 |
| Forthcoming platforms | 1 | 1 | 3 | - | 3 | - | - | 2 | 1 | 1 | - |

listed in Table 1 have also published renewable energy projects). Platforms generally offer investments in renewables energy projects from a mix of technologies, including solar photovoltaics, wind, biomass, hydropower and, in some cases, in energy technologies startups or energy related technological research and development projects. Investments in energy efficiency projects are much less common, apart from one German platform, Bettervest, which is almost totally concentrated on them. Some platforms are instead focused on a specific renewable technology, usually structuring their project scouting and communication strategy explicitly targeted on the renewable technology chosen. Clear examples are We Share Solar or Windcentrale in the Netherland, Mosaic and Sun Funder in the USA or Solar Schools in the United Kingdom whose respective descriptive slogans are as follows:

- "The Windcentrale enables households to produce their own sustainable energy with their own wind turbine" (Reitsma, 2014);
- "We are a crowdfunding platform for collective solar-energy initiatives", We Share Solar website (WeShareSolar, n.d.);
- "Turn sunshine into savings" or "The simplest home solar loans", Mosaic website (Mosaic, n.d.);
- "Financing solar beyond the grid", Sunfunder website (SunFunder, n.d.);
- "Powered by sunshine, funded by everyone", Solar Schools website (Solar Schools, n.d.).

## How Are Different Crowdfunding Models Applied?

The landscape of renewable energy platforms is still fragmented, the concept of crowdfunding in energy is not a commonplace and platforms differ among themselves in particular in terms of the crowdfunding model chosen.

There exists several types of crowdfunding models and several classifications have been produced by previous contributions (EU Commission, 2014; Massolution, 2015; Pais et al., 2014; World Bank, 2013). For the purpose of this review and analysis crowdfunding platforms are grouped in two overarching categories (Figure 1), which differentiate themselves on the basis of the relationship between those who provide financial resources and those that receive the funds (i.e. recipients):

1. Non-financial or donation crowdfunding, where individuals' contributions are not associated with a financial return and;
2. Financial or investing crowdfunding, where financial instruments are sold in relation to companies assets and/or financial performance.

*Figure 1. Overview of major crowdfunding models*
*Source: Author's elaboration.*
*Note: Other, less common models are recently gaining traction: 1. Hybrid models, platforms offering one or more crowdfunding models on a single platform; 2. Royalty-based model, where investors receive a royalty interest derived from intellectual property of the fundraising company.*

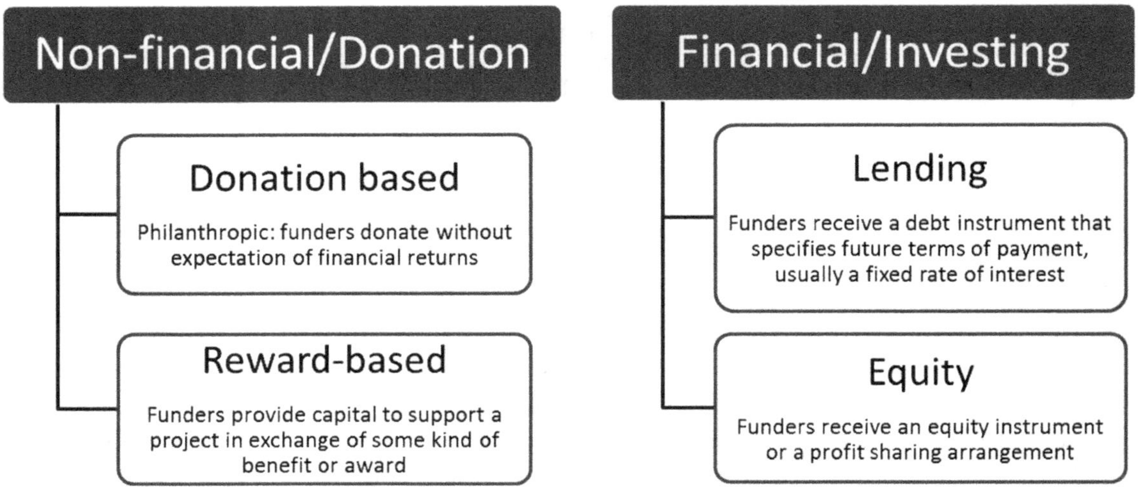


Non-financial crowdfunding can be pure calls for *donations* which are given without expectation of any financial returns or benefit, thus relying on altruistic motives. Typical donation campaigns are run for charitable or public interest causes. Another form of non-financial crowdfunding is the *reward based* model where individuals provide capital to support a project in exchange for some kind of benefit or award; such award could also be the product that will be produced with the capital collected (typical examples are musicians who raise funds to record an album in exchange of the copy of the album once released). Reward based crowdfunding has also been used extensively by businesses or innovators, often as a tool to assess demand for a new potential product or service by raising on crowdfunding platforms capital for pilot design and production. The main financial crowdfunding models are *lending* and *equity based* crowdfunding. With lending crowdfunding funders receive a debt instrument that specifies future terms of payment, usually a fixed rate of interest. Lending platforms can be peer-to-peer (i.e. lending among individuals) or peer-to-business (i.e. lending to support a business activity). With equity crowdfunding funders receive an equity instrument or a profit sharing arrangement. In this case the returns on the investment are dependent on the company's performance, thus potentially higher than the case of lending, but also accompanied with a higher risk.

In practice there are several variations of these basic models, but it is useful to use such categories to give a picture of how the crowdfunding tool has been applied to date within the energy sector. Over 80% of the active platforms are financial/investing platforms (twenty-four out of twenty 29 – see Table 3) with the remaining 20% dedicated to non-financial/donation models (Figure 2). Of the financial ones a good majority have adopted some sort of declination of the lending model (14 out of 24), some the equity/community share model (6 out of 24) and the remaining have adopted an hybrid model allowing investors to choose among different types of investing options (debt instruments, bonds, company equity or shares in local cooperatives).

*Table 3. Financial active platforms*

| Financial | | | |
|---|---|---|---|
| **Platform** | **Model** | **Investor's Benefit** | **Country** |
| Abundance Generation | Lending | % of revenue of profit generated by the projects | United Kingdom |
| DuurzaamInvesteren | Lending | % of revenue of profit generated by the projects | Netherlands |
| Econeers | Lending | % of revenue of profit generated by the projects | Germany |
| GreenVesting | Lending | % of revenue of profit generated by the projects | Germany |
| Lumo | Lending | % of revenue of profit generated by the projects, plus use of the electricity generated | France |
| Mosaic | Lending | % of revenue of profit generated by the projects | USA |
| We share Solar | Lending | % of revenue of profit generated by the projects | Netherlands |
| Lendosphere | Lending | % of revenue of profit generated by the projects (higher % returns for investors living in the project area) | France |
| Enerfip | Lending | % of revenue of profit generated by the projects | France |
| Bettervest | Lending | % of revenue of profit generated by the projects | Germany |
| GreenXmoney | Lending | % of revenue of profit generated by the projects | Germany |
| LeihDeinerUmweltGeld | Lending | % of revenue of profit generated by the projects | Germany |
| Collective Sun | Lending (for no profit projects) | % of revenue of profit generated by the projects | USA |
| SunFunder | Lending (for rural electrification) | % of revenue of profit generated by the projects | USA/Tanzania |
| Crowdener.gy | Equity (Community Shares) | Financial returns proportional to the share owned | Germany |
| Gen Community | Equity (Community Shares) | Financial returns proportional to the share owned, plus surplus income reinvested in the community | United Kingdom |
| Microgenius | Equity (Community Shares) | Financial returns proportional to the share owned | United Kingdom |
| WindCentrale | Equity (Community Shares) | Financial returns proportional to the share owned, plus use of the electricity generated | Netherlands |
| Village Power | Equity (Community Shares) | Financial returns proportional to the share owned | USA |
| Coopernico | Equity (Community Shares) | Financial returns proportional to the share owned | Portugal |
| Trillion Fund | Hyprid (Lending, Equity, Community Shares) | Returns vary in accordance to the financial tool chosen (bonds, community shares, equity, funds) | United Kingdom |
| Greencrowd | Hybrid (Lending, Equity) | Returns vary in accordance to the financial tool chosen (loan or quity) | Netherlands |
| Veolis | Hybrid (Lending, Reward) | Benefits vary: % of the revenue of profit generated by the projects (Lending) or rewards provided (Reward) | Switzerland |
| GridShare | Hybrid (Lending, Equity, Reward) | Benefits vary: % of the revenue of profit generated by the projects (Lending) or rewards provided (Reward) | USA |

*Figure 2. Number of energy crowdfunding platforms per model*
*Source: Author's elaboration*

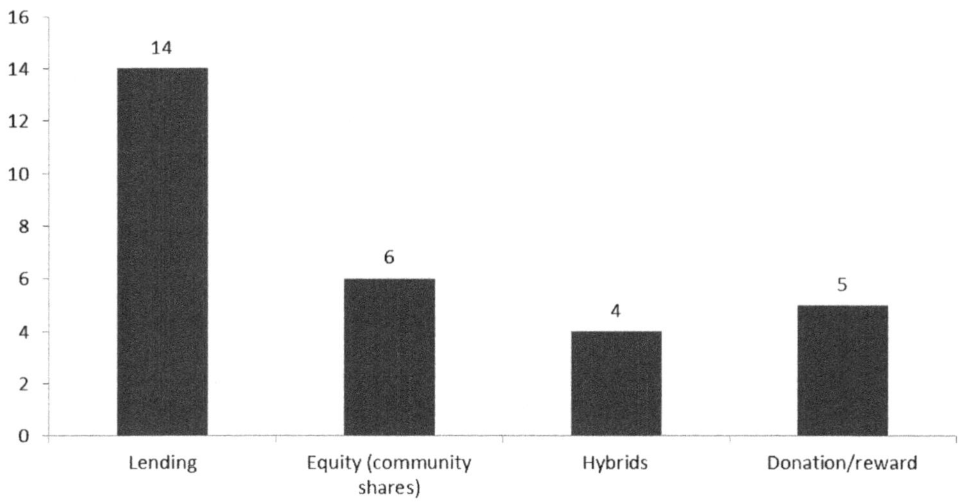


But let's have a closer look: Tables 3 and 4 show how the twenty-nine active energy crowdfunding platforms are differentiated across models.

The *lending platforms* in this context are all 'peer to business' meaning that, according to the definition above, some money is lent through the platform by the investor to support an energy related business or activity. Investments in renewable energy projects (and in few cases in energy efficiency projects) are solicited online by the platform providing the investor with a debt instrument which is then repaid and remunerated over time as a function of the revenues generated by the project. Payments of revenues are often administered by the platform itself which, in this case, acts as a sort of financial intermediary between the project developers and the wider investors (e.g. Abundance Generation, We Share Solar, Mosaic). In some cases, the remuneration is variable as a function of the revenues accruing from the sale of the electricity generated by the renewable project, which vary due to both the fluctuations in the

*Table 4. Non-financial platforms*

| Non-Financial | | | |
|---|---|---|---|
| **Platform** | **Model** | **Investor's Benefit** | **Country** |
| Milaap | Lending (for social value projects) | Loan without interests or repayment – high social value project | India |
| Re-Volv | Donation (plus Revolving Fund) | No monetary returns for investors. The lease payments are reinvested in a revolving fund, the Solar Seed Fund, to further finance community-based solar projects | USA |
| Clean Reach | Donation | No monetary returns for investors. Money collected are used for testing new technologies and organize thematic events | USA |
| Solar Schools | Donation | No monetary returns for investors. Money collected are used by a charity to install photovoltaic plants on UK schools | United Kingdom |
| Divvy | Donation/Reward | No monetary returns. Rewards provided | USA |

generation itself (e.g. solar, wind are intermittent and variable renewable power) and the variability of electricity selling price. In other cases the remuneration guaranteed is fixed.

All lending platforms' mission and communication strategy emphasize the ethical and environmental dimension of the initiative; a quick browsing of slogans on platforms' websites provides the best evidence for it:

- "Investments that build a better world", Abundance Generation (Adundance Investment, n.d.)
- "Working together for a better future. Crowdfunding for energy efficiency and green technologies", (Econeers, n.d.)
- "Enerfip, la plateforme de financement participatif pour la transition énergétique", Enerfip (n.d.)

However, some platforms have a stronger ethical/social connotation, either in the type of investments they offer or/and in the way they structure investors' involvement and returns. To give some examples:

- Collective Sun mainly supports solar projects development for non-profit entities, such as schools, churches, community centres and so on;
- SunFunder's mission is to unlock solar energy in emerging and developing countries to improve access to energy and rural electrification;
- Lendosphere offers higher returns to investors living in the area of the project, thus emphasizing the importance of the involvement local community in the project.

The *equity crowdfunding* platforms active in the energy sector generally offer a form of *community shares*, usually for investments in renewables projects: investors give money in exchange of shares of the legal entity making a specific investment (usually a company or a cooperative), as such benefiting from returns on the investment itself through dividend payments and acquiring rights to participate to the legal entity activities (which vary according to the form of the legal entity, e.g. generally higher in the case of cooperatives (SpaceTec Capital Partners, 2014)). In the case of renewable generation and energy efficiency revenues accrue respectively from the sale of the electricity produced and the savings on the electricity bills.

Most of these platforms have a strong focus on community based projects, with a strong geographical characterization and with the explicit aim of increasing and improving the participation of local communities and citizens in the renewable energy investments accruing in their territories. Platforms such as Gen Community, Microgenius, WindCentrale, Village Power and Crowdener.gy explicitly shape their mission as well as scout and structure their projects in order to empower communities by offering them a stake and voting rights in the legal entities (usually cooperatives) making the renewable energy investment. Gen Community's projects also encompass a sort of revolving fund[1] which allow to reinvest surplus income from the project into other activities of community interest. WindCentrale also offers to investors the use of the electricity generated by the wind farms thanks to a partnership with an energy supplier (Reitsma, 2014).

Finally *hybrid platforms* allow investors to choose among different types of investing options: debt instruments, bonds, company equity or shares in local cooperatives. Each of the four hybrid platforms listed have its own peculiarity. Veolis is a lending platform, which has also launched some reward based campaigns to support energy technologies startups or energy related R&D activities. Greencrowd is also structured as a lending platform, mainly raising from online investors loan to finance the construction phase of renewable energy projects; however, it also offers equity instruments to finance initial development costs and/or to fund the equity component of a renewable investment funded through project financing. In this context equity is offered as a higher risk but higher return investment. Grid share offers a combination of loan, equity and reward, within the same projects. However, from the evidence presented online it is difficult to gauge whether the platforms is fully operational as very limited money has been raised to date and none of the projects has been funded. Trillion Fund instead looks as a successful case of hybrid platforms (see also Section 2.3). On Trillion Fund investors can browse among several types of investments, including:

- Lending instruments such as bonds and funds (issued respectively by renewable energy company or investment funds) and direct loan to renewable energy projects.
- **Community Shares:** Shares in co-operatives or firms serving a community purpose.
- **Equity Investments:** Trillion Fund is the first platform to have launched a pure equity crowdfunding campaign, offering shares to capitalize Triodos Renewables, a renewable energy company developing projects across the UK.

In this case the platform acts as an intermediary, thus being responsible only for the crowdfunding campaign; payments of returns to the investment are then managed autonomously by the proponent of the project i.e. company, cooperative or investment fund.

Five of the active platforms are *non-financial/donation* (Table 4), four of which are strongly socially oriented, providing renewable energy projects to communities and no-profit in their respective countries, and one aiming at raising funds to support technological development in ocean energy (Clean Reach).

## How Have Energy Crowdfunding Platforms Performed to Date?

Such high variability of crowdfunding models adopted raises the question of how have the different platforms performed to date. Is there a model more successful than another? What is the success rate of the projects? Is success affected by the returns offered and/or by other factors such as e.g. the ethical/social mission of the projects? Let's begin by looking at the evidence gathered on projects published to date by each active platform (data summarized in Table 5). Platforms performances differ quite substantially, in terms of both number of projects and amount of money raised. This is certainly an indication of how young and fragmented this sector still is. However, it is possible to highlight some initial trends out of the gathered evidence.

When data are aggregated by country, focusing on the most active in the sector (Table 6), United Kingdom stand out as the market leader, in terms of number of projects and money raised (both total amount raised and average project size). This is not surprising, considering United Kingdom leadership in the wider European crowdfunding market (accounting for market share of over 74% (Wadrop et al., 2015)). Netherland follows, with the next highest figures in terms of money raised, then USA and Germany whose numbers fall in a similar range.

Table 7 instead presents data aggregated by crowdfunding model. Lending platforms have been publishing the highest number of projects to date, which could have been expected considering they account for about 60% of the active platforms. However, equity/community shares and hybrids platforms seem to have better performed in terms of money raised (both total amount and average project size). Average returns offered to investors are comparable between lending and hybrid platforms and slightly

*Table 5. Performance of active platforms*

| Platform | Model | Number of Projects | Success Rate | Average Returns | Money Raised € |
|---|---|---|---|---|---|
| Abundance Generation | Lending | 23 | 61% | 7.49% | 13,526,367 |
| Mosaic | Lending | 20 | NA | NA | 5,785,414 |
| LeihDeinerUmweltGeld | Lending | 19 | 89% | 5.25% | 4,029,050 |
| Econeers | Lending | 10 | 90% | 5.34% | 3,915,150 |
| DuurzaamInvesteren | Lending | 9 | 78% | 5.68% | 2,892,000 |
| Bettervest | Lending | 31 | 90% | 7.21% | 1,774,240 |
| Lendosphere | Lending | 16 | 94% | 5.42% | 1,446,990 |
| We share Solar | Lending | 8 | 63% | 5.75% | 1,075,574 |
| GreenVesting | Lending | 9 | 100% | 5.33% | 970,950 |
| GreenXmoney | Lending | 13 | 85% | 4.54 | 277,550 |
| Lumo | Lending | 6 | 67% | 3.00% | 185,000 |
| Enerfip | Lending | 1 | 100% | NA | 60,000 |
| Collective Sun | Lending (for no profit projects) | 4 | 75% | 6.20% | 254,769 |
| SunFunder | Lending (for rural electrification projects) | 34 | 100% | 3.52% | 986,238 |
| Trillion Fund | Hybrid (Lending, Equity, Community Shares) | 29 | 97% | 7.60% | 102,095,794 |
| Greencrowd | Hybrid (Lending, Equity) | 23 | 96% | 4.88% | 2,097,040 |
| Veolis | Hybrid (Lending, Reward) | 3 | 100% | 2.50% | 77,498 |
| GridShare | Hybrid (Lending, Equity, Reward) | 14 | 0% | 6.38% | 54,874 |
| WindCentrale | Equity (Community Shares) | 9 | NA | 3-8% | 14,300,000 |
| Village Power | Equity (Community Shares) | 13 | 62% | NA | 5,303,674 |
| Gen Community | Equity (Community Shares) | 2 | 100% | 7.00% | 1,351,516 |
| Microgenius | Equity (Community Shares) | 2 | 100% | NA | 1,436,458 |
| Coopernico | Equity (Community Shares) | 7 | 86% | NA | 267,600 |
| Crowdener.gy | Equity (Community Shares) | 5 | 80% | NA | 181,800 |
| Solar Schools | Donation | 65 | 82% | - | 691,976 |
| Re-Volv | Donation (plus Revolving Fund) | 3 | 100% | - | 106,448 |
| Divvy | Donation/Reward | 4 | 50% | - | 1-,423 |
| Clean Reach | Donation | 6 | 0% | - | 5,698 |
| Milaap | Lending (for social value projects) | 15 | NA | - | 338 |

Notes:

1. Success rate is the percentage of the published projects which get successfully funded.
2. Money raised figure is the result of the sum of money raised in each single project published online. The survey thus assume that figures published by platforms for each project are accurate and factual.
3. Money raised figures for Mosaic and WindCentrale are taken from websites and grey literature, as information and data of specific projects is not available online.
4. Trillion Fund figure for money raised presented is conservative as three successful 'Fund' projects are not added to the figure as data not available online.

*Table 6. Performance by country*

| Country | Number of Projects | Money Raised € | Average Raised per Project € | Average Returns by Country | Number of Platforms |
|---|---|---|---|---|---|
| France | 23 | 1,691,990 | 73,565 | 4.21% | 3 |
| Germany | 87 | 11,148,740 | 128,146 | 5.53% | 6 |
| USA | 98 | 12,406,537 | 126,597 | 6.29% | 8 |
| Netherlands | 49 | 20,364,614 | 415,604 | 5.58% | 4 |
| United Kingdom | 121 | 118,234,845 | 977,147 | 7.36% | 5 |

higher for equity/community shares projects (which is again expected, as usually equity is remunerated at higher rates than debt). Thus, these data seem to show that despite lending platforms are the mostly implemented, they tend to attract smaller size projects and raise lower amount of money overall.

However, a closer look reveals that the data presented above are strongly affected by the performance of two specific platforms: Trillion Fund in the United Kingdom and WindCentrale in the Netherlands. Indeed, Trillion Fund significantly outperforms to at least an order of magnitude all other energy crowdfunding platforms across the world (Table 5), thus being the main contributor of United Kingdom leadership in the sector. Similarly, WindCentrale is the largest platform in the Netherlands and the second biggest in Europe (Table 8), comparable only to Abundance Generation (second biggest platform in the United Kingdom - Table 5).

Moreover, disaggregating Trillion Fund performance data across their different project types (Table 9) it is quite striking to notice how the majority of the money has been raised through their lending projects,

*Table 7. Performance by crowdfunding model*

| Model | Number of Projects | Number of Funded Projects | Money Raised € | Average Raised per Project € | Average Returns |
|---|---|---|---|---|---|
| Lending | 203 | 153 | 37,179,291 | 243,002 | 5.39% |
| Equity (community shares) | 38 | 31 | 22,740,049 | 733,550 | 7.00% |
| Hybrids | 66 | 53 | 104,325,206 | 1,968,400 | 5.34% |
| Donation/reward | 93 | 58 | 814,883 | 14,050 | - |

Note: 'Number of projects' figure for Hybrid platforms does not include three Trillion Fund successful 'Fund' projects for which data on amount of money raised is not available online.

*Table 8. WindCentrale performance*

| WindCentrale | |
|---|---|
| Money raised € | 14,300,000 |
| Number of projects | 9 |
| Average return | 3-8% |
| Average raised per project € | 1,588,889 |

*Table 9. Trillion fund performance, by project type*

| Trillion Fund | Equity | Community Shares | Loan | Fund |
|---|---|---|---|---|
| Money raised € | 4,747,208 | 6,972,470 | 34,017,910 | 56,358,206 |
| Number of projects | 1 | 9 | 13 | 3 |
| Average return | 9.5 | 5.29% | 7.34% | 15.13% |
| Average raised per project € | 4,747,208 | 774,719 | 2,616,762 | 18,786,069 |

Note: Successful 'Fund' projects are in reality six, but three have not been included as data on amount of money raised is not available online.

i.e. loans and funds issued by renewable energy company or investment funds. Average returns for such type of investments are also higher than average lending platforms returns (just above 5%).

Moving Trillion Fund lending projects from the hybrid model group to the lending one (Table 10) leads to very different results than above in terms of energy crowdfunding models' performances: lending model is not only the more frequently applied by energy crowdfunding platforms, but it's also the best performing. Table 10 clearly shows how lending projects have been the highest in numbers, raised the largest amount of money (over 75% of the total raised to date) and for largest average project size. Returns offered are also the highest, although the 9.29% figure in Table 10 is slightly biased upper ward by the Trillion Fund's average returns to funds of 15.13%, without which average returns would stay around 6% (thus still slightly lower than equity/community shares' returns).

Compared to the financial platforms, donation/reward initiatives have raised much less money, but for a good number of projects overall. They sit in a fairly different category. In this contexts crowdfunding campaigns are smaller in size than in financial platforms as they are generally used to raise funds for rather small renewable energy projects, such as small/medium solar energy plants on schools roofs or solar for rural electrification projects (e.g. in the case of Milaap, the currently open campaigns would finance solar lamps for women empowerment in India and have set targets between a min of 80$ and a maximum of 250$).

The most successful of these platforms, e.g. Milaap (15 projects published in the first month of activity, August 2015) and Solar School (65 published projects, 53 of which successfully funded), offer very homogenous and replicable projects (i.e. solar lamps in the case of Milaap; solar projects on schools roofs mobilizing the interest of parents, local businesses, former students and wider local community in the case of Solar Schools) which allows them to raise money from a high number of similar small projects.

*Table 10. Performance by crowdfunding model – accounting for Trillion Fund lending projects*

| Model | Number of Projects | Money Raised € | Average Raised per Project € | Average Returns |
|---|---|---|---|---|
| Lending (with Trillion Fund's lending projects) | 219 | 127,555,407 | 768,406 | 9.29% |
| Equity (community shares) | 38 | 22,740,049 | 733,550 | 7.00% |
| Hybrids (no Trillion Fund's lending projects) | 40 | 13,949,090 | 348,727 | 4.58% |
| Donation/reward | 93 | 814,883 | 14,050 | - |

## How Does It Compare with the Wider Crowdfunding Sector?

Overall the energy crowdfunding sector has raised about €165millions since 2012, which accounts for only 0,75% of the crowdfunding funding volume cumulated worldwide roughly over the same period of time (i.e. almost €23billions) (Massolution, 2015). However, it is interesting to notice that, despite being a still quite small market, the size of energy crowdfunding campaigns is almost double than the average wider worldwide crowdfunding (Figure 3). And this pattern is consistent across all models.

Moreover, the energy crowdfunding campaigns are also on average more successful. Figure 4 shows much higher success rates[2] for energy crowdfunding projects than wider crowdfunding averages, and across all models. The only exception is lending based campaigns showing the highest success rate (around 90% and above), mainly due to the popularity of peer-to-peer microloans (which have the highest success rate across all campaigns). However, the peer-to-peer microloan model has not been used in the energy sector to date, thus such figure is not directly comparable with the success rate of lending projects in the energy sector, which are instead peer-to-business. When lending in the real estate market (which is instead peer-to-business) is taken as a more realistic benchmark, energy crowdfunding sector still outperform the average global trend. Similarly, donation/reward and equity/community shares projects are also very successful in the energy sector. Success rate of the latter is particularly striking if compared with wider equity projects success rate (which is expectable due to the fact that they tend to be focused on innovative start-up businesses offering highly risky and volatile returns) and, even more striking, with equity projects in the real estate sector (which is instead characterized by less risky investment with relatively stable and foreseeable returns, such as the energy sector). A discussion of some of the potential drivers of such higher success rate are presented in Section 4.

## SHAPING INVESTMENT DECISIONS

The review has highlighted how all energy crowdfunding platforms are characterized by a clear environmental and social mission both in the type of projects they publish and in their wider communication

*Figure 3. Comparing average crowdfunding campaign size, by model*
*Source: Author's elaboration, (Massolution, 2015).*

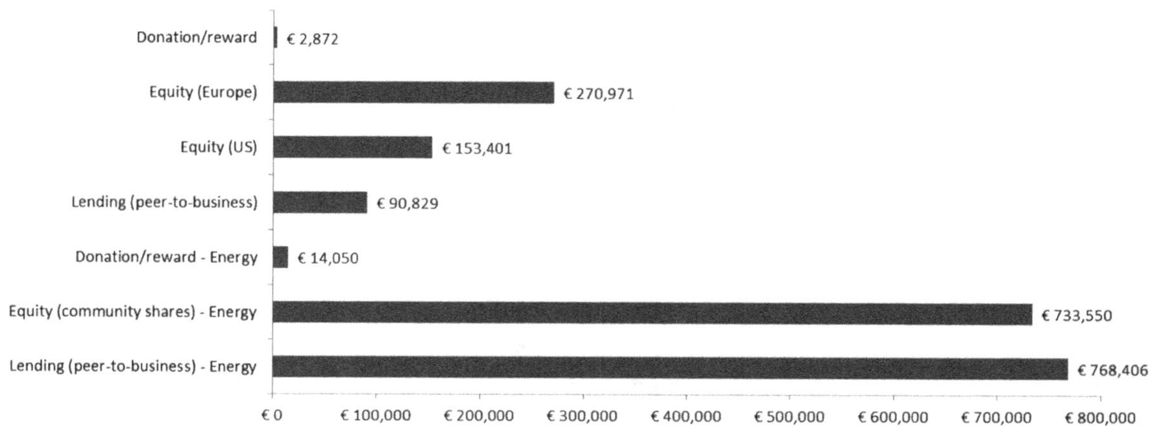

*Figure 4. Comparing average success rate, by model*
*Source: Author's elaboration, (Massolution, 2015).*

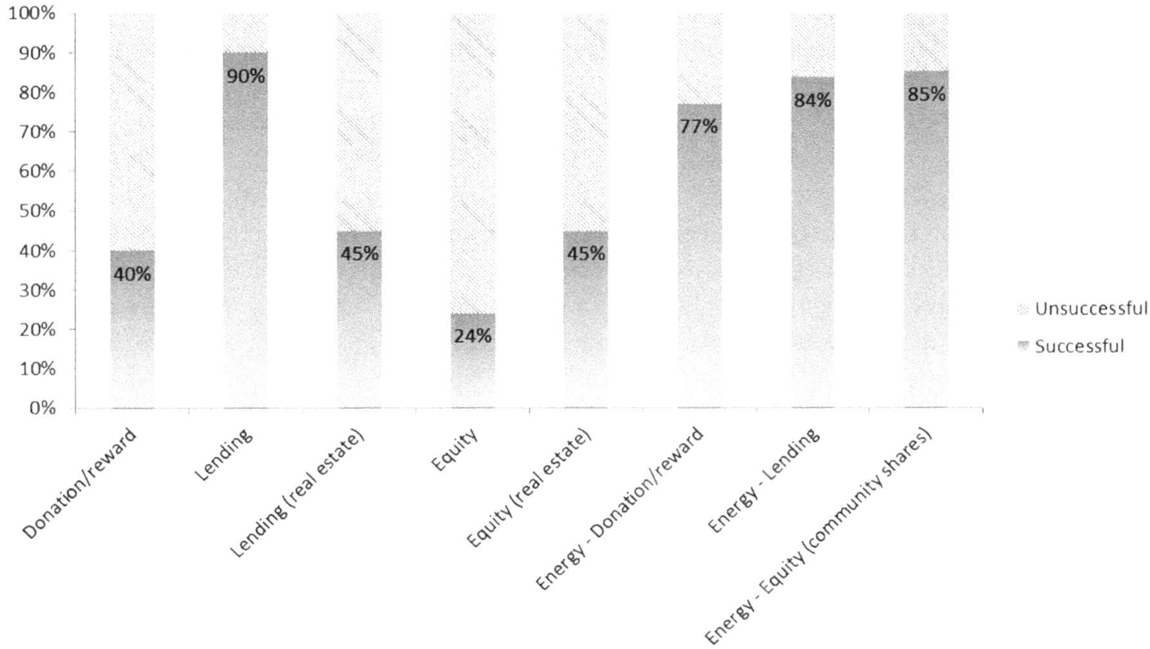


strategy. Some lending platforms have explicit focus on projects in developing countries for rural electrification or renewable projects for not-for-profit, and equity/community shares platforms emphasise the community and local nature of the investments they offer.

Indeed, crowdfunding in energy not only is mainly focused on clean and sustainable energy investments, but, as also previously noted, has also flourished within in a wider process of democratization of the energy sector, which has seen the development of several community energy and energy cooperative initiative across Europe and USA in particular (Kunze & Becker, 2014, ILO, 2013). These new approaches for renewable energy deployment bring forward forms of shared ownership of energy assets with the aim of maximising localization of economic returns and increasing communities' participation. Thus it would seem that the environmental, social/ethical and participatory characteristics of the investment are at least some of the drivers behind the decision to invest from the general public and the communities interested by the energy project (in particular in the case of community shares). However, this is definitely not the only driver. Most of the platforms are financial models, thus offering quite interesting monetary returns on the investments, on average in the 4% to 7% range.

It would be interesting to explore more in details what are the drivers behind investment decisions in energy crowdfunding as well as the investors' profiles. Which type of people and investors tend to approach investments on energy crowdfunding platforms? Are more the social and environmental commitments or the financial returns driving investment decisions? Unfortunately, data on investors profile and behaviour are owned by platforms under strict privacy controls and the sector is relatively young, thus publicly available analysis on this topic is still limited.

However, some initial evidence can be drawn from experience shared by some platforms. An analysis from a leading UK energy crowdfunding platform (Abundance Generation) points out that about 34% of their target investors are the so called 'ethical investors' (those that actively consider the impact of their investment activities, thus motivated by the 'green value' of the investment), but over 50% are 'financial planners' (experienced investors, who plan for their financial futures, thus less motivated by the sustainability cause and more by the returns on the investment and the possibility to choose among additional and novel investment options) (Abundance Generation, 2014). This UK experience seems to indicate that returns on the investments play a considerable role in making energy crowdfunding investments attractive. On the other hand, at the recent Energy Crowdfunding Conference (London, 5th November 2015) several platforms' representatives (among those Trine and Lendosphere) have emphasized the importance of communicating social and environmental impacts of investments to acquire investors and guarantee high success rates to their projects (Energy Crowdfunding Conference, 2015). Lendosphere also stated the success of their strategy of offering higher returns to people geographically closer to the project, which has to date allowed them to raise about 50% of the money among local people (Blais, 2015).

## TRENDS: LOOKING INTO THE FUTURE

The analysis of the evidence gathered on energy crowdfunding platforms pictures a very new, but quite dynamic sector. The crowdfunding tool has been applied in most of its forms, ranging from peer-to-business lending to pure donation platforms. The strong environmental mission is a common place, with all platforms focusing on clean energy projects and most of them explicitly shaping their project scouting and communication strategy emphasising the ethical dimension of the initiative.

Indeed, the use of crowdfunding in the energy sector has begun as a fairly niche application to grass root and community energy projects, with the explicit aim of increasing participation of citizens in renewable energy investment. First ground-breaking campaigns are those launched by platforms such as WindCentrale (Netherlands), which in 2013 managed to raise €1.3 million in less than 13 hours from 1700 households in the Netherlands to finance a community owned wind turbine (Reitsma, 2014, Bayar, 2013); or Abundance Generation (United Kingdom), which is allowing citizens to invest in solar or wind plants with as little as £5 investment since 2012 (Abundance Generation, 2014). More recently platforms such as Lendhosphere (France), LeihDeinerUmweltGeld (Germany), DuurzaamInvesteren (Netherlands) or Village Power (USA) have managed to successfully raise millions of euros in renewable energy projects with explicit community and local focus.

This local and social connotation is probably one of main elements guaranteeing higher success rate to energy crowdfunding campaigns than worldwide averages, in particular when compared to potentially less locally and socially characterized businesses such as real estate, where the driver to invest is likely to mainly be the financial returns offered. Evidence shows that funding from local population is more easily raised, and potentially lower returns is counterbalanced by the possibility of being involved in the project with some sort of participation or control rights (Yildiz, 2014; Dóci & Vasileiadou, 2015).

Evidence from forthcoming platforms also seems to indicate that the social and ethical advocacy of energy crowdfunding platforms will continue to drive the sector. The 2015 has seen the entry into the energy sector of the Indian micro lending platform Milaap as well as the announcement of several new platforms focused on raising funds for rural electrification projects in developing countries.

However, evidence presented in previous sections also shows how energy crowdfunding platforms, while maintaining their environmental and clean energy focus, have already been moving from niche, grass root initiatives into larger projects and collaboration with energy private sector and institutional finance. Results from a recent European survey among projects proponents (Figure 5) highlight how the majority of projects have been developed by private sector stakeholders, either limited companies or energy service companies (Klaes, 2015). In other words, crowdfunding has been used by energy project developers to finance new investments in the sector, despite with a participatory and shared ownership approach. The case of Trillion Fund is a clear indication of this trend: structured finance instruments such as energy loans and funds coming along community shares on the same platforms.

Thus, how is the energy crowdfunding sector likely to develop in the future? What are the potential conditions to scaling up? Are larger companies, institutional investors and finance going to play a role in the future, along with locally sourced and community energy projects?

More generally crowdfunding, despite beginning as a grass root phenomenon, it has quickly scaled up and moved into financing innovation and providing seed capital for innovators and startups, raising money in the billion euros range and being increasingly used by established businesses to finance growth and expansion activities (Lehner, 2013; Mollick, 2014). The energy sector is already showing signs of similar trend: from small/medium locally characterized energy projects to larger projects, from energy cooperatives to private sector led campaigns often launched to support the expansion of renewable energy companies rather than single projects (in particular in the United Kingdom). Trillion Fund performance to date highlights the important role of structured finance and more institutional investors for scaling up the sector: over 50% of the total funding volume of worldwide energy crowdfunding sector (about €165 millions) has been raised only by 16 loan and fund campaigns launched by Trillion Fund in the last couple of years. Whether grass root, community approaches will coexists with more structured finance instruments in the scaling up of the sector is an open question to which we will get an answer only in the years to come.

On the other hand, utilities are in need of developing new business models as a response to the ongoing energy sector transition which is challenging their historical role of market makers and leaders. Could crowdfunding become an interesting tool for utilities or large energy companies?

*Figure 5. Company structure of projects proponents*
*Source: Author's elaboration, (Klaes, 2015).*

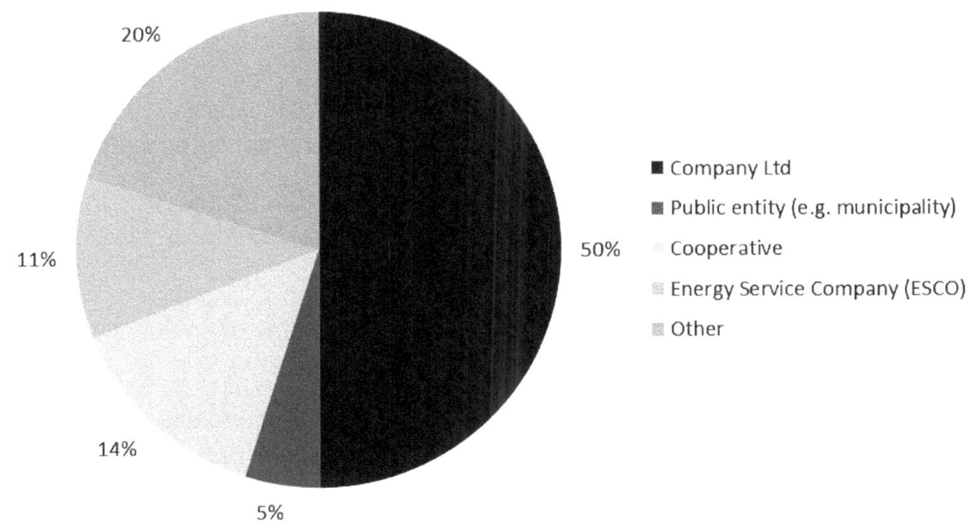

Some platforms are already collaborating with energy utilities as for example WindCentrale (which is partnering with the utility Greenchoice in order to offer to investors the use of the electricity generated by the wind turbine they own (Reitsma, 2014)) or Lendosphere (which had developed a project in partnership with Électricité de France (EDF) (Blais, 2015). Moreover, crowdfunding platforms are generally perceived to be a more democratic form of financing (Lehner, 2013), so they could for example become a tool for improving utilities' public image and accountability with final energy end users (i.e. their clients) or to fund (fully or partially) projects in their core business activities through a democratic, empowering and engaging financing tool. Indeed, crowdfunding, beside its primary role (of financing projects) has a strong marketing and communication function, which can favour its use by institutional actors, corporate and large companies in the energy sector for public advocacy and marketing activities. Some examples have already started to appear, as the recently announced partnership between the French oil company Total and the crowdfunding platform Babyloan to develop a crowdfunding platform dedicated to access to energy in developing countries (Total, 2015). As above, only the years to come will tell us whether such type of initiatives will be well received by the general public and will actually successfully become new business models for energy utilities and large energy companies along with other community led and bottom up approaches.

In conclusion, crowdfunding in energy is a novel sector, with large room from innovation, testing and standardization of business models (both for platforms themselves and projects launched on them). It is a highly innovative sector characterized by flexible and dynamic firm structures to keep the pace with fast market developments. Initially developed as a socio-technical practice in a niche, it is already showing signs of scaling up and is turning into a new business model, which can help in fostering and scaling up sustainable solutions in the future (Bocken et al., 2014, Vasileiadou et al.).

## ACKNOWLEDGMENT

Research funding from the Marie Curie Fellowship LocalEnergy (FP7-PEOPLE-2012-IEF, GA-331081) is gratefully acknowledged.

## REFERENCES

Abundance Generation. (2014). *Investments that build a better world to spend your money in*. Renewable Energy Crowdfunding Conference. Retrieved from http://www.recrowdfunding.eu/#intro-2015

Bayar, T. (2013). Dutch Wind Turbine Purchase Sets World Crowdfunding Record. *Renewable Energy Worl*. Available at: http://www.renewableenergyworld.com/articles/2013/09/dutch-wind-turbine-purchase-sets-world-crowdfunding-record.html

Blais, A. (2015). *Crowdlending for Renewable energies. A simple way to associate citizens*. Renewable Energy Crowdfunding Conference, London.

Bocken, N. M. P., Short, S. W., Rana, P., & Evans, S. (2014). A literature and practice review to develop sustainable business model archetypes. *Journal of Cleaner Production*, *65*, 42–56. doi:10.1016/j.jclepro.2013.11.039

Dóci, G., & Vasileiadou, E. (2015). "Let's do it ourselves" Individual motivations for investing in renewables at community level. *Renewable & Sustainable Energy Reviews*, *49*, 41–50. doi:10.1016/j.rser.2015.04.051

*Econeers*. (n.d.). Retrieved from: https://www.econeers.de/

*Enerfip*. (n.d.). Retrieved from: https://www.enerfip.fr/

EU Commission. (2014). *Crowdfunding: regulatory framework in EU member states and perspectives for the EU*. Luxembourg: Publications Office of the European Union.

Frankfurt School-UNEP & BNEF. (2015). *Global Trends in Renewable Energy Investment*. Available at: http://fs-unep-centre.org/sites/default/files/attachments/key_findings.pdf

Geels, F. W. (2013). The impact of the financial–economic crisis on sustainability transitions: Financial investment, governance and public discourse. *Environmental Innovation and Societal Transitions*, *6*, 67–95. doi:10.1016/j.eist.2012.11.004

Gray, M., Leaton, J., Schuwerk, R., & Fulton, M. (2015). *Coal: Caught in the EU Utility Death Spiral*. Carbon Tracker Initiative Report. Available at: http://www.carbontracker.org/report/eu_utilities/

ILO. (2013). *Providing clean energy access through cooperatives*. Geneva: International Labour Office Cooperative Unit Report. Green Jobs Programme.

*Indiegogo*. (n.d.). Retrieved from: https://www.indiegogo.com/

Kempener, R., Malhotra, A., & De Vivero, G. (2015). *The age of renewable energy power. Designing national roadmaps for a successful transformation power*. IRENA Power Sector Transformation Report. Available at: http://www.irena.org/DocumentDownloads/Publications/IRENA_PST_Age_of_Renewable_Power_2015.pdf

*Kickstarter*. (n.d.) Retrieved from: https://www.kickstarter.com/

Klaes, M. (2015). *Crowdfunding for Renewable Energy Projects: European Survey Results.Renewable Energy Crowdfunding Conference*.

Lehner, O. M. (2013). Crowdfunding social ventures: A model and research agenda. *Venture Capital*, *15*(4), 289–311. doi:10.1080/13691066.2013.782624

Massolution. (2015). *The crowdfunding industry report*. Massolution Report. Available at: www.massolution.com

Mccann, D. (2013). Solar Crowdfunding Heats Up. *CFO*, *29*, 9–9.

Mollick, E. (2014). The dynamics of crowdfunding: An exploratory study. *Journal of Business Venturing*, *29*(1), 1–16. doi:10.1016/j.jbusvent.2013.06.005

*Mosaic*. (n.d.). Retrieved from: https://joinmosaic.com/

Pais, I., Peretti, P., & Spinelli, C. (2014). *Crowdfunding. La via collaborativa all'imprenditorialita*. Milan: Egea.

Phimister, E., & Roberts, D. (2012). The Role of Ownership in Determining the Rural Economic Benefits of On-shore Wind Farms. *Journal of Agricultural Economics*, *63*(2), 331–360. doi:10.1111/j.1477-9552.2012.00336.x

Reitsma, H. D. W. (2014). *Accelerating the Energy Transition*. Renewable Energy Crowdfunding Conference. Retrieved from http://www.recrowdfunding.eu/#intro-2015

Renewable Energy Crowdunding Conference. (2014). *Proceedings of the Renewable Energy Crowdfunding Conference*.

Seyfang, G., Hielscher, S., Hargreaves, T., Martiskainen, M., & Smith, A. (2014). A grassroots sustainable energy niche? Reflections on community energy in the UK. *Environmental Innovation and Societal Transitions*, *13*, 21–44. doi:10.1016/j.eist.2014.04.004

*Solar Schools*. (n.d.). Retrieved from: http://www.solarschools.org.uk/

Solarplaza. (2014a). *Renewable Energy Crowdfunding World Map (Update).* Renewable Energy Crowdfunding Conference. Available at: http://www.recrowdfunding.eu/news-updates/2014/8/20/renewable-energy-crowdfunding-world-map

Solarplaza. (2014b). *Top 5 Renewable Energy Crowdfunding Platforms.* Available at: http://www.solarplaza.com/channels/archive/11313/top-5-renewable-energy-crowdfunding-platforms/

Solarplaza. (2015). *Renewable Energy Crowdfunding. Overview and map.* Renewable Energy Crowdfunding Conference. Available at: http://www.recrowdfunding.eu/news-updates/2015/9/14/tracking-renewable-energy-crowdfunding

Spacetec Capital Partners. (2014). Crowdfunding innovative ventures in Europe. the financial ecosystem and regulatory landscape. Prepared for the European Commission, DG Communications Networks.

*Sun Funder*. (n.d.). Retrieved from: http://sunfunder.com/

The Economist. (2013). *How to lose half a trillion euros. Europe's electricity providers face an existential threat.* Available online at: http://www.economist.com/news/briefing/21587782-europes-electricity-providers-face-existential-threat-how-lose-half-trillion-euros

Total. (2015). *Babyloan and Total Team Up to Create the First Ever Crowdfunding Platform Dedicated to Access to Energy.* Total Press Release. Available at: http://www.total.com/en/media/news/press-releases/babyloan-and-total-team-create-first-ever-crowdfunding-platform-dedicated-access-energy

Van Der Schoor, T., & Scholtens, B. (2015). Power to the people: Local community initiatives and the transition to sustainable energy. *Renewable & Sustainable Energy Reviews*, *43*, 666–675. doi:10.1016/j.rser.2014.10.089

Vasileiadou, E., Huijben, J. C. C. M., & Raven, R. P. J. M. (2015). Three is a crowd? Exploring the potential of crowdfunding for renewable energy in the Netherlands. *Journal of Cleaner Production*. doi:10.1016/j.jclepro.2015.06.028

Wadrop, R., Zhang, B., Raghavendra, R., & Gray, M. (2015). Moving mainstream: The European alternative finance benchmarking report. Cambridge Centre for Alternative Finance Report. University of Cambridge.

*We Share Solar*. (n.d.). Retrieved from: http://wesharesolar.com/

World Bank. (2013). *Crowdfunding's potential for the developing world. Info Dev Growing Innovation Report*. Washington, DC: The World Bank.

Yildiz, Ö. (2014). Financing renewable energy infrastructures via financial citizen participation – The case of Germany. *Renewable Energy*, *68*, 677–685. doi:10.1016/j.renene.2014.02.038

## ENDNOTES

1 A revolving fund is a legal 'vehicle' e.g. a company or a cooperative, which funds local energy investments (by which is also repaid) whose returns are further invested in local carbon reduction investments and community activities.

2 The success rate is the percentage of the published projects which get successfully funded.

# Chapter 16
# A Crowd-Funder Value (CFV) Framework for Crowd-Investment:
## A Roadmap for Entrepreneurial Success in the Contemporary Society

**Payal Jain**
*University of Northampton, UK*

## ABSTRACT

*Crowd-funding is used by business and social entrepreneurs to fund projects that impact society in many ways. Although crowd-funders fund and support entrepreneurial projects, stakeholders are less aware of crowd-funder motivations and behaviours towards products/services, which impact on the success of these projects. The purpose of this research, therefore, is two-fold: a) development of crowd-funder value framework for understanding crowd-funder motivations; and b) provision of robust theoretical basis to the construct of 'crowd-funding'. Different social constructs, such as social identity, and symbolic, crowd and relational social capital related to crowd-funding are discussed based on symbolic convergence theory of communication and social identity theory. This framework will act as a roadmap to understand how crowd-funder motivations impact and create value for individuals, organisations and society, and inform how social, environmental and economic value and impact can be maximised through crowd-funding initiatives.*

## INTRODUCTION

Financiers, financial markets and financial innovation play an important role in social and environmental value creation, with recent financial innovations including - crowd-funding - that helps in democratising finance, enables small investors to become venture capitalists and creates social and environmental value along with economic value for the welfare of their communities (Shiller, 2013; Van Staveren, 2013).

DOI: 10.4018/978-1-5225-0568-6.ch016

Crowd-funding is used as the main informal capital resource that can be used as seed capital or early stage capital, where angel capital (i.e. private capital from business individuals) or venture capital may not be available (Bains, Wooder, & Guzman, 2014; Tomczak & Brem, 2013; Valančienė & Jegelevičiūtė, 2014). Entrepreneurs, crowd-funders and online platforms utilise different combinations of human, financial, social, economic and cultural resources to benefit society in various ways. For instance, VolunteerMatch provides human resources for individuals and organisations in need (Gerber & Hui, 2013), thus, filling the gap related to human capital in society.

With the introduction of Jumpstart Our Business Startups (JOBS) Act in the US supporting crowd-funding mechanisms and techniques, crowdfunding is increasingly gaining importance in business and society (Dushnitsky & Maron, 2013; Stemler, 2013). There are many national associations of crowdfunding operating in different countries, such as the UK Crowd-funding Association, National Crowdfunding Association of India, National Crowd-funding Association of Canada, The National Crowd-funding Association (U.S.) and the European Crowd-Funding Network (Kshetri, 2015a). Governments of many countries have implemented crowd-funding regulatory mechanisms that vary from country to country, aim to promote social and commercial entrepreneurship (Harrison, 2013; Kshetri, 2015b; Tomczak & Brem, 2013) and are providing financial support for achieving the bigger social, economic, and environmental aims for benefitting the wider communities, thereby, promoting crowd-funding ecosystems in their societies (Kshetri, 2015b; Valančienė & Jegelevičiūtė, 2014). However, although all forms of crowd-funding are supported by governments, equity and lending based crowd-funding are more strictly regulated than donation and reward based crowd-funding in most countries, such as China, India, US and UK (Cumming & Johan, 2013; Kshetri, 2015b; Lehner, 2013). Nevertheless, crowd-funding market continues to grow leaps and bounds, for instance, the UK market for crowd-funding was estimated at £200 m in 2012 and was expected to rise to £300 m in 2013 (Harrison, 2013).

Recent market trends have shown that crowdfunding is increasingly being considered and utilised as a way to fund social and business ventures by business and social entrepreneurs alike (Groves, 2015a; Lehner, 2013). Crowdfunding is still a new concept, especially in the social entrepreneurship field as compared to the traditional forms of investment models, used by entrepreneurs to fund their businesses and social ventures (Lehner & Nicholls, 2014). Business and social entrepreneurs utilise various different business models to create new ventures that impact society at different levels (Seelos & Mair, 2005; Teece, 2010). These business and social entrepreneurs that lack economic capital are increasingly utilising crowdfunding techniques and methods to fund their projects (Acs, Boardman, & McNeely, 2011; Ibrahim & Verlyantina, 2012a; Jiao, 2011) as there are very few individuals with access to funds that operate informally and provide venture-capital to other entrepreneurs (Van Staveren, 2013). Although there are informal venture capitalists that support crowd-funding, driven by social, economic and/or personal motives, few of them share these investment opportunities with friends and families, despite the fact that the costs and risks associated with crowdfunding projects are relatively low than the traditional capital venture projects (Ordanini, Miceli, Pizzetti, & Parasuraman, 2011; Shiller, 2013). The traditional approaches to raise economic capital through loans are considered costlier and difficult to access (Lehner & Nicholls, 2014; Mollick, 2014), which is encouraging social entrepreneurs to benefit from crowdfunding that allow peer to peer lending, equity, rewards or donation, to fund new social ventures aimed at radical change and positive impact on society (Dushnitsky & Maron, 2013; Kshetri, 2015b). Since micro businesses also face similar restrictions to get access to economic capital due to lack of financial history (Ibrahim & Verlyantina, 2012b), crowdfunding serves as a powerful tool for entrepreneurs contemplating to start-up new projects.

Crowdfunding may not replace old systems of financing projects, but is an attractive financing option for entrepreneurs (Feder, 2013a) as it seeks to cater to the needs of present society in unique ways and enable them to scale-up social impact in society (Lehner & Nicholls, 2014). *Crowd-funding* and *crowd-sourcing* are both closely related social constructs and rely on large crowds to participate in entrepreneurial initiatives by providing finance in small amounts to novice and experienced entrepreneurs. Enablers of crowd-funding activity is similar to that of crowd-sourcing and include 7Ps (Burkett, 2011; Dushnitsky & Maron, 2013; Gerber & Hui, 2013; Kappel, 2009; Lehner, 2013; Ordanini et al., 2011; Prpić, Shukla, Kietzmann, & McCarthy, 2015; Nagalakshmi & Sudhakar, 2013; Zheng, Li, Wu, & Xu, 2014). These are:

- **Pool:** Large number of people required to participate in crowd0funding projects to make it successful.
- **Platform:** Presence of web technology and/or a marketplace.
- **Preference:** Crowd-funders must be motivated and interested in the product/service they would invest in.
- **Product:** Crowd-funders provide feedback regarding the product or service in some way.
- **Promotion:** Crowd-funders actively or passively promote the products/services they invest in.
- **Promise:** Crowd-funders are connected and committed to the product.
- **Payment:** Crowd-funders are paid in some form through social, economic or environmental benefits.

Direct and indirect crowd-funding approaches are used by entrepreneurs when they want to reach out to a set of pre-defined crowd audiences directly through appeal for funds on their own platforms, such as a marketplace or website; and indirectly through intermediary platforms for an open call for funds to general public and crowds (Burkett, 2011), thus, expanding their horizons. Crowd-funding platforms (or service agents) are the organisations that help entrepreneurs or project proponents to gain crowdfunding from the crowd-investors, act as intermediaries and help in balancing the demand and supply of the products/services funded through crowd-investment (Ordanini et al., 2011). Crowdfunding can be understood as a two-sided market on a common platform with the crowd-investors on the subsidy side, an intermediary platform provider that provides a common platform for both investors and investees and the fundraiser on the entrepreneur side (Belleflamme, Lambert, & Schwienbacher, 2014; Kshetri, 2015b; Tomczak & Brem, 2013). Crowd-investors pay no fees to the agent, whilst entrepreneurs pay a particular amount of commission, depending upon the amount of funds raised through their crowd-funding platforms (Gerber & Hui, 2013; Tomczak & Brem, 2013), thus, creating value and network effects through enabling interactions among groups and individuals (Belleflamme et al., 2014), which indicates how valuable the project is to the crowd-investors.

Although crowdfunding is being criticised for a tool to become-rich-quickly investment option and fraudulent projects (Gobble, 2012), many benefits of utilising crowdfunding mechanisms to the crowd-investor include accessing tangible and intangible capitals, such as social capital and economic capital (Dushnitsky & Maron, 2013; Mollick, 2014; Zheng et al., 2014), which are not only essential for the operation of enterprises, but also help them achieve success. Although social capital of entrepreneur is found to be critical to the success of crowd-funded projects (Colombo, Franzoni, & Rossi-Lamastra, 2014; Lehner, 2014), multidimensional view of social capital that includes relational, structural and cognitive social capital will impact crowd-funding cross-culturally even more (Zheng et al., 2014). Structural social capital refers to the social interaction, cognitive social capital refers to the shared visions and views of the members of the group and relational social capital refers to the presence of trust between the members of the group (Nahapiet & Ghoshal, 1998; Tsai & Ghoshal, 1998). Since relational, structural and cognitive dimensions of social capital impact upon the value creation within and outside the firms (Tsai & Ghoshal, 1998) and the interaction of crowd-investors results in crowd capital that can be utilised (Prpić et al., 2015), it is important for individuals involved in crowd-funding to utilise social and crowd capital for not only the benefit of crowd-investors, entrepreneurs and crowd-funding platforms, but also the wider public. Crowd-funding platforms can enable crowd-investors belonging to different social groups to interact and come together for a common cause, which can help leverage social capital for utilisation by entrepreneurs in crowd-funded projects (Zheng et al., 2014). Additionally, since crowd-investors involved in crowdfunded projects contribute through time, effort and labour, it is critical to identify and understand the role of stakeholders involved, including crowd-investors and entrepreneurs, to improve the overall crowd-funding process and experience (Valančienė & Jegelevičiūtė, 2014).

Entrepreneurs, however, lack understanding of the motivation and preferences (social and economic rewards) of their existing and potential backers (Dushnitsky & Maron, 2013), which is why this research is essential and timely. Also, since potential and existing backers struggle with the lack of trust for crowdfunded entrepreneurial activity (Colombo et al., 2014; Gerber & Hui, 2013), it is the responsibility of entrepreneurs and campaigners to create value for all those involved that invest in crowd-funding through their projects in different ways. Since crowd-funding is derived from crowd-sourcing, it is a novice concept, especially in developing countries like India (Gerber & Hui, 2013; T. Nagalakshmi And A. Sudhakar, 2013). This is why a theoretical basis of crowd-funding is important and timely to help a comprehensive exploration of the construct of *crowd-funding* by researchers and practitioners in future. The theoretical basis of crowdfunding research will also enable crowd-investors and entrepreneurs to influence approaches to crowd-funding that can be attuned closely to the unmet needs and wants of the crowd-funders and members of society, and provide effective and viable solutions for future entrepreneurs seeking to engage with crowd-investors. Therefore, as further research on crowd-investors is essential for understanding crowd-investor perceptions and motivations about crowd-funding and how crowd-funding relates to social reality, the section below investigates the theoretical basis upon which the construct of crowd-funding is related to society.

## THEORETICAL UNDERPINNINGS OF THE SOCIAL CONSTRUCT OF CROWD-FUNDING

Crowdfunding research related to understanding the perceptions of crowd-investors is inspired by *social constructivist* approach as it allows multiple realities to co-exist and develop a common understanding of the concept, idea, motivations or perceptions of stakeholders (Benson, 2001; Lindgren & Packendorff, 2009). It helps in co-constructing reality in social, economic, political and cultural contexts, and utilise various qualitative or quantitative methods of research for understanding the reality subjectively (Sánchez & Loredo, 2009). Sociological crowd theories seem to form the basis of crowdfunding including - symbolic convergent theory of communication and social identity theory – is explored and discussed below in relation to crowdfunding. Convergence culture that refers to the participation of the users in the production and distribution of the product, results in collective intelligence where diverse members of a group help in the creation of value (Aitamurto, 2011). Since collective intelligence is present in the collective nature of the public or crowd – it applies to both the concepts: 'crowd-sourcing' and 'crowd-funding'(Jian & Shin, 2014; Prpić et al., 2015; T. Nagalakshmi And A. Sudhakar, 2013) as crowds are considered to be the source of crowd intelligence (Aitamurto, 2011; Lehner & Nicholls, 2014; Rossi, 2014). Convergent theories are based on collective efforts and collective input from a group of people that share a common viewpoint (Allport, 1962; Prpić et al., 2015), which is why the symbolic convergent theory (SCT) of communication is closely related to crowd-funding and its mechanisms.

SCT advocates that the group works in tandem to co-create and co-construct reality that impacts upon the behaviour, beliefs, attitudes, motivations and perceptions of the group and stakeholders involved (Bormann, Cragan, & Shields, 2003). SCT utilises symbols and language to facilitate interaction and understanding of group dynamics, whist impacting upon the communication flow within the group, between groups and outside the boundaries of groups in the public, thus enabling shared meanings to emerge that influences the perception and actions of individuals in the group (Allport, 1962; Dougherty, Mobley, & Smith, 2010). Sharing information between members also leads to the development of symbolic capital, resulting in identity formation of the individual (McArthur, 2015) that influences the perceptions of stakeholders about themselves. Symbolic and cultural capital creates economic value through resource exchange in crowd-funded projects, whilst social capital of entrepreneurs contributes towards the success of the crowd-funded projects (Lehner, 2014), thus, indicating that value creation can be enhanced through collective intelligence and effective utilisation of capitals and resources available to the entrepreneurs. Symbolic exchange between the members of community result in the creation of symbolic value or symbolic realities (Bormann et al., 2003; Ordanini et al., 2011) that help crowd-funders to relate to their preferred brand. Furthermore, crowd-investors that tend to associate themselves with a particular brand that offers them symbolic benefits, help them share common values and/or beliefs, thus, leading them to share identity with the brand (Ordanini et al., 2011). Brand-community members also create shared value, irrespective of the financial support from the users, and are particularly motivated by the learning, identification, and status that identify the members with the brand they associate with (Ordanini et al., 2011). This indicates that the symbolic convergent theory of communication, underpins the construct of 'crowd-funding' as crowd-investors tend to bond with the brand that it supports, and are, therefore, influenced by the its decisions.

In addition to symbolic convergent theory, social identity theory (SIT) bears important implications for crowd-funding. Identity, social or professional, is crucial for crowd-investors and the proponents of the projects alike, and crowdfunding process is contributing to the perceptions and building of social and professional identity (Aitamurto, 2011). Social identity refers to the perception of self in relation to other members of the group to which they may already belong or want to belong to, such as social and cultural groups, for instance - nationality, class, family, culture, and religion - that influence group behaviour and the values of the individuals and the groups (Hauge, 2006). Although the collective effort of the group is important, it is the social identity of individuals in the group that influences group behaviours and these groups possess the potential to become powerful entities, which is why social identity achievement for these groups to gain power through their collectiveness is crucial for achieving social change in society (Jian & Shin, 2014; Reicher, Haslam, Spears, & Reynolds, 2012). Additionally, these crowd-investors tend to help the brand in all its stages: design, development distribution and consumption (Ordanini et al., 2011), thus, highlighting and manifesting the ownership crowd-investors possess when associated with particular projects that depict their social identities. This indicates that *identity* is central to crowd-investors who associate themselves with a particular brand, image or cause and want to become a part of the project, and therefore, end up supporting these projects (Gerber & Hui, 2013; Zhou, Wu, Zhang,

& Xu, 2013). It is, therefore, important to explore the role played by social identity in crowd-investor motivations as it impacts upon the investment choices of crowd-funders.

Social identity Theory (SIT) seems to provide a robust basis for crowd-funder value framework development and facilitates our understanding of the interaction of group members. SIT also forms the basis of cognitive and motivational forces that exists in and between the groups, and discusses the group dynamics and individual self-organisation in a way that joins members of the group with an element of empathy playing a role in and between groups (DeTurk, 2001; Reicher et al., 2012). Social identity of the crowd-funders, therefore, may influence attitudes, behaviours and values of the group and may reflect in the perceptions and attitudes of the crowd-investors not only in a particular group, but also other crowd-funded projects and the public (Jian & Shin, 2014). For instance, crowdfunded projects, such as Sella Band and Kapipal, crowd-investors shared identity in with the projects or proponents in a way that motivated them to invest or donate money for social, personal or economic reasons (Ordanini et al., 2011). According to Porter and Kramer (2006, 2011) shared value is created when the venture or activity benefits both the company and society. Shared value, therefore, refers to the value created by the company when it comes in contact with its stakeholder and the environment (Manning, 2010; Porter, Hills, Pfitzer, Patscheke, & Hawkins, 2012; Porter & Kramer, 2006). This value could be social, financial or environmental that seeks to benefit the company as well as the wider society (Weitzman, 1986). Shared values help in reducing social exclusion and allow heterogeneous members of society to interact i.e. utilise social capital (Strike, 1999), which is confirmed by Putnam (1993) that social capital helps in reducing social inequalities and social exclusion existing within the social structures in society (Manning, 2010; Putnam, 1993). Therefore, social constructs that influence crowd-funder motivations, such as identity and capitals (social, relational, symbolic, and crowd) are inter-related, impact the entrepreneurial activity and are at the very heart of value creation (social, cultural, symbolic, and economic). This indicates that we must explore motivations of crowd-funders in detail to decipher the underlying phenomenon that drives crowd-funding.

## MOTIVATIONS OF CROWD-FUNDERS

The role of crowd-funders in improving socio-economic conditions of society is paramount as it is up to the crowd-investors to accord importance, provide development and distribution options and finally, facilitate consumption for the creation of value for organisations, crowd-funding platforms, crowd-investors, entrepreneurs and the public (Ordanini et al., 2011). Improving end-user experience and sustaining the crowd-funded projects are the two main objectives that can be achieved by understanding the motivations of crowd-investors (Gerber & Hui, 2013). Crowd-funded research has better acceptance rates and improved perception of the product in the market due to the level of participation of crowd-investors at all the stages: planning, development and distribution of products (Rossi, 2014). Motivations of crowd-funders are also dependent upon the field of activity in which the project is being crowd-funded, this includes journalism, where fun, belief in freedom of content, self-esteem, understanding, altruism and image are more important for the members of the community (Jian & Shin, 2014). Although, it has been found that crowd-funder motivations are not static and may change over time (Jian & Shin, 2014), it is essential for entrepreneurs to understand the motivations, values, and preferences of their crowd-funders to attract, engage and positively contribute to all the activities related to crowd-funding so as to ensure success of crowd-funded projects (Dushnitsky & Maron, 2013; Vezzoli, Ceschin, Diehl, & Kohtala, 2015).

There are a number of reasons that motivate crowd-funders to invest in crowd-funding projects. The quality of rewards influence crowd-funders to invest in particular projects whilst ignoring others (Hörisch, 2015). Social and environmental value creation are also a deciding factor for some crowd-funders that are interested in community and social cohesion (Perilleux, 2015). Online philanthropic behaviour influences crowd-funders to donate due to the feelings of empathy, sympathy, network and community building, including social identity due to the charitable cause of crowd-funding projects. Online consumer behaviour, such as the ease of online transfer of funds also motivate crowd-funders to donate funds more easily (Gerber & Hui, 2013; Rossi, 2014). Entrepreneurs that possess good credit score are chosen by crowd-funders to invest money in for their projects (Gerber & Hui, 2013). Online peer production, where the crowd-funders increase their interaction or learning process by coming in contact with the members of free online community, also impacts upon the crowd-funders choice of investment in particular projects (Cordova, Dolci, & Gianfrate, 2013; Gerber & Hui, 2013). Increased socialisation, new learning processes, public attention, knowledge and information sharing, social approval and maximisation of community benefit through cooperation result in shared value creation (Bröring et al., 2015; Gerber & Hui, 2013; Rossi, 2014). Additionally, as the motivations of crowd-investor include: reward-collection, self-promotion, social benefits, direct compensation, learning, support others and causes, involvement in community projects and philanthropy (Bröring et al., 2015; Gerber & Hui, 2013; Ordanini et al., 2011), these are also the primary reasons for existing and potential crowd-investors to get involved in projects. Becoming a part of project is a personal incentive for some crowd-investors that facilitate shared beliefs and values exhibited by campaigns (Feder, 2013b). Thus, there are many inter-related factors that influence crowd-investor perceptions and behaviours.

The role of social media, public relations, bloggers, news channels in creating awareness and raising interest can never be less emphasised. Customers that become crowd-funders tend to play different roles depending upon their involvement, such as participator, investor, information sources, social networker, entrepreneur, promoter, agents and financer (Ordanini et al., 2011). Likewise, the role of crowd-investors is no longer limited to that of investors as they also act as promotion and marketing managers for their investments, which is why they engage in virtual communities for collective value creation (Rossi, 2014). As knowledge transfer and knowledge-share between and within groups is facilitated utilising social capital in crowd-funding based social platforms, such as Kickstarter, or third-party social platforms, such as Facebook (Zheng et al., 2014), entrepreneurs utilise social media platforms, such as facebook, twitter, and internet; and utilise media tools and techniques to promote their investments and become successful (Gerber & Hui, 2013). It must, therefore, be emphasised that these social media platforms adopt similar approach to entrepreneurship, which is to understand behaviours, practices and beliefs of their users (Yoo, Choi, Choi, & Rho, 2014) that enable their businesses to run effectively and successfully, therefore, allowing scaling up social and environmental impact in society. Entrepreneurs utilise tools such as project monitoring and tutorials, make audios and videos, and exploit social media platforms such as Youtube and Facebook to reach out to their potential investors (Feder, 2013b; Gerber & Hui, 2013; Ibrahim & Verlyantina, 2012b). It is, therefore, crucial for entrepreneurs to not only gain an in-depth understanding of what underpins shared economy, but also make efforts to undertake strategy-driven marketing campaigns for success of the projects (Dushnitsky & Maron, 2013).

Many strategies are suggested to motivate crowd-funders to invest, and to think strategically and take strategic actions to attract them (Hörisch, 2015). This may include designing and implementing a marketing campaign, such as a compelling video campaign that emphasises the rewards, perks, financial and social aims to excite crowd-funders to invest and attach themselves with the team, product or idea

(Dushnitsky & Maron, 2013). In addition, project narratives must be published on the project website, along with using text messaging to engage with their crowd-investors. Since quality of the project and geography also play a role in the success of the project (Mollick, 2014), entrepreneurs must ensure and be more sensitive about the challenges faced by the crowd-investors in trusting and lending money to crowdfunded projects online. The choice of a large number of entrepreneurs, between running a project as for-profit or non-profit entrepreneurial activity, also affect the likelihood of reaching funding goals (Pitschner & Pitschner-Finn, 2014) as Belleflamme et al (2013) found that projects established as non-profit entities were more successful. This resulted in affecting potential shared value (social, economic and environmental) creation and impact in society. Engaging with crowd-funders at all level of crowd-funding project is key and will not only encourage knowledge and mind share, but also lead to the development of pool base and spark communication of common interest, thus, building momentum as early as possible (Bröring et al., 2015; Dushnitsky & Maron, 2013). As crowd-investor motivations are reflected in the typology of crowd-funding models that these investors choose to adopt, select and invest in (Jian & Shin, 2014), various crowd-funding models, are therefore, elaborated discussed in detail below.

## MODELS OF CROWD-FUNDING

Crowd-funding models have been used in diverse fields, including entertainment, social entrepreneurship (Lehner, 2013), environment (Hörisch, 2015), renewable energy (Groves, 2015a; Vasileiadou, Huijben, & Raven, 2015), entertainment, music, fashion and recording industry (Kappel, 2009; Tomczak & Brem, 2013), and social finance (Perilleux, 2015). One of the fastest growing crowd-funded fields include real estate, where $1 bn has been invested in 2014 (Cooper, 2015). Crowd-funding models can be divided into two basic forms: ex post facto and ex ante: in ex post facto - crowd-investors are offered a product/service in exchange of funds at the time of investment, whilst in ex ante - the funds offered by the crowd-investors are used for the realisation of the product/service - so that a particular outcome can be achieved (Kappel, 2009). As most of the crowdfunding is ex ante, this model of funding is becoming increasingly popular as it allows investors to become involved in crowd-funded projects early on and receive monetary benefit in return, as is the case with US and European entertainment, fashion, music and recording industry in which fans assume the role of financiers of their artists so that more records can be produced and sold in the market (Kappel, 2009; Tomczak & Brem, 2013). In ex ante, crowd-investors not only play the role of funders but also entrepreneurs that utilise their internal and external social capital and networks to increase the sale of the products in order to increase the impact of their investments (Colombo et al., 2014). For instance, Kickstarter, a crowd-funding platform, displays information on the number of crowd-investors backing up a particular project and the amount of money collected, thus amassing $1 billion in March 2014, thereby, encouraging other investors to support crowd-funded projects more (Colombo et al., 2014; Meer, 2014).

There are two types of ex ante crowdfunding models being utilised in the music and recording industry: the Betting model and the Investment model. In the Betting model, fans receive the rate of return depending upon the album sales that leads them to play the role of advertiser and promoter of the records/products they support, model presently being utilised by the UK based company: Slicethepie (Kappel, 2009; Ordanini et al., 2011). In the Investment model, profits generated through the sales of the records are split in a particular ratio between the investors, artists and the company, model presently being utilised by a German company: SellaBond and a UK company: Bandstocks (Kappel, 2009; Ordanini et al., 2011). These models are driven by moral and social character that fans exhibit for their artists as much as it is driven by profit, and there are many state and federal laws that regulate the functioning of these companies (Kappel, 2009)., which is why it is important to understand the value these models offer for organisations, talented individuals (such as artists and writers) and the members of society. Therefore, different models of crowd-funding are adopted by entrepreneurs to fund their projects and offer value propositions to crowd-funders, crowd-funding platforms and the general public (Ibrahim & Verlyantina, 2012b; Porter & Kramer, 2011).

Entrepreneurs that are interested in additional funding or alternative funding consider crowd-funding to raise capital (Rossi, 2014; Valančienė & Jegelevičiūtė, 2014). Crowd-investors not only integrate resources available to them through their networks to create value, but also act as agents to realise projects that would otherwise either not be activated or carried out due to lack of social or economic value (Ordanini et al., 2011). Since there is ambiguity and inconsistency in the way crowd-funding models are being understood and adopted by entrepreneurs for funding their projects (Dushnitsky & Maron, 2013; Harrison, 2013; Hörisch, 2015; Lehner, 2013), therefore, this section below intends to provide a broader understanding of the crowd-funding models presently being used in society:

- **Equity Investment:** Crowd-funders are motivated by social, financial and intrinsic aims and expect a financial return on investment over time (Cordova et al., 2013; Dushnitsky & Maron, 2013). This is considered as real investment with high risks involved and is used to generate additional money for projects (Rossi, 2014). This model is followed by many entrepreneurs looking to invest money in entrepreneurial projects. For instance, Triodos, a renewable company, is owned by 5,000 shareholders and boosted capacity generation by 50% utilises crowd-funding for raising equity (Earley, 2015). Similarly, Trampoline, a finance company, aims to commercialise a software program: SONAR – a product of the London based technology company- that enhances employee's skills by mapping relationships inside the organisation. Trampoline invite bigger investments from individuals with higher net worth, and those who invest become shareholders in the company (Ordanini et al., 2011). Similarly, Crowdcube, a crowd-funding platform for venture capitalists, run a funding campaign to achieve economic value for itself and its crowd-funders. It is regulated by the UK Financial Conduct Authority and is one of the leading crowd-funding platforms in the UK (Parker, 2014; Tomczak & Brem, 2013). This type of funding managed to raise £2.9m in 2012, £ 28m in 2013, and expected to have raised £84m in 2014 in the UK (Baeck, Collins, & Zhang, 2014; Gill, 2015). Also, SellaBand, a German company, allows music artists to record a professional music album funded by the SellaBand community once an artist reaches a minimum investment target(Belleflamme, Lambert, & Schwienbacher, 2013; Burkett, 2011). Sellaband crowd-funders receive some return on their investment in the form of revenue, and to this date, more than 40 artists have recorded their music albums and receive a share of the revenues and a copy of the limited edition of the CD (Harrison, 2013; Ordanini et al., 2011).
- **Peer-to-Peer (P2P) Lending:** Crowd-funders are motivated by financial, social and intrinsic aims, and funds are provided in the form of loan to the entrepreneurs for funding their projects. Crowd-funders may expect a rate of interest on their investments depending upon their motivations (Dushnitsky & Maron, 2013; Rossi, 2014). This type of lending is flexible and may be equity, debt, profit sharing or hybrid (Macaulay, 2015). P2P helps in increasing the liquidity into the market, is regulated by the British Financial Conduct Authority and has received government support in the form of £30 m of investment from the Department of Business, Innovation and

Skills (Harrison, 2013). Crowd-funded projects are likely to attract more funding from crowd-investors when they possess good credit scores (Gerber & Hui, 2013; Macaulay, 2015). For instance, Funding circle is one of the leading peer-to-peer lending platforms that supports small and medium sized enterprises (SMEs) in the UK (Cooper, 2015; Macaulay, 2015).

- **Donation:** Crowd-funders are interested in funding these projects as they facilitate increased social participation, and these projects are mostly philanthropic in nature (Lehner, 2013). Crowd-funders are not paid back in equity, perks or monetarily that leads entrepreneurs to feel a sense of responsibility towards crowd-investors (Aitamurto, 2011; Dushnitsky & Maron, 2013). These types of crowd-funding is seen in the traditional non-profit and non-governmental organisations (Lehner, 2013). Donations are based principally on altruistic motivations and intrinsic values than the instrumental ones leading to social change (Dushnitsky & Maron, 2013; Rossi, 2014), allowing the relationship between the donor and the entrepreneur to strengthen (Aitamurto, 2011). An example of donation based crowdfunding is: Kapipal, an Italian company that specialised in Web 2.0 and that allows individuals to act as Kapipalists and investors to act as contributors. Kapiplists create a public or private fundraising page for personal fundraising that includes raising money for medical treatment, birthday gifts and wedding lists, and every contribution from contributors is considered a donation (Ordanini et al., 2011). Besides, donation based crowd-funding can be contribution of time in addition to economic capital as well. For instance, enviroCar, a tool that allows sensor data to be collected from the vehicles of the city in order to reduce traffic related issues, such as pollution, noise and improper traffic lights, was crowdfunded by volunteers and has been very successful (Bröring et al., 2015).
- **Crowd-Sponsoring:** Is used to re-advertise a product/service of an organisation and is used to fund non-profit ventures that has no economic return (Tomczak & Brem, 2013). It is different than donation as in crowd-sponsoring, nothing is expected in return. This type of crowd-funding is in its initial stages and project sponsors are interested in funding individuals that possess expertise, skills in the project and have similar vision as regards the potential of the crowd-funded projects (Zheng et al., 2014). Although project sponsors receive rewards and perks, they not only donate funds, but are also involved in all the stages of project (Bröring et al., 2015; Vasileiadou et al., 2015). For instance, Solar Impulse project, is a crowd-sponsored project that aims to make solar powered airplanes, in which sponsors of the project include: Deutsche Bank and the general public (Tomczak & Brem, 2013).
- **Reward (Also Called Pre-Purchase or Pre-Selling):** Crowd-funders pre-purchase a product/service and are motivated by rewards, perks or social aims that follows (Dushnitsky & Maron, 2013). It does not provide a return on investment, or share in business and projects. They, however, can receive a nominal return in the form of the product/service that allows demand to be generated and regulated, and provides market validation, thus reducing the risk of product failure (Harrison, 2013; Rossi, 2014). Some projects also provide voting rights to the crowd-funders under this model (Zheng et al., 2014).
- **Collecting from Friends, Family, and Followers (CFFF) (Also Called 5Fs: Founders, Family, Friends, Fools, and Fans):** Crowd-funders raise money from close family members, friends and social circles, thus, lowering the initial rate of interest that is required in the start-up phase (Harrison, 2013). Fools in 5F's refers to individuals taking risk with their investment (Dushnitsky & Maron, 2013; Harrison, 2013). The presence of trust (i.e. relational social capital) is found to be key in this type of crowd-funding model and particularly seen in countries where family-friends

bond is closely tied, such as China, (Zheng et al., 2014). Kickstarter, a platform based on crowd-funding, has raised more than $1 bn from crowd-funding emphasises that money raised from CFFF has substantially contributed to the success of many crowd-funded projects (Dushnitsky & Maron, 2013).

- **Social Crowd-Funding Finance:** This is a fairly new type of crowd-funding model, where crowd-funders are particularly interested in social, cultural and environmental value creation, scaling social impact investments and provide interest free loans to social entrepreneurs (Lehner & Nicholls, 2014). For instance, Trillion Fund, is the biggest UK' social crowd-funding finance service, providing support to social entrepreneurs in the renewable energy sector in the world (Groves, 2015b). Similarly, Socrowd is a Belgian based crowd-funding platform, and have provided its membership to 700 people. This organisation is considered a social enterprise in itself that is active in social finance (Perilleux, 2015). Social crowd-funding model helps in creating direct and open relationship between the lenders and borrowers as both are aware of community and social benefits that arise from the crowd-funded projects, and therefore, are likely to fulfil their respective responsibilities of a crowd-investor and crowd-investee (Aitamurto, 2011; Perilleux, 2015).

Since, many organisations employ hybrid models of crowd-funding, such as Razoo.com that focuses on US public schools and combine donation based crowd-funding model with peer-to-peer lending crowd-funding model (Dushnitsky & Maron, 2013), it can be argued that crowd-funding, therefore, allows different combinations of models to co-exist in entrepreneurial projects and facilitate economic and social value creation. Moreover, since crowd-investors may directly enable the production and sale of products/services, crowd-investors impact upon every stage of the design, development and distribution of crowdfunded products and/or services (Ordanini et al., 2011), thus, ensuring greater flexibility on the participation in the crowd-funding process. Crowd-investors may involve themselves actively and passively in the crowd-funded projects depending upon their interest and the level of involvement in each project. Active crowd-investors are very involved in projects they invest in, and may engage in business activities, such as product development and testing, providing suggestions and feedback on the product, and promotion of the product/service (Rossi, 2014; Zheng et al., 2014), thereby, impacting upon the outcome and success of the crowd-funded projects. Passive crowd-investors may be interested in the rewards and recognition that their investments provide, but are not much involved with the entrepreneurs and/or the sale, purchase or promotion of the product (Zheng et al., 2014). Horisch (2015) suggests that active crowd-funding includes equity based crowdfunding, whilst passive crowd-funding includes reward and lending based crowd-funding models. Upon the success of the crowd-funded projects, entrepreneurs pay a platform usage fee and payment processing fee (Colombo et al., 2014; Dault, 2014; Gerber & Hui, 2013). Crowd-funding platforms utilise two types of models during the facilitation of crowd-funding (Colombo et al., 2014; Dault, 2014; Gerber & Hui, 2013), these are:

- **All-or-Nothing:** All the funds generated are returned to crowd-investors if the project goals is not met, for instance, used by the crowd-funding platform, Kickstarter.
- **All-or-More:** All the funds that are generated are kept by the entrepreneurs, irrespective of the success of the project, such as the crowd-funding platform RocketHub.

## A CROWD-FUNDER VALUE THEORETICAL FRAMEWORK

This section seeks to propose a Crowd-Funder Value (CFV) theoretical framework that is grounded in the sociological crowd theories, and will provide a theoretical understanding of crowd-investor perspective. Crowd-funder value framework, primarily, aims to increase our understanding of how crowd theories are related to crowd-funders; and how research on crowd-investors may impact future entrepreneurial activity and research directly and indirectly. It is argued that such an integrated framework is crucial for primary in-depth understanding of crowd-investor motivations and perceptions for funding projects that will help in setting up and running business and social ventures. Stakeholders will also gain an insight into sociological constructs related to crowd-funding. As crowd-funding is itself embedded within the society as much as business, this framework will add to the existing interdisciplinary evidence based by combining sociology and business in this research. Furthermore, since there is a dearth of research on the perceptions of crowd-funders and the factors that influence their selection of projects, whether depending upon crowd-funding model or crowd-funding platform, at individual level, this comprehensive framework will provide an in-depth understanding of crowd-funder influences on the overall activity of crowd-funding and how it impacts upon the overall well-being of society.

Figure 1 demonstrates a holistic overview of the crowd-funder value framework and intends to show the relationships between values and motivations of crowd-investors, thus, highlighting how crowd-investor motivations contribute towards the creation of societal values (i.e. social, environmental, economic and cultural) in society. Social media also plays a significant role in influencing the motivations of crowd-investors to invest in crowd-funding projects at initial, intermediary and later stages of crowd-funding experience of crowd-investors (Gerber & Hui, 2013; Ibrahim & Verlyantina, 2012a; Padgett, 2014). New crowd-investors are influenced by the formal and informal communication channels to understand and invest in crowd-funding projects, whilst in later stages, the former experience of crowd-investor's guide them to invest in some projects and avoiding others, depending upon their motivations at that

*Figure 1. A holistic crowd-funder value (CFV) theoretical framework*

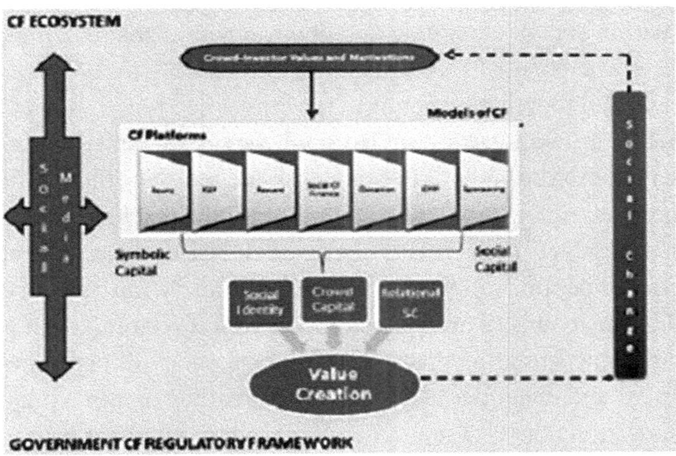

time (Colombo et al., 2014; Zachary & Mishra, 2013; Zheng et al., 2014). Crowd-investor motivations, therefore, are susceptible to change according to their experiences, knowledge and information that they gain throughout their investment process (Lehner, 2014; Yoo et al., 2014).

The motivations and preferences of a crowd-investor guide him/her to select a crowd-funding model that appeals to his/her intrinsic (such as moral and philanthropic) and extrinsic values (such as economic and material) and is influenced by either his/her own earlier experiences of crowdfunding or through other communication channels (i.e. social media or friends and family) (Dushnitsky & Maron, 2013; Valančienė & Jegelevičiūtė, 2014). It can be argued that social capital and symbolic capital are essential to the development of crowd capital, social identity and relational social capital (that refers to the trust within and between groups). Social identity is formed by individuals in a group, in this case crowd-funding projects, when crowd-investors interact with each other irrespective of the social or cultural groups they belong to, which is why crowd-funding platforms are helpful in bridging the communication gap between crowd-investors belonging to heterogeneous groups (Hauge, 2006; Heere et al., 2013; Reicher et al., 2012; Tomczak & Brem, 2013). Social capital allows interaction to take place online or offline between crowd-funding members that result in developing crowd capital and relational social capital at the same time (Frydrych, Bock, Kinder, & Koeck, 2014; Gerber & Hui, 2013; Yoo et al., 2014). Symbolic capital also contributes through the choice of language and non-verbal communication that helps in the development of crowd capital and relational social capital at the same time as well (Heere et al., 2013; McArthur, 2015; Nakassis & Searle, 2013). Although unpleasant communication between crowd-funded members could affect their relationship, possessing similar aim and vision to invest in a particular project may help overcome the challenges that inhibit crowd-funders to co-create value. Since it has been found that crowd-funders and entrepreneurs co-create value (Colombo et al., 2014; Gummerus, 2013), crowd-funders also play a central role in the value creation process and decision-making process related to crowd-funding activity, in addition to entrepreneurs (Ordanini et al., 2011). It can, therefore, be contested that the elements that enable crowd-investors to relate to crowd-funding and crowd-funding projects include social identity, and the presence of crowd capital, relational social capital and symbolic capital, that impact upon the selection of a particular crowd-funding project by crowd-investors to invest in. The presence of trust (i.e. relational social capital) between the crowd-funders, therefore, will lead to the change in the perceptions, attitudes, values, motivations and preferences of the crowd-funders as well as other stakeholders, such as entrepreneurs, sponsors and crowd-funding platforms, thereby, resulting in the value creation within social, economic and environmental spheres, thus, a step forward towards enabling social change in society.

Crowd-funding is a useful tool that responds to the unmet needs and wants of the public by providing investment opportunities to crowd-investors and through the support and promotion of the entrepreneurial projects. This framework, therefore, intends to emphasise and reinforce the creation of shared value among crowd-funding stakeholders. Since value creation is essential for achieving sustainability in social and economic systems in society (Acs et al., 2011; Gummerus, 2013), the contribution of crowd-funding to achieving value creation for organisations, individuals and society can be seen through multiple projects being undertaken within the realms of crowd-funding in varied sectors. It can, therefore, be posited that stakeholders involved in crowd-funding processes may enhance building of positive social identity, social capital, crowd capital and engender trust through providing support to individual crowd-investors at the project and organisational levels (such as crowd-funding platform level). In addition to providing support, government may help crowd-funders access information that may be useful and essential for crowd-investors in making wise investment decisions at individual level. This framework, therefore, simplifies our understanding of preferences and motivations of crowd-funders, so that efforts are made by entrepreneurs, government, policy-makers, investors, sponsors, scholars and crowd-investors themselves to include crowd-investors at every stage of the product development and testing process, so that shared value is generated throughout the crowd-funding process, thus, bringing crowd-funders closer to the crowd-funded products and projects.

## A Classic Way to Maximise Social Value and Impact through Crowd-Funding

Social enterprises lead in the creation of social value and work towards identifying and tackling social problems of society in many ways (J. Austin, Stevenson, & Wei-skillern, 2006). The primary aim of social enterprises is social and environmental value creation and maximisation, whilst their secondary aim is profit maximisation in order to reinvest their profits back into solving social problems of the community (Desai & Ács, 2007; Jiao, 2011). Social enterprises strive to achieve triple bottom line i.e. the creation of social, environmental and economic value for the welfare of the community (Cornelius & Wallace, 2013; Peloza & Shang, 2011; Ruebottom, 2011). This is why social entrepreneurs are considered as social change makers, due to their capability of creating social and environmental value despite resource challenges (J. Austin et al., 2006; Ridley-Duff & Bull, 2011). The success of these enterprises are attributable to their presence in public, private and third sectors and within their local communities (Acs et al., 2011; J. Austin et al., 2006; Dacanay, 2012). Social entrepreneurs are as much as intrinsically motivated as extrinsically to achieve social aims that they continuously look for opportunities that help them become self-sustainable in the long term (Acs et al., 2011; J. E. Austin & Seitanidi, 2012; Jiao, 2011; Porter & Kramer, 2006; Ruebottom, 2011).

Utilising social entrepreneurship model that advocates creation of enterprises primarily for the benefit of society, societal value could not only be created, but possibly maximised. Motivations of crowd-investors to invest in social enterprises vary from social, psychological, cognitive to financial, and may contribute towards the success of social enterprises (Lehner, 2013). Crowd-funding models are being used by social enterprises, some of which provide social finance to entrepreneurs for the creation of social and environmental value in society (Perilleux, 2015). For instance, Trillion Fund, a fastest growing social enterprise in the energy sector, develops projects on renewable energy and utilises crowd-funding to create economic, social and environmental value for individuals, organisations and society as a whole (Earley, 2015; Groves, 2015a). Since social enterprises have a huge potential to positively impact society, it can, therefore, be argued that if social entrepreneurship and crowd-funding model, such as social crowd-funding finance, are combined together, it may be possible to maximise and add societal value for the overall development and welfare of society.

Additionally, the framework discussed above also demonstrated that by adopting social crowd-funding finance model, it may be possible to engage community members and help them become crowd-funders or entrepreneurs, leading to increased community and social cohesion. Moreover, as crowd-funders, the members of society may not only propose and consume a product, but may become the lead users, thereby, leading the product innovation through their expertise and timely anticipation and identification of the future consumer needs (Ordanini et al., 2011). Most importantly, by scaling up investment in social crowd-funding finance model, the power to run and change a society will be transferred back to the crowd-investors and the public that contribute in various crowd-funding models (debt, equity, donation, reward, sponsoring, CFFF and social crowd-funding finance) directly and indirectly in the first place, thus, leading to the possibility of democratic social change in society. The potential of crowd-funding as a powerful tool to systematic social change through community involvement and engagement must not, therefore, be underestimated.

## FUTURE RESEARCH AND RECOMMENDATIONS

There are many gaps in the literature related to crowdfunding, some of these gaps are mentioned below. Firstly, in order to understand the motivations of a crowd-investor, it is essential to understand the construct of crowd-funding through the application of social constructionist paradigmatic approach to research. Since, the crowd-investors may possess distinct understanding of crowdfunding, and its application, and may possess common values that these projects exhibit, social constructivism approach to future research will contribute towards exploration, and prediction of motivation, preferences and behaviours of crowd-investors (Hennink, Hutter, & Bailey, 2011). Secondly, there is a need to conduct an in-depth study on social and economic conditions that facilitate or hinder crowd-funding activities to take place in order to ease the investment process for both the crowd-funders and the entrepreneurs in the present society. Thirdly, investigation into the impact of social identity and symbolic capital on crowd-funding and crowd-funder motivations in a real world setting. This research will be useful for understanding crowd-funders in a particular society. Fourthly, empirical exploration and research on the theoretical basis of symbolic convergence theory of communication and social identity theory in relation to crowd-funding will help us expand our knowledge and information base to understand the perceptions and motivations of crowd-investors in detail. Most importantly, further empirical research into the social crowd-funding finance model and social entrepreneurship related to crowd-funding will increase our limited understanding of the social, economic and environmental impact and the factors that influence or hinder these crowd-funded social projects.

## CONCLUSION

Crowd-funders play a central role in enabling an entrepreneurial project to be funded, supported and promoted in the community. From the product development and testing through to distribution, crowd-funders are involved in all the key stages actively and passively. Therefore, it is imperative for all stakeholders, including entrepreneurs, sponsors, crowd-funding platforms, and government to understand the perceptions and motivations of crowd-funders to be able to increase their involvement, engagement and responsibly respond to their needs and wants. The crowd-funder value framework posited the need to reflect and consider the motivations and perceptions of crowd-funders during crowd-funding projects at all stages due to the impact crowd-funders have on entrepreneurial activity and the value of their predictions of future societal needs. This research emphasises the need to understand theoretical basis of crowd-funding research by elaborating on the social identity theory and symbolic convergence theory of communication. These theories are not only inter-related but also prove an in-depth understanding of the process of crowd-funder engagement or disengagement. This research argues that social capital and symbolic capital are essential for the creation of crowd capital and social capital, along with building of relational social capital (i.e. trust), which leads to the creation of societal values, such as social, economic, environmental and culture for the benefit of individuals, organisations and society, thus, leading to positive social change.

## REFERENCES


Acs, Z. J., Boardman, M. C., & McNeely, C. L. (2011). The social value of productive entrepreneurship. *Small Business Economics*, *40*(3), 785–796. doi:10.1007/s11187-011-9396-6

Aitamurto, T. (2011). The Impact of Crowdfunding on Journalism. *Journalism Practice*, *5*(4), 429–445. doi:10.1080/17512786.2010.551018

Allport, F. H. (1962). A structuronomic conception of behavior: Individual and collective. *Journal of Abnormal and Social Psychology*, *64*(1), 3–30. doi:10.1037/h0043563 PMID:13860640

Austin, J., Stevenson, H., & Wei-skillern, J. (2006). Social and Commercial Entrepreneurship : Same, Different, or Both? *Entrepreneurship Theory and Practice*, *3*(1), 1–22. doi:10.1111/j.1540-6520.2006.00107.x

Austin, J. E., & Seitanidi, M. M.Value Creation Spectrum and Collaboration Stages. (2012). Collaborative Value Creation: A Review of Partnering Between Nonprofits and Businesses: Part I. Value Creation Spectrum and Collaboration Stages. *Nonprofit and Voluntary Sector Quarterly*, *41*(5), 726–758. doi:10.1177/0899764012450777

Baeck, P., Collins, L., & Zhang, B. (2014). *Understanding Alternative Finance*. London.

Bains, W., Wooder, S., & Guzman, D. R. M. (2014). Funding biotech start-ups in a post-VC world. *Journal of Commercial Biotechnology*, *20*(1), 10–27. doi:10.5912/jcb628

Belleflamme, P., Lambert, T., & Schwienbacher, A. (2013). Individual crowdfunding practices. *Venture Capital*, *15*(4), 313–333. doi:10.1080/13691066.2013.785151

Belleflamme, P., Lambert, T., & Schwienbacher, A. (2014). Crowdfunding: Tapping the right crowd. *Journal of Business Venturing*, *29*(5), 585–609. doi:10.1016/j.jbusvent.2013.07.003

Benson, G. D. (2001). Science Education from a Social Constructivist Position : A Worldview. *Studies in Philosophy and Education*, *20*(5), 443–452. doi:10.1023/A:1012035404449

Bormann, E. G., Cragan, J. F., & Shields, D. C. (2003). Defending symbolic convergence theory from an imaginary Gunn. *The Quarterly Journal of Speech*, *89*(4), 366–372. doi:10.1080/0033563032000160990

Bröring, A., Remke, A., Stasch, C., Autermann, C., Rieke, M., & Möllers, J. (2015). enviroCar: A Citizen Science Platform for Analyzing and Mapping Crowd-Sourced Car Sensor Data. *Transactions in GIS*, *19*(3), 362–376. doi:10.1111/tgis.12155

Burkett, E. (2011). A Crowdfunding Exemption? Online Investment Crowdfunding and U.S. Securities Regulation. *Transactions: The Tennessee Journal of Business, 13*(2004), 63–106.

Colombo, M. G., Franzoni, C., & Rossi-Lamastra, C. (2014). Internal social capital and the attraction of early contributions in crowdfunding. *Entrepreneurship: Theory and Practice*, *39*(1), 75–100. doi:10.1111/etap.12118

Cooper, J. (2015). Technology: Crowd-funding - Playing the game. *The Banker*. The Financial Times Limited.

Cordova, A., Dolci, J., & Gianfrate, G. (2013). The Bearable Lightness of Crowdfunding : Evidences from Technology Projects. *4th European Conference on Corporate R&D and Innovation CONCORDi-2013*. http://doi.org/ doi:<ALIGNMENT.qj></ALIGNMENT>10.1016/j.sbspro.2015.04.872

Cornelius, N., & Wallace, J. (2013). Capabilities, urban unrest and social enterprise: Limits of the actions of third sector organisations. *International Journal of Public Sector Management*, *26*(3), 232–249. doi:10.1108/IJPSM-11-2008-0043

Cumming, D., & Johan, S. (2013). Demand-driven securities regulation: Evidence from crowdfunding. *Venture Capital*, *15*(4), 361–379. doi:10.1080/13691066.2013.847635

Dacanay, M. L. (2012). *Social Enterprises and the Poor Enhancing Social Entrepreneurship and Stakeholder Theory*. Academic Press.

Dault, A. (2014). Crowdfunding Indie Theatre: Understanding the Costs. *Canadian Theatre Review*, *160*, 64–67. doi:10.3138/ctr.160.012

Desai, S., & Ács, Z. J. (2007). *A theory of destructive entrepreneurship* (No. 085). Jena Economic Research Paper.

DeTurk, S. (2001). Intercultural empathy: Myth, competency, or possibility for alliance building? *Communication Education*, *50*(4), 374–384. doi:10.1080/03634520109379262

Dougherty, D. S., Mobley, S. K., & Smith, S. E. (2010). Language Convergence and Meaning Divergence: A Theory of Intercultural Communication. *Journal of International and Intercultural Communication*, *3*(2), 164–186. doi:10.1080/17513051003611628

Dushnitsky, G., & Maron, D. (2013). Crowd Monogamy. *Business Strategy Review*, *24*(4), 24–26. doi:10.1111/j.1467-8616.2013.00990.x

Earley, K. (2015). UK market ripe for investment. *Renewable Energy Focus*, *16*(1), 22–23. doi:10.1016/S1755-0084(15)30019-3

Feder, T. (2013). Scientists experiment with crowdfunding. *Physics Today*, *66*(4), 23. [REMOVED HYPERLINK FIELD] doi:10.1063/PT.3.1942

Frydrych, D., Bock, A., Kinder, T., & Koeck, B. (2014). Exploring entrepreneurial legitimacy in reward-based crowdfunding. *Venture Capital*, *16*(3), 247–269. doi:10.1080/13691066.2014.916512

Gerber, E. M., & Hui, J. (2013). Crowdfunding : Motivations and Deterrents for Participation. *ACM Transactions on Computer-Human Interaction*, *20*(6), 32. doi:10.1145/2530540

Gill, D. E. (2015, April). Government intervention in risk capital. *Venture Capital: An International Journal of Entrepreneurial Finance*, 1–16. http://doi.org/<ALIGNMENT.qj></ALIGNMENT>10.1080/13691066.2015.1021029

Gobble, M. A. M. (2012). Everyone is a venture capitalist: The new age of crowdfunding. *Research-Technology Management*, *55*(4), 4.

Groves, J. (2015). Trillion Fund seeks to disrupt the paradigm. *Renewable Energy Focus*, *16*(2), 32–33. doi:10.1016/S1755-0084(15)30048-X

Gummerus, J. (2013). Value creation processes and value outcomes in marketing theory: Strangers or siblings? *Marketing Theory*, *13*(1), 19–46. doi:10.1177/1470593112467267

Harrison, R. (2013). Crowdfunding and the revitalisation of the early stage risk capital market: Catalyst or chimera? *Venture Capital*, *15*(4), 283–287. doi:10.1080/13691066.2013.852331

Hauge, Å. L. (2006). Identity and Place: A Critical Comparison of Three Identity Theories. *Architectural Science Review*, *50*(1), 44–51. doi:10.3763/asre.2007.5007

Heere, B., Walker, M., Gibson, H., Thapa, B., Geldenhuys, S., & Coetzee, W. (2013). The power of sport to unite a nation: The social value of the 2010 FIFA World Cup in South Africa. *European Sport Management Quarterly*, *13*(4), 450–471. doi:10.1080/16184742.2013.809136

Hennink, M., Hutter, I., & Bailey, A. (2011). *Qualitative Research Methods*. Sage Publications; doi:10.1016/B978-0-12-799943-2.00010-0

Hörisch, J. (2015). Crowdfunding for Environmental Ventures: An empirical analysis of the influence of environmental orientation on the success of crowdfunding initiatives. *Journal of Cleaner Production*, 1–10. doi:10.1016/j.jclepro.2015.05.046

Ibrahim, N., & Verlyantina, N. (2012). The Model of Crowdfunding to Support Small and Micro Businesses in Indonesia Through a Web-based Platform. *Procedia Economics and Finance*, *4*, 390–397. doi:10.1016/S2212-5671(12)00353-X

Jian, L., & Shin, J. (2015). Motivations behind donors' contributions to crowdfunded journalism. *Mass Communication & Society*. doi:10.1080/15205436.2014.911328

Jiao, H. (2011). A conceptual model for social entrepreneurship directed toward social impact on society. *Social Enterprise Journal*, *7*(2), 130–149. doi:10.1108/17508611111156600

Kappel, T. (2009). Ex Ante Crowdfunding and the Recording Industry: A Model for the U.S. *Loyola of Los Angeles Entertainment Law Review*, *29*(3), 375–386.

Kshetri, N. (2015). Success of Crowd-based Online Technology in Fundraising : An Institutional Perspective. *Journal of International Management*, *21*(2), 100–116. doi:10.1016/j.intman.2015.03.004

Lehner, O. M. (2013). Crowdfunding social ventures: A model and research agenda. *Venture Capital*, *15*(4), 289–311. doi:10.1080/13691066.2013.782624

Lehner, O. M. (2014, July). The formation and interplay of social capital in crowdfunded social ventures. *Entrepreneurship & Regional Development*, 1–22. http://doi.org/<ALIGNMENT.qj></ALIGNMENT>10.1080/08985626.2014.922623

Lehner, O. M., & Nicholls, A. (2014). Social finance and crowdfunding for social enterprises: A public–private case study providing legitimacy and leverage. *Venture Capital*, *16*(3), 271–286. doi:10.1080/13691066.2014.925305

Lindgren, M., & Packendorff, J. (2009). Social constructionism and entrepreneurship: Basic assumptions and consequences for theory and research. *International Journal of Entrepreneurial Behaviour & Research*, *15*(1), 25–47. doi:10.1108/13552550910934440

Macaulay, C. R. (2015). Capitalism's renaissance? The potential of repositioning the financial "meta-economy.". *Futures*, *68*, 5–18. doi:10.1016/j.futures.2014.10.016

Manning, P. (2010). Putnam and radical socio-economic theory. *International Journal of Social Economics*, *37*(3), 254–269. doi:10.1108/03068291011018794

McArthur, E. (2015). Many-to-many exchange without money: Why people share their resources. *Consumption Markets & Culture*, *18*(3), 239–256. doi:10.1080/10253866.2014.987083

Meer, J. (2014). Effects of the price of charitable giving: Evidence from an online crowdfunding platform. *Journal of Economic Behavior & Organization*, *103*, 113–124. doi:10.1016/j.jebo.2014.04.010

Mollick, E. (2014). The dynamics of crowdfunding: An exploratory study. *Journal of Business Venturing*, *29*(1), 1–16. doi:10.1016/j.jbusvent.2013.06.005

Nagalakshmi & Sudhakar. (2013). Conceptual Study And Awareness Of Crowd Sourcing. *Indian Streams Research Journal, 3*(8), 1–8. Retrieved from http://www.isrj.net/UploadedData/3056.pdf

Nahapiet, J., & Ghoshal, S. (1998). Social Capital, Intellectual Capital, and the Organizational Advantage. *Academy of Management Review*, *23*(2), 242. doi:10.2307/259373

Nakassis, C. V., & Searle, L. G. (2013). Introduction: Social value projects in post-liberalisation India. *Contributions to Indian Sociology*, *47*(2), 169–183. doi:10.1177/0069966713482962

Ordanini, A., Miceli, L., Pizzetti, M., & Parasuraman, A. (2011). Crowd-funding: Transforming customers into investors through innovative service platforms. *Journal of Service Management*, *22*(4), 443–470. doi:10.1108/09564231111155079

Padgett, B. L. (2014). Crowd funding: A case study at the intersection of social media and business ethics. *Journal of the International Academy for Case Studies*, *20*(3), 61–67.

Parker, S. C. (2014). Crowdfunding, Cascades and Informed Investors. *IZA Discussion Paper, No. 7994, 125*(3), 432–435. http://doi.org/<ALIGNMENT.qj></ALIGNMENT>10.1016/j.econlet.2014.10.001

Peloza, J., & Shang, J. (2011). How can corporate social responsibility activities create value for stakeholders? A systematic review. *Journal of the Academy of Marketing Science*, *39*(1), 117–135. doi:10.1007/s11747-010-0213-6

Perilleux, A. (2015). When Social Enterprises Engage in Finance: Agents of Change in Lending Relationships, a Belgian Technology. *Strategic Change*, *24*(3), 285–300. doi:10.1002/jsc.2009

Pitschner, S., & Pitschner-Finn, S. (2014). Non-profit differentials in crowd-based financing: Evidence from 50,000 campaigns. *Economics Letters*, *123*(3), 391–394. doi:10.1016/j.econlet.2014.03.022

Porter, M. E., Hills, G., Pfitzer, M., Patscheke, S., & Hawkins, E. (2012). Measuring Shared Value How to Unlock Value by Linking Social and Business Results. In *Conference Report*. Retrieved from http://www.fsg.org/tabid/191/ArticleId/740/Default.aspx?srpush=true

Porter, M. E., & Kramer, M. R. (2006). Strategy and Society: The Link Between Competitive Advantage and Corporate Social Responsibility. *Harvard Business Review*, *85*(12). PMID:17183795

Porter, M. E., & Kramer, M. R. (2011). Creating Shared Value. *Harvard Business Review*, *89*(1-2), 62–77.

Prpić, J., Shukla, P. P., Kietzmann, J. H., & McCarthy, I. P. (2015). How to work a crowd: Developing crowd capital through crowdsourcing. *Business Horizons*, *58*(1), 77–85. doi:10.1016/j.bushor.2014.09.005

Putnam, B. R. D. (1993). Social Capital and Public Life. *The American Prospect*.

Reicher, S. D., Haslam, S. A., Spears, R., & Reynolds, K. J. (2012). A social mind : The context of John Turner 's work and its influence. *European Review of Social Psychology*, *23*(1), 344–385. doi:10.1080/10463283.2012.745672

Ridley-Duff, R., & Bull, M. (2011). *Understanding social enterprise: Theory and practice*. Sage Publications.

Rossi, M. (2014). The New Ways to Raise Capital: An Exploratory Study of Crowdfunding. *International Journal of Financial Research*, *5*(2), 8. doi:10.5430/ijfr.v5n2p8

Ruebottom, T. (2011). Counting social change : Outcome measures for social enterprise. *Social Enterprise Journal*, *7*(2), 173–182. doi:10.1108/17508611111156628

Sánchez, J. C., & Loredo, J. C. (2009). Constructivisms from a genetic point of view: A critical classification of current tendencies. *Integrative Psychological & Behavioral Science*, *43*(4), 332–349. doi:10.1007/s12124-009-9091-1 PMID:19306117

Seelos, C., & Mair, J. (2005). Social entrepreneurship: Creating new business models to serve the poor. *Business Horizons*, *48*(3), 241–246. doi:10.1016/j.bushor.2004.11.006

Shiller, R. J. (2013). Capitalism and Financial Innovation. *Financial Analysts Journal*, *69*(1), 1–23. doi:10.2469/faj.v69.n1.4

Stemler, A. R. (2013). The JOBS Act and crowdfunding: Harnessing the power-and money-of the masses. *Business Horizons*, *56*(3), 271–275. doi:10.1016/j.bushor.2013.01.007

Strike, K. A. (1999). Can Schools Be Communities? The Tension Between Shared Values and Inclusion. *Educational Administration Quarterly*, *35*(1), 46–70. doi:10.1177/00131619921968464

Teece, D. J. (2010). Business Models, Business Strategy and Innovation. *Long Range Planning*, *43*(2-3), 172–194. doi:10.1016/j.lrp.2009.07.003

Tomczak, A., & Brem, A. (2013). A conceptualized investment model of crowdfunding. *Venture Capital*, *15*(4), 335–359. doi:10.1080/13691066.2013.847614

Tsai, W., & Ghoshal, S. (1998). Social Capital and Value Creation : The Role of Intrafirm Networks. *Academy of Management Journal*, *41*(4), 464–476. doi:10.2307/257085

Valančienė, L., & Jegelevičiūtė, S. (2014). Crowdfunding for Creating Value: Stakeholder Approach. *Procedia: Social and Behavioral Sciences*, *156*(April), 599–604. doi:10.1016/j.sbspro.2014.11.248

Van Staveren, I. (2013). Caring Finance Practices. *Journal of Economic Issues*, *47*(2), 419–426. doi:10.2753/JEI0021-3624470215

Vasileiadou, E., Huijben, J. C. C. M., & Raven, R. P. J. M. (2015). Three is a crowd? Exploring the potential of crowdfunding for renewable energy in the Netherlands. *Journal of Cleaner Production*, *3*, 1–14. doi:10.1016/j.jclepro.2015.06.028

Vezzoli, C., Ceschin, F., Diehl, J. C., & Kohtala, C. (2015). New Design Challenges to Widely Implement "Sustainable Product-Service Systems.". *Journal of Cleaner Production*, *97*, 1–12. doi:10.1016/j.jclepro.2015.02.061

Weitzman, M. L. (1986). The share economy conquering stagflation.pdf. *Industrial & Labor Relations Review*, *39*(2), 285–290. doi:10.1177/001979398603900210

Yoo, J., Choi, S., Choi, M., & Rho, J. (2014). Why people use Twitter: Social conformity and social value perspectives. *Online Information Review*, *38*(2), 265–283. doi:10.1108/OIR-11-2012-0210

Zachary, R., & Mishra, C. S. (2013). Research on Angel Investments: The Intersection of Equity Investments and Entrepreneurship. *Entrepreneurship Research Journal*, *3*(2), 160–170. doi:10.1515/erj-2013-0044

Zheng, H., Li, D., Wu, J., & Xu, Y. (2014). The role of multidimensional social capital in crowdfunding: A comparative study in China and US. *Information & Management*, *51*(4), 488–496. doi:10.1016/j.im.2014.03.003

Zhou, Z., Wu, J. P., Zhang, Q., & Xu, S. (2013). Transforming visitors into members in online brand communities: Evidence from China. *Journal of Business Research*, *66*(12), 2438–2443. doi:10.1016/j.jbusres.2013.05.032

## ADDITIONAL READING

Bains, W., Wooder, S., & Guzman, D. R. M. (2014). Funding biotech start-ups in a post-VC world. *Journal of Commercial Biotechnology*, *20*(1), 10–27. doi:10.5912/jcb628

Bull, M., & Crompton, H. (2006). Business practices in social enterprises. *Social Enterprise Journal*, *2*(1), 42–60. doi:10.1108/17508610680000712

Defourny, J., & Nyssens, M. (2008). Social enterprise in Europe: Recent trends and developments. *Social Enterprise Journal*, *4*(3), 202–228. doi:10.1108/17508610810922703

Dey, P., & Steyaert, C. (2010). The politics of narrating social entrepreneurship. *Journal of Enterprising Communities: People and Places in the Global Economy*, *4*(1), 85–108. doi:10.1108/17506201011029528

Doong, S.-C., Fung, H.-G., & Wu, J.-Y. (2011). Are social, financial, and human capital value enhancing? Evidence from Taiwanese firms. *International Review of Economics & Finance*, *20*(3), 395–405. doi:10.1016/j.iref.2010.11.018

Emerson, J. (2003). The Blended Value Proposition: Integrating social and financial returns. *California Management Review*, *45*(4), 35–52. doi:10.2307/41166187

Gobble, M. A. M. (2012). Everyone is a venture capitalist: The new age of crowdfunding. *Research-Technology Management*, *55*(4), 4.

McWade, W. (2012). The Role for Social Enterprises and Social Investors in the Development Struggle. *Journal of Social Entrepreneurship*, *3*(1), 96–112. doi:10.1080/19420676.2012.663783

Millar, R., & Hall, K. (2013). Social Return on Investment (SROI) and Performance Measurement. *Public Management Review*, *15*(6), 923–941. doi:10.1080/14719037.2012.698857

Pirson, M. (2012). Social entrepreneurs as the paragons of shared value creation? A critical perspective. *Social Enterprise Journal*, *8*(1), 31–48. doi:10.1108/17508611211226575

Pitschner, S., & Pitschner-Finn, S. (2014). Non-profit differentials in crowd-based financing: Evidence from 50,000 campaigns. *Economics Letters*, *123*(3), 391–394. doi:10.1016/j.econlet.2014.03.022

Porter, M. E., Hills, G., Pfitzer, M., Patscheke, S., & Hawkins, E. (2012). Measuring Shared Value How to Unlock Value by Linking Social and Business Results. In *Conference Report*. Retrieved from http://www.fsg.org/tabid/191/ArticleId/740/Default.aspx?srpush=true

Ridley-Duff, R., & Bull, M. (2011). *Understanding social enterprise: Theory and practice*. Sage Publications.

Wilson, D., & Bull, M. F. (2013). SROI in practice: The Wooden Canal Boat Society. *Social Enterprise Journal*, *9*(3), 315–325. doi:10.1108/SEJ-03-2013-0013

## KEY TERMS AND DEFINITIONS

**Crowd Capital:** Refers to the social, economic or cultural resources available to a particular group that can be accessed and utilised by its members when required.

**Crowd-Funder Value:** The value of engagement and involvement of crowd-funders at all stages of crowd-funding due to the potential positive contribution to the investment process.

**Crowd-Funding Platforms:** A third party service for enabling crowd-funding activity to happen by bringing together crowd-funders and crowd-funders (i.e. entrepreneurs).

**Social Enterprise:** An organisation that aims to benefit society by undertaking community projects, employing jobless, and looking after vulnerable populations.

**Social Identity:** The perception of self in relation to others when belonging to a particular group, and the feeling of similarities and differences in relation to others.

**Social Relational Capital:** The strength in the bond shared with others, and refers to building trust gradually.

**Social Value:** The social benefit that arises by employing community resources and engaging the members of the community.

**Symbolic Capital:** Refers to the non-verbal communication and pointers indicating that a person may belong to a specific group, by way of its associations to non-tangible things, such as brand identity.

# Compilation of References


*About MyFootballClub*. (n.d.). Retrieved August 20, 2015 from http://www.myfootballclub.co.uk/about-myfootballclub

*About the D.R.E.A.M.S Framework from TrustMan by BlaBlaCar.* (n.d.). Retrieved from www.betrustman.com/#about-framework

Abundance Generation. (2014). *Investments that build a better world to spend your money in*. Renewable Energy Crowd-funding Conference. Retrieved from http://www.recrowdfunding.eu/#intro-2015

Accredited Investor. (n.d.). *Securities and Exchange Commission*. Retrieved 30 August 2015 from http://www.sec.gov/accreditedinvestor

Acs, Z. J., Boardman, M. C., & McNeely, C. L. (2011). The social value of productive entrepreneurship. *Small Business Economics*, *40*(3), 785–796. doi:10.1007/s11187-011-9396-6

AFME. (2015). *Bridging the growth gap*. Boston Consulting Group.

Ahlers, G. K. C., Cumming, D. J., Guenther, C., & Schweizer, D. (2012). *Signaling in equity crowdfunding*. Retrieved May 15, 2015, from http://ssrn.com/abstract=2161587

Aitamurto, T. (2011). The Impact of Crowdfunding on Journalism. *Journalism Practice*, *5*(4), 429–445. doi:10.1080/17512786.2010.551018

Algesheimer, R., Dholakia, U. M., & Herrmann, A. (2005). The social influence of brand community: Evidence from European car clubs. *Journal of Marketing*, *69*(3), 19–34. doi:10.1509/jmkg.69.3.19.66363

Allison, T. H., Davis, B. C., Short, J. C., & Webb, J. W. (2015, January). Crowdfunding in a prosocial microlending environment: Examining the role of intrinsic versus extrinsic cues. *Entrepreneurship Theory and Practice,* 53-73.

Allport, F. H. (1962). A structuronomic conception of behavior: Individual and collective. *Journal of Abnormal and Social Psychology*, *64*(1), 3–30. doi:10.1037/h0043563 PMID:13860640

AMA. (n.d.). *Definition of Marketing*. Retrieved from: https://www.ama.org/AboutAMA/Pages/Definition-of-Marketing.aspx

Amabile, T. M. (1988). A model of creativity and innovation in organizations. *Research in Organizational Behavior*, *10*, 123–167.

Ambani, P. (2015). *Four ways crowdsourcing will change our lives in 2015*. Retrieved on October 10, 2015 from http://www.virgin.com/entrepreneur/four-ways-crowdsourcing-will-change-our-lives-in-2015

Anderson, S., & Cavanagh, J. (2005). *Field Guide to the Global Economy* (2nd ed.). New Press.

Arvidsson, A. (2008). The ethical economy of customer co-production. *Journal of Macromarketing*, *28*(4), 326–338. doi:10.1177/0276146708326077

Austin, J. E., & Seitanidi, M. M. Value Creation Spectrum and Collaboration Stages. (2012). Collaborative Value Creation: A Review of Partnering Between Nonprofits and Businesses: Part I. Value Creation Spectrum and Collaboration Stages. *Nonprofit and Voluntary Sector Quarterly*, *41*(5), 726–758. doi:10.1177/0899764012450777

Austin, J., Stevenson, H., & Wei-skillern, J. (2006). Social and Commercial Entrepreneurship : Same, Different, or Both? *Entrepreneurship Theory and Practice*, *3*(1), 1–22. doi:10.1111/j.1540-6520.2006.00107.x

Baeck, P., Collins, L., & Zhang, B. (2014). *Understanding Alternative Finance*. London.

Bagozzi, R. P., & Dholakia, U. M. (2006a). Open source software user communities: A study of participation in Linux user groups. *Management Science*, *52*(7), 1099–1115. doi:10.1287/mnsc.1060.0545

Bagozzi, R. P., & Dholakia, U. M. (2006b). Antecedents and purchase consequences of customer participation in small group brand communities. *International Journal of Research in Marketing*, *23*(1), 45–61. doi:10.1016/j.ijresmar.2006.01.005

Bagozzi, R. P., & Lee, K. H. (2002). Multiple routes for social influence: The role of compliance, internalization, and social identity. *Social Psychology Quarterly*, *65*(3), 226–247. doi:10.2307/3090121

Bains, W., Wooder, S., & Guzman, D. R. M. (2014). Funding biotech start-ups in a post-VC world. *Journal of Commercial Biotechnology*, *20*(1), 10–27. doi:10.5912/jcb628

Bakan, J. (2004). *The Corporation: The Pathological Pursuit of Profit and Power*. New York: Free Press.

Barkawi, W. (2015). *The Dynamic New Trends Of Entrepreneurship*. Retrieved on October 10, 2015 from http://www.entrepreneur.com/article/247319

Bayar, T. (2013). Dutch Wind Turbine Purchase Sets World Crowdfunding Record. *Renewable Energy Worl*. Available at: http://www.renewableenergyworld.com/articles/2013/09/dutch-wind-turbine-purchase-sets-world-crowdfunding-record.html

Bell, D. (1974). *The Coming of Post-Industrial Society*.New York: Harper Colophon Books.

Belleflamme, P., Lambert, T., & Schwienbacher, A. (2012). *Crowdfunding: Tapping the Right Crowd*. CORE Discussion Paper No. 2011/32.

Belleflamme, P., Lambert, T., & Schwienbacher, A. (2013). Individual crowdfunding practices. *Venture Capital*, *15*(4), 313–333. doi:10.1080/13691066.2013.785151

Belleflamme, P., Lambert, T., & Schwienbacher, A. (2014). Crowdfunding: Tapping the right crowd. *Journal of Business Venturing*, *29*(5), 585–609. doi:10.1016/j.jbusvent.2013.07.003

Bennett, L., Chin, B., & Jones, B. (2014). Crowdfunding. *New Media & Society*, 1–8.

Benson, G. D. (2001). Science Education from a Social Constructivist Position : A Worldview. *Studies in Philosophy and Education*, *20*(5), 443–452. doi:10.1023/A:1012035404449

Berger, A. N., & Udell, G. F. (1998). The Economics of Small Business Finance. The Roles of Private Equity and Debt Markets in the Financial Growth Cycle. *Journal of Banking & Finance*, *22*, 613. doi:10.1016/S0378-4266(98)00038-7

Berlingen, F., Gauthey, M. A., De Grave, A., Filippova, D., Guedira, A., Léonard, A., & Tordo, M. (2015). *Société collaborative, la fin des hiérarchies*. Paris: Rue de l'Echiquier.

Betts, B. (2010). Bringing the factory home. *Engineering & Technology*, *5*(8), 56–58. doi:10.1049/et.2010.0813

Beugré, C. D., & Das, N. (2013). Limited capital and new venture creation in emerging economies: A model of crowd-capitalism. *SAM Advanced Management Journal.*, *78*(3), 21–27.

Bhidé, A. (1992). Bootstrap Finance: The Art of Start-Ups. *Harvard Business Review*, *70*(6), 109. PMID:10122688

Blais, A. (2015). *Crowdlending for Renewable energies. A simple way to associate citizens*. Renewable Energy Crowdfunding Conference, London.

Bocken, N. M. P., Short, S. W., Rana, P., & Evans, S. (2014). A literature and practice review to develop sustainable business model archetypes. *Journal of Cleaner Production*, *65*, 42–56. doi:10.1016/j.jclepro.2013.11.039

Booth, P. (2015). Crowdfunding: A spimatic application of digital fandom. *New Media & Society*, *17*(2), 149–166. doi:10.1177/1461444814558907

Bormann, E. G., Cragan, J. F., & Shields, D. C. (2003). Defending symbolic convergence theory from an imaginary Gunn. *The Quarterly Journal of Speech*, *89*(4), 366–372. doi:10.1080/0033563032000160990

Botsman, R. (2013). *The Sharing Economy Lacks A Shared Definition*. Retrieved from: www.fastcoexist.com/3022028/the-sharing-economy-lacks-a-shared-definition

Bradford, C. S. (2012). Crowdfunding and the Federal Securities Laws. *Columbia Business Law Review*, *1*, 1–150.

Bradshaw, S., Bowyer, A., & Haufe, P. (2010). The intellectual property implications of low-cost 3D printing. *ScriptEd*, *7*(1), 5–31.

Braet, O., Spek, S., & Pauwels, C. (2013). Crowdfunding the movies: A Business Analysis of Crowdfinanced Moviemaking in Small Geographical Markets. *Journal of Media Business Studies*, *10*(1), 1–23. doi:10.1080/16522354.2013.11073557

Bretschneider, U., Knaub, K., & Wieck, E. (2014). Motivations for crowdfunding: what drives the crowd to invest in start-ups?*Twenty Second European Conference on Information Systems.*

Bretschneider, U., Knaub, K., & Wieck, E. (2014). Motivations for crowdfunding: What drives the crowd to invest in start-ups?*Twenty Second European Conference on Information Systems.*

Bröring, A., Remke, A., Stasch, C., Autermann, C., Rieke, M., & Möllers, J. (2015). enviroCar: A Citizen Science Platform for Analyzing and Mapping Crowd-Sourced Car Sensor Data. *Transactions in GIS*, *19*(3), 362–376. doi:10.1111/tgis.12155

Brown, T. (2010). *Blog post*. Retrieved from http://designthinking.ideo.com/?p=482

Burkett, E. (2011). A Crowdfunding Exemption? Online Investment Crowdfunding and U.S. Securities Regulation. *Transactions: The Tennessee Journal of Business, 13*(2004), 63–106.

Burtch, G., Ghose, A., & Wattal, S. (2012). *An empirical examination of the antecedents and consequences of investment patterns in crowd-funded Markets*. Retrieved May 15, 2015, from http://www.krannert.purdue.edu/academics/mis/workshop/2012/crowdfunding%20journalism.pdf

Butler, B. (2014). *IDC: Cloud will be $107B industry by 2017*. Retrieved from: www.networkworld.com/article/2175333/cloud-computing/idc--cloud-will-be--107b-industry-by-2017.html

Buysere, K. D., Gajda, O., Kleverlaan, R., & Marom, D. (2012). *A framework for European crowdfunding*. Retrieved May 14, 2015, from www.crowdfundingframework.eu

Calvino, I. (1957). *The Baron in the Trees*. Turin: Einaudi.

Camarinha-Matos, L., & Afsarmanesh, H. (2006). Collaborative networks: value creation in a knowledge society. *PROLAMAT 2006 Conference Proceedings- Knowledge Enterprise: Intelligent Strategies in Product Design, Manufacturing, and Management*, (pp. 26-40). Springer.

Carpini, M. X. D. (2000). *Gen.Com: Youth civic engagement and the new information environment. Departmental Paper.* University of Pennsylvania, Annenberg School for Communication.

Chateau, B., & Rossetti di Valdalbero, D. (Eds.). (2011). *World and European Energy and Environment Transition Outlook*. Luxembourg: European Commission.

Chen, P., Strickler, Y., & Adler, C. (2012). *Kickstarter Is Not a Store*. Retrieved on October 10, 2015 from https://www.kickstarter.com/blog/kickstarter-is-not-a-store

Chesbrough, H. (2003). *Open Innovation: The new imperative for creating and profiting from technology*. Boston: Harvard Business School Press.

Chesbrough, H. W. (2006). *Open innovation: The new imperative for creating and profiting from technology*. Boston: Harvard Business Press.

Chiu, C. M., Hsu, M. H., & Wang, E. T. G. (2006). Understanding knowledge sharing in virtual communities: An integration of social capital and social cognitive theories. *Decision Support Systems*, *42*(3), 1872–1888. doi:10.1016/j.dss.2006.04.001

Cholakova, M., & Clarysse, B. (2015, January). Does the possibility to make equity investments in crowdfunding projects crowd out reward-based investments?. *Entrepreneurship Theory and Practice,* 145-172.

CIM. (n.d.). *What is Marketing?* Retrieved from: http://www.cim.co.uk/more/getin2marketing/what-is-marketing/

Colombo, M. G., Franzoni, C., & Rossi-Lamastra, C. (2014). Internal social capital and the attraction of early contributions in crowdfunding. *Entrepreneurship: Theory and Practice*, *39*(1), 75–100. doi:10.1111/etap.12118

Conte, A. (2014). *Efficiency of R&D Spending at national and regional level*. Luxembourg: EC, Joint Research Centre.

Cooper, J. (2015). Technology: Crowd-funding - Playing the game. *The Banker*. The Financial Times Limited.

Cordova, A., Dolci, J., & Gianfrate, G. (2013). The Bearable Lightness of Crowdfunding : Evidences from Technology Projects. *4th European Conference on Corporate R&D and Innovation CONCORDi-2013*. http://doi.org/ doi:<ALIGNMENT.qj></ALIGNMENT>10.1016/j.sbspro.2015.04.872

Cornelius, N., & Wallace, J. (2013). Capabilities, urban unrest and social enterprise: Limits of the actions of third sector organisations. *International Journal of Public Sector Management*, *26*(3), 232–249. doi:10.1108/IJPSM-11-2008-0043

Costa, D. L., & Kahn, M. E. (2002). *Civic engagement and community heterogeneity: An economist's perspective.* Presentation on Conference on Social Connectedness and Public Activism. Harvard University.

Cova, B., & Cova, V. (2012). On the road to prosumption: Marketing discourse and the development of consumer competencies. *Consumption Markets & Culture*, *15*(2), 149–168. doi:10.1080/10253866.2012.654956

Crane, A., & Ruebottom, T. (2011). Stakeholder theory and social identity: Rethinking stakeholder identification. *Journal of Business Ethics*, *102*(S1), 77–87. doi:10.1007/s10551-011-1191-4

Creative Funding for Creative Media. (2014). *Future Media Lab. & European Crowdfunding Network*. Retrieved June 14, 2015, from http://www.eurocrowd.org/wp-content/blogs.dir/12/files/2014/01/FML-ECN-EMMA-crowdfunding-guide-final.pdf

Crowdcube. (2014). *Pizza Rossa campaign*. Retrieved from https://www.crowdcube.com/investment/pizza-rossa-ltd-12705

Crowdfunding Buzz. (2015). *Pmi in Uk Raccoglie più di £2,8 Milioni con Equity Crowdfunding di cui 1 Milione da un solo Investitore*. Retrieved from http://www.crowdfundingbuzz.it/pmi-in-uk-raccoglie-piu-di-28-milioni-con-equity-crowdfunding-di-cui-1-milione-da-un-solo-investitore-e-record/

*Crowdfunding Industry Report*. (2013). Retrieved May 14, 2015, http://www.buzzacott.co.uk/getattachment/a4e47539-1038-40c9-b90d-8709e3f4e4f/crowdfunding-industry-report-1.pdf.aspx

Crowdfunding's Potential for the Developing World. (2013). infoDev, Finance and Private Sector Development Department. Washington, DC: World Bank.

Crowdsourcing.org (2015). *Global Crowdfunding Market to Reach $34.4B in 2015, Predicts Massolution's 2015CF Industry Report*. Retrieved from: www.crowdsourcing.org/editorial/global-crowdfunding-market-to-reach-344b-in-2015-predicts-massolutions-2015cf-industry-report/45376

Cumming, D., & Johan, S. (2013). Demand-driven securities regulation: Evidence from crowdfunding. *Venture Capital*, *15*(4), 361–379. doi:10.1080/13691066.2013.847635

Dacanay, M. L. (2012). *Social Enterprises and the Poor Enhancing Social Entrepreneurship and Stakeholder Theory*. Academic Press.

Dault, A. (2014). Crowdfunding Indie Theatre: Understanding the Costs. *Canadian Theatre Review*, *160*, 64–67. doi:10.3138/ctr.160.012

Davidson, R., & Poor, N. (2015). The barriers facing artists' use of crowdfunding platforms: Personality, emotional labor, and going to the well one too many times. *New Media & Society*, *17*(2), 289–307. doi:10.1177/1461444814558916

Davies, R. (2014). *Civic crowdfunding as a marketplace for participation in urban development*. Center for Work, Technology and Organizations, Stanford University. Retrieved June 18, 2015, from http://ipp.oii.ox.ac.uk/sites/ipp/files/documents/IPP2014_Davies.pdf

Davies, R. (2015). Three provocations for civic crowdfunding. *Information Communication and Society*, *18*(3), 342–355. doi:10.1080/1369118X.2014.989878

Dawson, T. (2015). *Want To Prove Your Product? Validate On Indiegogo!* Retrieved from Indiegogo Blog: https://go.indiegogo.com/blog/2015/07/validate-your-product-on-indiegogo.html

De Meo, P., Ferrara, E., Fiumara, G., & Provetti, A. (2014). On Facebook, most ties are weak. *Communications of the ACM*, *57*(11), 78–84. doi:10.1145/2629438

Dees, G. (1998). *The Meaning of Social Entrepreneurship*. Kansas City: Kauffman Foundation and Stanford University.

Desai, S., & Ács, Z. J. (2007). *A theory of destructive entrepreneurship* (No. 085). Jena Economic Research Paper.

DeTurk, S. (2001). Intercultural empathy: Myth, competency, or possibility for alliance building? *Communication Education*, *50*(4), 374–384. doi:10.1080/03634520109379262

Diamond, J. (1997). *Guns, Germs and Steel: The Fates of Human Societies*. New York: W.W. Horton & Co.

*Digital Single Market*. (2015). Retrieved May 26, 2015, from http://ec.europa.eu/digital-agenda/en/digital-single-market

Dimitrov, D., Schreve, K., & de Beer, N. (2006). Advances in three dimensional printing– state of the art and future perspectives. *Rapid Prototyping Journal*, *12*(3), 136–147. doi:10.1108/13552540610670717

Dóci, G., & Vasileiadou, E. (2015). "Let's do it ourselves" Individual motivations for investing in renewables at community level. *Renewable & Sustainable Energy Reviews*, *49*, 41–50. doi:10.1016/j.rser.2015.04.051

Dougherty, D. S., Mobley, S. K., & Smith, S. E. (2010). Language Convergence and Meaning Divergence: A Theory of Intercultural Communication. *Journal of International and Intercultural Communication*, *3*(2), 164–186. doi:10.1080/17513051003611628

Dovidio, J. F., Gaertner, S. L., Pearson, A. R., & Riek, B. M. (2005). Social identities and social context: Social attitudes and personal well-being. *Social Identification in Groups: Advances in Group Processes*, *22*, 231–260. doi:10.1016/S0882-6145(05)22009-X

Duhigg, C., & Baraboza, D. (2012, January 5). In China, Human Costs Are Built Into an iPad. *The New York Times*. Retrieved from http://www.nytimes.com

Dushnitsky, G., & Maron, D. (2013). Crowd Monogamy. *Business Strategy Review*, *24*(4), 24–26. doi:10.1111/j.1467-8616.2013.00990.x

Earley, K. (2015). UK market ripe for investment. *Renewable Energy Focus*, *16*(1), 22–23. doi:10.1016/S1755-0084(15)30019-3

*EC-European Commission* (2014a), *Communication, Towards a circular economy: A zero waste programme for Europe.* COM(2014)398, 2 July.

EC-European Commission. (2010). Europe 2020: A strategy for smart, sustainable and inclusive growth. *COM*, *2020*, 3.

EC-European Commission. (2013a). *Global Europe 2050.* Luxembourg: European Union.

EC-European Commission. (2013b). *Accessibility Based Business Models for Peer-to-Peer Markets*. Brussels: European Commission.

EC-European Commission. (2013c). Innovation Union Competitiveness Report 2013[December.]. *SWD*, *505*, 2.

EC-European Commission. (2013d). *Guide to social innovation*. Brussels: European Commission.

EC-European Commission. (2014b). *EU research in Foresight*. Brussels: European Commission.

EC-European Commission. (2014c). *The 2015 Ageing Report*. Brussels: European Commission.

EC-European Commission. (2014d). *Scoping Paper of Horizon 2020, Societal Challenge 'Europe in a changing world - inclusive, innovative and reflective societies*. Brussels: European Commission.

EC-European Commission. (2014e). *Communication Enhancing and focusing EU international cooperation in research and innovation: A strategic approach*. COM(2012)497, 14 September.

EC-European Commission. (2015a). Communication on steps towards completing Economic and Monetary Union. *COM*, *600*, 21.

EC-European Commission. (2015b). Communication, Closing the loop - An EU action plan for the Circular Economy. *COM*, *614*, 2.

EC-European Commission. (2016). *Open Innovation 2.0*. Retrieved from https://ec.europa.eu/digital-single-market/en/open-innovation-20

*Econeers*. (n.d.). Retrieved from: https://www.econeers.de/

Ellemers, N., Kortekaas, P., & Ouwerkerk, J. W. (1999). Self-categorisation, commitment to the group and group self-esteem as related but distinct aspects of social identity. *European Journal of Social Psychology*, *29*(2-3), 371–389. doi:10.1002/(SICI)1099-0992(199903/05)29:2/3<371::AID-EJSP932>3.0.CO;2-U

Ellen MacArthur Foundation. (2012), Towards the Circular Economy: Economic and business rationale for an accelerated transition. Isle of Weight, UK: Author.

Emerson, J., & Twersky, F. (1996). *New Social Entrepreneurs: The Success, Challenges, and Lessons of Non-Profit Enterprise Creation*. Roberts Foundation.

*Enerfip*. (n.d.). Retrieved from: https://www.enerfip.fr/

EU Commission. (2014). *Crowdfunding: regulatory framework in EU member states and perspectives for the EU*. Luxembourg: Publications Office of the European Union.

European Commission Report. (2013). *3D printing*. Retrieved January 9, 2015, from http://ec.europa.eu/digital-agenda/futurium/en/content/3d-printing

European Union. (2014). *Global trends to 2030 – Can the EU meet the challenges ahead?* ESPAS report. Brussels: European Union.

Eurostat. (2015). *Statistics by theme*. Eurostat website.

Fabricant, R. (2013). Meet Your New R&D Team: Social Entrepreneurs. *Harvard Business Review*. Retrieved from: https://hbr.org/2013/03/meet-your-new-rd-team-social-e

*Fan Funded Music*. (n.d.). Retrieved July 12, 2015, from https://www.sellaband.com/en/pages/about_us

Federal Reserve Bank. (n.d.). *Consumer Credit*. Retrieved 1 August 2015 from http://www.federalreserve.gov/economicdata&research/statisticalreleaseandhistoricaldata/consumercredit

Feder, T. (2013). Scientists experiment with crowdfunding. *Physics Today*, *66*(4), 23. [REMOVED HYPERLINK FIELD] doi:10.1063/PT.3.1942

Frankfurt School-UNEP & BNEF. (2015). *Global Trends in Renewable Energy Investment*. Available at: http://fs-unep-centre.org/sites/default/files/attachments/key_findings.pdf

*Freddie Mac*. (n.d.). Retrieved 15 September 2015 from http://www.freddiemac.com/economic&housingresearch/mortgageoriginations

Freeman, R. E., Harrison, J. S., Wicks, A., Parmar, B. L., & De Colle, S. (2010). Stakeholder theory – The state of the Art. Cambridge.

Frydrych, D., Bock, A., Kinder, T., & Koeck, B. (2014). Exploring entrepreneurial legitimacy in reward-based crowdfunding. *Venture Capital*, *16*(3), 247–269. doi:10.1080/13691066.2014.916512

Galak, J., Small, D., & Stephen, A. T. (2011). Microfinance decision making: A field study of prosocial lending. *JMR, Journal of Marketing Research*, *48*(SPL), 130–137. doi:10.1509/jmkr.48.SPL.S130

Galuszka, P., & Bystov, V. (2014). The rise of fanvestors: A study of a crowdfunding community. *First Monday*, *19*(5). doi:10.5210/fm.v19i5.4117

Garnefeld, I., Iseke, A., & Krebs, A. (2012). Explicit incentives in online communities: Boon or bane? *International Journal of Electronic Commerce*, *17*(1), 11–37. doi:10.2753/JEC1086-4415170101

Geels, F. W. (2013). The impact of the financial–economic crisis on sustainability transitions: Financial investment, governance and public discourse. *Environmental Innovation and Societal Transitions*, *6*, 67–95. doi:10.1016/j.eist.2012.11.004

Gerber, E. M., Hui, J. S., & Kuo, P. Y. (2013). *Crowdfunding: why People are motivated to post and fund Projects on crowdfunding platforms*. Retrieved May 16, 2015, from http://www.juliehui.org/wp-content/uploads/2013/04/CSCW_Crowdfunding_Final.pdf

Gerber, E. M., Hui, J. S., & Kuo, P. Y. (2013). *Crowdfunding: Why people are motivated to post and fund Projects on crowdfunding platforms*. Retrieved May 16, 2015, from http://www.juliehui.org/wp-content/uploads/2013/04/CSCW_Crowdfunding_Final.pdf

Gerber, E. M., & Hui, J. (2013). Crowdfunding : Motivations and Deterrents for Participation. *ACM Transactions on Computer-Human Interaction*, *20*(6), 32. doi:10.1145/2530540

Gerber, E. M., & Hui, J. S. (2013). Crowdfunding: Motivations and deterrents for participation. *ACM Transactions on Computer-Human Interaction. 20*(6). *Article*, *34*(December), 1–32.

Gill, D. E. (2015, April). Government intervention in risk capital. *Venture Capital: An International Journal of Entrepreneurial Finance*, 1–16. http://doi.org/<ALIGNMENT.qj></ALIGNMENT>10.1080/13691066.2015.1021029

Gobble, M. A. M. (2012). Everyone is a venture capitalist: The new age of crowdfunding. *Research-Technology Management*, *55*(4), 4.

Gould, R. W. (2012). Open Innovation and Stakeholder involvement. *Journal of Technology Management & Innovation*, *7*(3).

Graeber, D. (2011). *Debt: The First 5,000 Years*. New York: Melville House.

Granovetter, M. S. (1973). The strength of weak ties. *American Journal of Sociology*, *78*(6), 1360–1380. doi:10.1086/225469

Gray, B. (2014). *Brady Capital Research Nov 24, 2014 II. Economic Abundance: Where the Long Tail Meets the Blue Ocean*. Retrieved from: http://www.linkedin.com/pulse/20141124150237-10448698-ii-where-the-long-tail-meets-the-blue-ocean?trk=mp-reader-card

Gray, M., Leaton, J., Schuwerk, R., & Fulton, M. (2015). *Coal: Caught in the EU Utility Death Spiral*. Carbon Tracker Initiative Report. Available at: http://www.carbontracker.org/report/eu_utilities/

Greenberg, M. D., & Gerber, E. M. (2014). Learning to fail: Experiencing public failure online through crowdfunding. In *Proceedings of the SIGCHI Conference on Human Factors in Computing Systems*, (pp. 581-590).

Grell. (2015). *The Kingii Crowd-First Tactics & Go-to-Market Strategy with Crowdfunding*. Retrieved form Crowdfund Insider: http://www.crowdfundinsider.com/2015/08/72662-the-kingii-crowd-first-tactics-go-to-market-strategy-with-crowdfunding/

Grepper, R. (2014). *COOLEST COOLER: 21st Century Cooler that's Actually Cooler*. Retrieved from: https://www.kickstarter.com/projects/ryangrepper/coolest-cooler-21st-century-cooler-thats-actually

Groves, J. (2015). Trillion Fund seeks to disrupt the paradigm. *Renewable Energy Focus*, *16*(2), 32–33. doi:10.1016/S1755-0084(15)30048-X

GSMA Intelligence (2015). *Data*. GSMA Intelligence website.

*Guinness World Records 2015 : Most Crowdfunded Projects*. (n.d.). Retrieved August 25, 2015, from http://www.guinnessworldrecords.com/2015/preview/files/assets/basic-html/page17.html

Gummerus, J. (2013). Value creation processes and value outcomes in marketing theory: Strangers or siblings? *Marketing Theory*, *13*(1), 19–46. doi:10.1177/1470593112467267

Harrison, R. (2013). Crowdfunding and the revitalisation of the early stage risk capital market: Catalyst or chimera? *Venture Capital*, *15*(4), 283–287. doi:10.1080/13691066.2013.852331

Hars, A., & Ou, S. (2002). Working for free? Motivations for participating in open-source projects. *International Journal of Electronic Commerce*, *6*(3), 25–39.

Hauge, Å. L. (2006). Identity and Place: A Critical Comparison of Three Identity Theories. *Architectural Science Review*, *50*(1), 44–51. doi:10.3763/asre.2007.5007

Heere, B., Walker, M., Gibson, H., Thapa, B., Geldenhuys, S., & Coetzee, W. (2013). The power of sport to unite a nation: The social value of the 2010 FIFA World Cup in South Africa. *European Sport Management Quarterly*, *13*(4), 450–471. doi:10.1080/16184742.2013.809136

Hemer, J. (2011). *A snapshot on crowdfunding*. Working papers firms and region, No. R2/2011. Retrieved May 16, 2015, from http://hdl.handle.net/10419/52302

Hennink, M., Hutter, I., & Bailey, A. (2011). *Qualitative Research Methods*. Sage Publications; doi:10.1016/B978-0-12-799943-2.00010-0

Higginbottom, S. (2015). Everyone's Betting on Blockchain because We Need Technology That Can Scale. *Fortune*. Retrieved from: http://fortune.com/2015/05/29/blockchain-big-because-scale/?iid=sr-link1

Hollow, M. (2013). Crowdfunding and civic society in Europe: A Profitable partnership? *Open Citizenship*, *4*, 68–73.

Hörisch, J. (2015). Crowdfunding for Environmental Ventures: An empirical analysis of the influence of environmental orientation on the success of crowdfunding initiatives. *Journal of Cleaner Production*, 1–10. doi:10.1016/j.jclepro.2015.05.046

Hui, J. S., Greenberg, M. D., & Gerber, E. M. (2014). Understanding the role of community in crowdfunding work. In *Proceedings of the 17th ACM conference on Computer supported cooperative work & social computing*, (pp. 62-74). ACM.

Hui, J. S., Gerber, E. M., & Greenberg, M. D. (2012). Easy money? The demands of crowdfunding work. *Northwestern University Segal Design Institute Technical Report*, *12*, 4.

Huizingh, E. K. R. (2011). Open innovation: State of the art and future perspectives. *Technovation*, *31*(1), 2–9. doi:10.1016/j.technovation.2010.10.002

Hull, P. (2013). Talk about your idea, it won't get stolen. *Forbes Magazine*. Retrieved from: http://www.forbes.com/sites/patrickhull/2013/06/13/talk-about-your-idea-it-wont-get-stolen/#2688ac5972d1

Ibrahim, N., & Verlyantina, N. (2012). The Model of Crowdfunding to Support Small and Micro Businesses in Indonesia Through a Web-based Platform. *Procedia Economics and Finance*, *4*, 390–397. doi:10.1016/S2212-5671(12)00353-X

ILO. (2013). *Providing clean energy access through cooperatives*. Geneva: International Labour Office Cooperative Unit Report. Green Jobs Programme.

*Indiegogo Terms and Rules*. (n.d.). Retrieved August, 2015 from http://www.indiegogo.com

Indiegogo. (2014). *Kinoma Campaign*. Retrieved from https://www.indiegogo.com/projects/kinoma-create#/

*Indiegogo*. (n.d.). Retrieved from: https://www.indiegogo.com/

Innoventures, S., & Wyman, O., & the Anthemis Group. (2015). *The Fintech 2.0 Paper: Rebooting financial services.* Retrieved from: http://santanderinnoventures.com/wp-content/uploads/2015/06/The-Fintech-2-0-Paper.pdf

Isler, S. T. (2014). Modern finansmanda yeni bir dönem: Kitlesel fonlama – Crowdfunding. *Gösterge.* Retrieved May 10, 2015, from http://www.tspb.org.tr/tr/Portals/0/aim__gosterge_guz_2014_02.pdf

Jeffery, N. (2009). *Stakeholder Engagement: A Road Map to Meaningful Engagement.* Doughty Centre, Cranfield School of Management.

Jian, L., & Shin, J. (2015). Motivations behind donors' contributions to crowdfunded journalism. *Mass Communication & Society*. doi:10.1080/15205436.2014.911328

Jiao, H. (2011). A conceptual model for social entrepreneurship directed toward social impact on society. *Social Enterprise Journal*, *7*(2), 130–149. doi:10.1108/17508611111156600

John Paul, I. I. (1991). *Pontifical Council for Justice and Peace*. Academic Press.

Juncker, J.-C. (2014). *Speech: Time for Action – Statement in the European Parliament plenary session ahead of the vote on the College.* speech/14/1525, Strasbourg, 22 October.

Juncker, J-C. (2015). *Speech: State of the Union 2015 – Time for honesty, unity and solidarity.* speech/15/5614, Strasbourg, 9 September.

Kappel, T. (2009). Ex Ante Crowdfunding and the Recording Industry: A Model for the U.S.?. *Loyola of Los Angeles Entertainment Law Review*, 375.

Kappel, T. (2009). Ex Ante Crowdfunding and the Recording Industry: A Model for the U.S. *Loyola of Los Angeles Entertainment Law Review*, *29*(3), 375–386.

Katainen, J. (2015). *Speech: Circular Economy: boosting business, reducing waste.* Speech at the Circular economy stakeholder conference, Brussels, 25 June.

Kayser, O. (2009). *Ashoka to Partner with Hystra to Increase Corporate Engagement in Scaling Up Social Innovations.* Retrieved from https://www.ashoka.org/press/5809

Kempener, R., Malhotra, A., & De Vivero, G. (2015). *The age of renewable energy power. Designing national roadmaps for a successful transformation power.* IRENA Power Sector Transformation Report. Available at: http://www.irena.org/DocumentDownloads/Publications/IRENA_PST_Age_of_Renewable_Power_2015.pdf

Keushnigg, C. (2004). Venture Capital Backed Growth. *Journal of Economic Growth*, *9*(2), 239–261. doi:10.1023/B:JOEG.0000031428.35711.fc

Khatib, F., DiMaio, F., Cooper, S., & Kazmierczyk, M. (2011). Crystal structure of a monomeric retroviral protease solved by protein folding game players in *Nature Structural &. Molecular Biology*, *18*, 1175–1177.

*Kickstarter Terms and Rules*. (n.d.). Retrieved August, 2015 from http://www.kickstarter.com

Kickstarter. (2014). *Coolest Cooler Campaign*. Retrieved from: https://www.kickstarter.com/projects/ryangrepper/coolest-cooler-21st-century-cooler-thats-actually/description

*Kickstarter*. (n.d.) Retrieved from: https://www.kickstarter.com/

*Kickstarter 2014 : By The Numbers*. (n.d.). Retrieved August 25, 2015, from https://www.kickstarter.com/year/2014/data?ref=yir2014

Kirby, E., & Worner, S. (2014). *Crowdfunding: An infant industry growing fast.* Staff Working Paper of the International Organisation of Securities Commissions Research Department. Retrieved May 14, 2015, from http://www.memofin.fr/uploads/library/pdf/Crowd-funding-An-Infant-Industry-Growing-Fast[1].pdf

Klaes, M. (2015). *Crowdfunding for Renewable Energy Projects: European Survey Results.Renewable Energy Crowdfunding Conference.*

Korten, D. (2001). *When Corporations Rule the World.* Kumarian Press, Inc and Berrett-Koehler Publishers, *Inc.*

Kotler, P. (1986). The prosumer movement: A new challenge for marketers. *Advances in Consumer Research. Association for Consumer Research (U. S.), 13*(1), 510–513.

Kotler, P. (2012). *Marketing Management.* Pearson Education.

Kozinets, R. V. (1999). E-tribalized marketing?: The strategic implications of virtual communities of consumption. *European Management Journal, 17*(3), 252–264. doi:10.1016/S0263-2373(99)00004-3

Kraut, R. E., & Resnick, P. (2012). *Building successful online communities: Evidence-based social design.* Cambridge, MA: MIT Press.

Kshetri, N. (2015). Success of Crowd-based Online Technology in Fundraising : An Institutional Perspective. *Journal of International Management, 21*(2), 100–116. doi:10.1016/j.intman.2015.03.004

Kuppuswamy, V., & Bayus, B. L. (2014). *Crowdfunding Creative Ideas: The Dynamics of Project Backers in Kickstarter.* UNC Kenan-Flagler Research Paper. Retrieved June 27, 2015, from http://funginstitute.berkeley.edu/wp-content/uploads/2013/11/Crowdfunding_Creative_Ideas.pdf

Kuti, M., & Madarász, G. (2014). Crowdfunding. *Public Finance Quarterly, 3*, 355–366.

Lakhani, K. R., & Wolf, R. G. (2005). Why hackers do what they do: Understanding motivation and effort in free/open source software projects. In J. Feller, B. Fitzergerald, S. Hissam, & K. R. Lakhani (Eds.), *Perspectives on Free and Open Source Software* (pp. 3–27). Cambridge, MA: MIT Press.

Lambert, T., & Schwienbacher, A. (2010). *An empirical analysis of crowdfunding.* Retrieved May 10, 2015, from http://www.crowdsourcing.org/document/an-empirical-analysis-of-crowdfunding-/2458

Landes, D. (1969). *The Unbound Prometheus.* Cambridge, UK: Cambridge University Press.

Lego. (n.d.). *Lego Ideas.* Retrieved from: https://ideas.lego.com/

Lehner, O. M. (2014, July). The formation and interplay of social capital in crowdfunded social ventures. *Entrepreneurship & Regional Development*, 1–22. http://doi.org/<ALIGNMENT.qj></ALIGNMENT>10.1080/08985626.2014.922623

Lehner, O. M. (2013). Crowdfunding social ventures: A model and research agenda. *Venture Capital, 15*(4), 289–311. doi:10.1080/13691066.2013.782624

Lehner, O. M., & Nicholls, A. (2014). Social finance and crowdfunding for social enterprises: A public–private case study providing legitimacy and leverage. *Venture Capital, 16*(3), 271–286. doi:10.1080/13691066.2014.925305

Leonard, A. (2007). *The Story of change.* Retrieved from http://storyofstuff.org/wp-content/uploads/movies/scripts/SoChange_Annotated_Script.pdf

Leonard, A. (2007). *The Story of stuff.* Retrieved from http://storyofstuff.org/wp-content/uploads/movies/scripts/Story%20of%20Stuff.pdf

Levy, S. J. (1959). Symbols for sale. *Harvard Business Review, 37*(4), 117–124.

Light, J. (2014, August 27). Nonbank Mortgage Originators Bounce Back. *Wall Street Journal.*

Lindgren, M., & Packendorff, J. (2009). Social constructionism and entrepreneurship: Basic assumptions and consequences for theory and research. *International Journal of Entrepreneurial Behaviour & Research, 15*(1), 25–47. doi:10.1108/13552550910934440

Lin, Y., Boh, W. F., & Goh, K. H. (2014). How different are crowdfunders? Examining archetypes of crowdfunders and their choice of projects. *Organizational Communication & Information Systems, 1*, 13309.

Lyons, T. S. (2012). *Understanding Social Entrepreneurship: The Relentless Pursuit of Mission in an ever changing world.* New York: Routledge.

Macaulay, C. R. (2015). Capitalism's renaissance? The potential of repositioning the financial "meta-economy.". *Futures, 68*, 5–18. doi:10.1016/j.futures.2014.10.016

Macht, S. A., & Weatherston, J. (2014). The benefits of online crowdfunding for fund-seeking business ventures. *Strategic Change, 23*(1-2), 1–14. doi:10.1002/jsc.1955

Mael, F. A., & Ashforth, B. E. (1992). Alumni and their alma mater: A partial test of the reformulated model of organizational identification. *Journal of Organizational Behavior, 13*(2), 103–123. doi:10.1002/job.4030130202

Manning, P. (2010). Putnam and radical socio-economic theory. *International Journal of Social Economics, 37*(3), 254–269. doi:10.1108/03068291011018794

Market Wired. (2013). *Pebble Announces $15M in Funding and the Next Stage of Its Open Platform.* Retrieved from http://www.marketwired.com/press-release/pebble-announces-15m-in-funding-and-the-next-stage-of-its-open-platform-1792035.htm

Marom, D., & Sade, O. (2014). *Are the life and death of an early stage venture indeed in the power of the tongue? Lessons from online crowdfunding pitches.* Retrieved May 28, 2015, from http://www.researchgate.net/publication/256060198

Marom, D., Robb, A., & Sade, O. (2015), *Gender Dynamics in Crowdfunding (Kickstarter): Evidence on Entrepreneurs, Investors, Deals and Taste Based Discrimination.* Retrieved October 10, 2015 from http://ssrn.com/abstract=2442954

Martinez-Canas, R., Ruiz-Palomino, P., & Pozo-Rubio, R. (2012). crowdfunding and social networks in the music industry: Implications for entrepreneurship. *International Business & Economics Research Journal, 11*(13), 1471.

Massolution. (2015). *Real Estate Crowdfunding – Industry Report.* Author.

Massolution. (2015). *The crowdfunding industry report.* Massolution Report. Available at: www.massolution.com

*Massolution's 2013 CF Industry Report.* (2013). Retrieved May 15, 2015, from http://www.compromisoempresarial.com/wp-content/uploads/137356857-Massolution-2013CF-Excerpt-Revised-04182.pdf

*Massolution's 2015 CF Industry Report.* (2015). Retrieved May 10, 2015, from http://www.crowdsourcing.org/editorial/global-crowdfunding-market-to-reach-344b-in-2015-predicts-massolutions-2015cf-industry-report/45376

Mazzucato, M. (2013). *The Entrepreneurial State: debunking public vs. private sector myths.* London: Anthem Press.

McArthur, E. (2015). Many-to-many exchange without money: Why people share their resources. *Consumption Markets & Culture, 18*(3), 239–256. doi:10.1080/10253866.2014.987083

Mccann, D. (2013). Solar Crowdfunding Heats Up. *CFO, 29*, 9–9.

McCarthy, J. E. (1964). *Basic Marketing. A Managerial Approach.* Homewood, IL: Irwin.

McGowan, H. (2014). *Jobs are Over: The future is income generation*. Retrieved on October 10 from https://www.linkedin.com/pulse/20140720200547-22726740-jobs-are-over-the-future-is-income-generation-part-4?trk=mp-details-rr-rmpost

McKinsey Company. (2014). *The Trillion Dollar Convergence: Capturing the Next Wave of Growth in Alternative Investments*. New York: McKinsey Financial Services Practice.

Meer, J. (2014). Effects of the price of charitable giving: Evidence from an online crowdfunding platform. *Journal of Economic Behavior & Organization*, *103*, 113–124. doi:10.1016/j.jebo.2014.04.010

Meier, P. (2014). *iRevolutions. Results of the Crowdsourced Search for Malaysia Flight 370 (Updated)*. Retrieved from: https://irevolutions.org/2014/03/15/results-of-the-crowdsourced-flight-370-search/

Merchant, N. (2012). Let your ideas go. *Harvard Business Review*. Retrieved from: https://hbr.org/2012/06/let-your-ideas-go

Meric, J., Bouiass, K., & Maque, I. (2015). More than three's a crowd... in the best interest of companies!: Crowdfunding as zeitgeist or ideology? *Society and Business Review*, *10*(1), 23–39. doi:10.1108/SBR-09-2014-0042

Meyer, B. (2011). Macroeconomic modelling of sustainable development and the links between the economy and the environment. Study for the European Commission.

Milosevic, M. (2015). *Crowdfunding as an innovation booster factor and its dependence on information systems*. Retrieved June 10, 2015, from https://hal-paris1.archives-ouvertes.fr/hal-01158238/document

Milway, K. (2014). How social entrepreneurs can have the most impact. *Harvard Business Review*. Retrieved from: https://hbr.org/2014/05/how-social-entrepreneurs-can-have-the-most-impact/

Mitra, D. (2012). The role of crowdfunding in entrepreneurial finance. *Delhi Business Review*, *13*(2), 67–72.

Moedas, C. (2015a). *Speech: Trust in Europe: The contribution of the social sciences and humanities*. Speech at the Trust Conference, Brussels, 29 October.

Moedas, C. (2015b). *Speech: Open Innovation, Open Science, Open to the World*. Speech/15/5243, Brussels, 22 June.

Moedas, C. (2015c). *Speech: The EU approach to science diplomacy*. Speech at the European Institute, Washington, 1 June.

Mollick, E. R., & Kuppuswamy, V. (2014). *After the Campaign: Outcomes of Crowdfunding*. UNC Kenan-Flagler Research Paper No. 2376997. Retrieved from: http://papers.ssrn.com/sol3/papers.cfm?abstract_id=2376997

Mollick, E. (2014). The dynamics of crowdfunding: An exploratory study. *Journal of Business Venturing*, *29*(1), 1–16. doi:10.1016/j.jbusvent.2013.06.005

Morlock, G. S. (2010). Personal fabrication. *Communications of the ACM*, *53*(10), 14–15. doi:10.1145/1831407.1831414

*Mosaic*. (n.d.). Retrieved from: https://joinmosaic.com/

Moules, J. (2013). *Big companies show they are in with the in crowdfunding*. Retrieved form Financial Times: http://www.ft.com/intl/cms/s/0/1234e5f6-30e5-11e3-b478-00144feab7de.html#axzz3jHbGvQfG

Muller, M., Geyer, W., Soule, T., & Wafer, J. (2014). Geographical and organizational commonalities in enterprise crowdfunding. CSCW Technology and Information Workers.

*Musicraiser Terms and Rules*. (n.d.). Retrieved August, 2015 from http://www.musicraiser.com

Nagalakshmi & Sudhakar. (2013). Conceptual Study And Awareness Of Crowd Sourcing. *Indian Streams Research Journal*, *3*(8), 1–8. Retrieved from http://www.isrj.net/UploadedData/3056.pdf

Nahapiet, J., & Ghoshal, S. (1998). Social Capital, Intellectual Capital, and the Organizational Advantage. *Academy of Management Review*, *23*(2), 242. doi:10.2307/259373

Nakassis, C. V., & Searle, L. G. (2013). Introduction: Social value projects in post-liberalisation India. *Contributions to Indian Sociology*, *47*(2), 169–183. doi:10.1177/0069966713482962

Nash, R., & Beardsley, E. (2015). *The Future of Finance Part One, The Rise of New Shadow Bank*. Goldman Sachs.

Nash, R., & Beardsley, E. (2015). *The Future of Finance Part One, The Rise of the New Shadow Bank*. New York: Goldman Sachs.

Nekaj, E. L. (2014). *Forget the sharing economy. It's time for the crowd economy*. Retrieved on October 10, 2015 from http://www.virgin.com/entrepreneur/forget-the-sharing-economy-its-time-for-the-crowd-economy

Niederer, P. (n.d.). *Simple Equity Crowdfunding Fundability Evaluation Tool Revisited*. Retrieved from http://www.paulniederer.com/2015/06/a-simple-equity-crowdfunding-fundability-evaluation-tool-revisited/

Oliveira, P. (2014). *Patient Innovation: Crowdsourcing Healthcare Solutions - Crowdsourcing Week Global 2014*. Retrieved from: https://www.youtube.com/watch?v=sblJKVFv3_s

Ordanini, A., Miceli, L., Pizzetti, M., & Parasuraman, A. (2011). Crowdfunding: Transforming customers into investors through innovative service platforms. *Journal of Service Management*, *22*(4), 443–470. doi:10.1108/09564231111155079

Oreskes, N., & Conway, E. (2010). *The Merchants of Doubts*. London: Bloomsburry.

Padgett, B. L. (2014). Crowd funding: A case study at the intersection of social media and business ethics. *Journal of the International Academy for Case Studies*, *20*(3), 61–67.

Pais, I., Peretti, P., & Spinelli, C. (2014). *Crowdfunding. La via collaborativa all'imprenditorialita*. Milan: Egea.

Parker, S. C. (2014). Crowdfunding, Cascades and Informed Investors. *IZA Discussion Paper, No. 7994, 125*(3), 432–435. http://doi.org/<ALIGNMENT.qj></ALIGNMENT>10.1016/j.econlet.2014.10.001

*Patreon Terms and Rules*. (n.d.). Retrieved August, 2015 from http://www.patreon.com

Pearl, B. (2011). *Britta Riley: A garden in my apartment*. Retrieved from http://opensource.com/life/11/12/britta-riley-garden-my-apartment

Peloza, J., & Shang, J. (2011). How can corporate social responsibility activities create value for stakeholders? A systematic review. *Journal of the Academy of Marketing Science*, *39*(1), 117–135. doi:10.1007/s11747-010-0213-6

Perez, Y. B. (2015). *Santander: Blockchain Tech Can Save Banks $20 Billion a Year*. Retrieved from: www.coindesk.com/santander-blockchain-tech-can-save-banks-20-billion-a-year/

Perilleux, A. (2015). When Social Enterprises Engage in Finance: Agents of Change in Lending Relationships, a Belgian Technology. *Strategic Change*, *24*(3), 285–300. doi:10.1002/jsc.2009

Phimister, E., & Roberts, D. (2012). The Role of Ownership in Determining the Rural Economic Benefits of On-shore Wind Farms. *Journal of Agricultural Economics*, *63*(2), 331–360. doi:10.1111/j.1477-9552.2012.00336.x

Pitschner, S., & Pitschner-Finn, S. (2014). Non-profit differentials in crowd-based financing: Evidence from 50,000 campaigns. *Economics Letters*, *123*(3), 391–394. doi:10.1016/j.econlet.2014.03.022

*PledgeMusic Terms and Rules*. (n.d.). Retrieved August, 2015 from http://www.pledgemusic.com

Porter, M. E., Hills, G., Pfitzer, M., Patscheke, S., & Hawkins, E. (2012). Measuring Shared Value How to Unlock Value by Linking Social and Business Results. In *Conference Report*. Retrieved from http://www.fsg.org/tabid/191/ArticleId/740/Default.aspx?srpush=true

Porter, M. E., & Kramer, M. R. (2006). Strategy and Society: The Link Between Competitive Advantage and Corporate Social Responsibility. *Harvard Business Review*, *85*(12). PMID:17183795

Porter, M. E., & Kramer, M. R. (2011). Creating Shared Value. *Harvard Business Review*, *89*(1-2), 62–77.

Potter, D. (2014). Why the Future of Social Innovation Is Open. *Stanford Social Innovation Review*. Retrieved from: http://ssir.org/articles/entry/why_the_future_of_social_innovation_is_open

Prahalad, C. K., & Ramaswamy, V. (2004). Co-creation experiences: The next practice in value creation. *Journal of Interactive Marketing*, *18*(3), 5–14. doi:10.1002/dir.20015

Prpić, J., Shukla, P. P., Kietzmann, J. H., & McCarthy, I. P. (2015). How to work a crowd: Developing crowd capital through crowdsourcing. *Business Horizons*, *58*(1), 77–85. doi:10.1016/j.bushor.2014.09.005

*Pubslush Terms and Rules*. (n.d.). Retrieved August, 2015 from http://www.pubslush.com

Putnam, B. R. D. (1993). Social Capital and Public Life. *The American Prospect*.

PwC. (2015). *Peer Pressure - How peer-to-peer Lending Platforms are Transforming the Consumer Lending Industry*. Retrieved on October 10, 2015 from https://www.pwc.com/us/en/consumer-finance/publications/assets/peer-to-peer-lending.pdf

Pyburn, G. (2014). *Crowdfunding (Equity-Based) In Asia-Pacific: Legal and Regulatory Ambiguities without an Analogous us Jobs Act Exemption*. Academic Press.

Rayner, T. (2011). *The Coalition of the willing*. Retrieved from http://coalitionofthewilling.org.uk/

Reicher, S. D., Haslam, S. A., Spears, R., & Reynolds, K. J. (2012). A social mind : The context of John Turner 's work and its influence. *European Review of Social Psychology*, *23*(1), 344–385. doi:10.1080/10463283.2012.745672

REIT. (2014). *Market Caps*. Washington, DC: U.S. REIT Industry.

Reitsma, H. D. W. (2014). *Accelerating the Energy Transition*. Renewable Energy Crowdfunding Conference. Retrieved from http://www.recrowdfunding.eu/#intro-2015

Renewable Energy Crowdunding Conference. (2014). *Proceedings of the Renewable Energy Crowdfunding Conference*.

Report, C. F. Regulatory Framework in EU Member States and Perspectives for the EU. (2014). *European Commission*. Retrieved July 10, 2015, from http://capacity4dev.ec.europa.eu/crowdfunding/document/crowdfunding-regulatory-framework-eu-member-states-and-perspectives-eu-en-version

Review of Crowdfunding Regulation. (2014). *Interpretations of existing regulation concerning crowdfunding in Europe, North America and Israel*. European Crowdfunding Network. Retrieved May 25, 2015, from http://www.eurocrowd.org/files/2014/12/ECN-Review-of-Crowdfunding-Regulation-2014.pdf

Ridley-Duff, R., & Bull, M. (2011). *Understanding social enterprise: Theory and practice*. Sage Publications.

Ries, E. (2015). *Only on Kickstarter: THE LEADER'S GUIDE by ERIC RIES*. Retrieved on October 10, 2015 from https://www.kickstarter.com/projects/881308232/only-on-kickstarter-the-leaders-guide-by-eric-ries/description

Ritzer, G., & Jurgenson, N. (2010). Production, consumption, prosumption: The nature of capitalism in the age of the digital prosumer. *Journal of Consumer Culture*, *10*(1), 13–36. doi:10.1177/1469540509354673

Rossetti di Valdalbero, D. (2013). *Europe: Towards a socio-ecological transition*. Brussels: Belgian royal Academy of Sciences.

Rossetti di Valdalbero, D. (2014). *The nexus between man and technology*. Brussels: Belgian royal Academy of Sciences.

Rossetti di Valdalbero, D. (2015). *Quelle Europe en 2050*. Brussels: Belgian Royal Academy of Sciences.

Rossi, M. (2014). The New Ways to Raise Capital: An Exploratory Study of Crowdfunding. *International Journal of Financial Research*, *5*(2), 8. doi:10.5430/ijfr.v5n2p8

Rowan, D. (2015). *FireChat will become a totally free mobile carrier*. Retrieved from www.wired.co.uk/news/archive/2015-08/05/firechat-full-mobile-carrier-free

Ruebottom, T. (2011). Counting social change : Outcome measures for social enterprise. *Social Enterprise Journal*, *7*(2), 173–182. doi:10.1108/17508611111156628

Ruef, M. (2002). Strong ties, weak ties and islands: Structural and cultural predictors of organizational innovation. *Industrial and Corporate Change*, *11*(3), 427–449. doi:10.1093/icc/11.3.427

Safner, R. (2015). *Crowdfunding expressive works, crowding out copyright?*. Job Market Paper.

Sakamoto, M., & Nakajima, T. (2013). Micro-crowdfunding: Achieving a sustainable society through economic and social incentives in micro-level crowdfunding.*12th International Conference on Mobile and Ubiquitous Multimedia*. doi:10.1145/2541831.2541838

Sánchez, J. C., & Loredo, J. C. (2009). Constructivisms from a genetic point of view: A critical classification of current tendencies. *Integrative Psychological & Behavioral Science*, *43*(4), 332–349. doi:10.1007/s12124-009-9091-1 PMID:19306117

Sandlund, J. (2013). *The Rise of Meaningful Investing*. Retrieved from http://www.thecrowdcafe.com/rise-of-meaningful-investing/

Schinas, M. (2012). *The EU in 2030*. Brussels: Centre for European Studies.

Schumpeter, J. (1942). *Capitalism, Socialism, and Democracy*. New York: Harper & Row.

Schwienbacher, A., & Larralde, B. (2010). Crowdfunding of small entrepreneurial ventures. In *Handbook of Entrepreneurial Finance*. Oxford University Press.

Seedrs. (2015). *Domain Chanzy campaign*. Retrieved from https://learn.seedrs.com/seedrs-welcomes-domaine-chanzy-as-first-ipo-campaign/

Seelos, C., & Mair, J. (2005). Social entrepreneurship: Creating new business models to serve the poor. *Business Horizons*, *48*(3), 241–246. doi:10.1016/j.bushor.2004.11.006

Seforis Consortium. (2013). *The State of Social Entrepreneurship, Key facts and figures*. Retrieved from http://www.seforis.eu/upload/reports/1._Key_Facts_and_Figures_of_Social_Entrepreneurship.pdf

*Sellaband Terms and Rules*. (n.d.). Retrieved August, 2015 from http://www.sellaband.com

Serres, M. (1992). *Le contrat naturel*. Paris: Flammarion.

Seyfang, G., Hielscher, S., Hargreaves, T., Martiskainen, M., & Smith, A. (2014). A grassroots sustainable energy niche? Reflections on community energy in the UK. *Environmental Innovation and Societal Transitions*, *13*, 21–44. doi:10.1016/j.eist.2014.04.004

Shen, K. (2012). Identification vs. self-verification in virtual communities (VC): Theoretical gaps and design implications. In P. Maret & C. Morr (Eds.), *Virtual Community Building and the Information Society: Current and Future Directions* (pp. 208–236). IGI Global. doi:10.4018/978-1-60960-869-9.ch011

Shiller, R. J. (2013). Capitalism and Financial Innovation. *Financial Analysts Journal*, *69*(1), 1–23. doi:10.2469/faj.v69.n1.4

*Solar Schools*. (n.d.). Retrieved from: http://www.solarschools.org.uk/

Solarplaza. (2014a). *Renewable Energy Crowdfunding World Map (Update)*. Renewable Energy Crowdfunding Conference. Available at: http://www.recrowdfunding.eu/news-updates/2014/8/20/renewable-energy-crowdfunding-world-map

Solarplaza. (2014b). *Top 5 Renewable Energy Crowdfunding Platforms*. Available at: http://www.solarplaza.com/channels/archive/11313/top-5-renewable-energy-crowdfunding-platforms/

Solarplaza. (2015). *Renewable Energy Crowdfunding. Overview and map*. Renewable Energy Crowdfunding Conference. Available at: http://www.recrowdfunding.eu/news-updates/2015/9/14/tracking-renewable-energy-crowdfunding

Sony. (n.d.). *First Flight platform*. Retrieved from: https://first-flight.sony.com/

South Africa, F. C. B. (n.d.). *Giant Flag Project*. Retrieved from: https://www.giantflag.co.za/

Spacetec Capital Partners. (2014). Crowdfunding innovative ventures in Europe. the financial ecosystem and regulatory landscape. Prepared for the European Commission, DG Communications Networks.

*Star Citizen: Stats*. (n.d.). Retrieved August 25, 2015, from https://robertsspaceindustries.com/funding-goals

Stemler, A. R. (2013). The JOBS Act and crowdfunding: Harnessing the power-and money-of the masses. *Business Horizons*, *56*(3), 271–275. doi:10.1016/j.bushor.2013.01.007

Stiver, A., Barroca, L., Minocha, S., Richards, M., & Roberts, D. (2015). Civic crowdfunding research: Challenges, opportunities, and future agenda. *New Media & Society*, *17*(2), 249–271. doi:10.1177/1461444814558914

Strickler, Y. (2014). *The History of #1 UPDATED*. Retrieved on October 10, 2015 from https://www.kickstarter.com/blog/the-history-of-1-updated

Strike, K. A. (1999). Can Schools Be Communities? The Tension Between Shared Values and Inclusion. *Educational Administration Quarterly*, *35*(1), 46–70. doi:10.1177/00131619921968464

*Sun Funder*. (n.d.). Retrieved from: http://sunfunder.com/

Svendsen, A. (1998). *The Stakeholder Strategy: Profiting from Collaborative Business Relationships*. San Francisco: Berrett-Koehler Publishers, Inc.

Swart, R. (2014). *Why Crowdfunding Is a Must-Have Marketing Tool for Brands*. Retrieved from Marketing Profs: http://www.marketingprofs.com/opinions/2014/26060/why-crowdfunding-is-a-must-have-marketing-tool-for-brands#ixzz3jA6ZbNcv

Taxfoundation. (2015). *U.S. Federal Tax Rates*. Retrieved from http://www.taxfoundation.org

Teece, D. J. (2010). Business Models, Business Strategy and Innovation. *Long Range Planning*, *43*(2-3), 172–194. doi:10.1016/j.lrp.2009.07.003

The Economist. (2013). *How to lose half a trillion euros. Europe's electricity providers face an existential threat*. Available online at: http://www.economist.com/news/briefing/21587782-europes-electricity-providers-face-existential-threat-how-lose-half-trillion-euros

The Network. (2013, May 29). *Cisco's Visual Networking Index Forecast Projects Nearly Half the World's Population Will Be Connected to the Internet by 2017.* Retrieved from https://newsroom.cisco.com/press-release-content?type=webcontent&articleId=1197391

The S&P500 PE Ratio. (n.d.). In BloombergBusiness. Retrieved 1 September 2015 from http://www.bloomberg.com

*The White House Presidential Innovation Fellows.* (n.d.). Retrieved from https://www.whitehouse.gov/innovationfellows

*The whole truth about MegaTotal.pl.* (n.d.). Retrieved July 12, 2015, from http://www.megatotal.pl/pl/page/about

*The year in Kickstarter 2014.* (n.d.). Retrieved August 25, 2015 from https://www.kickstarter.com/year/2014?ref=2014datayir#stats-4

Times Realty News. (2015). *TRN Realty Crowdfunding Report 2015.* Author.

Tomczak, A., & Brem, A. (2013). A conceptualized investment model of crowdfunding. *Venture Capital, 15*(4), 335–359. doi:10.1080/13691066.2013.847614

Total. (2015). *Babyloan and Total Team Up to Create the First Ever Crowdfunding Platform Dedicated to Access to Energy.* Total Press Release. Available at: http://www.total.com/en/media/news/press-releases/babyloan-and-total-team-create-first-ever-crowdfunding-platform-dedicated-access-energy

Tozzi, J. (2013). *P&G Taps Crowdfunding Site to Scout for Startup Brands.* Retrieved on October 10, 2015 http://www.bloomberg.com/bw/articles/2013-01-29/p-and-g-taps-crowdfunding-site-to-scout-for-startup-brands

*Trampoline Systems.* (n.d.). Retrieved June 15, 2015, from http://crowdfunding.trampolinesystems.com/index.html

Tsai, W., & Ghoshal, S. (1998). Social Capital and Value Creation : The Role of Intrafirm Networks. *Academy of Management Journal, 41*(4), 464–476. doi:10.2307/257085

*Unbound Terms and Rules.* (n.d.). Retrieved August, 2015 from http://www.unbound.co.uk

*United States Census Bureau.* (n.d.). Retrieved from www.census.gov/population/international/data/worldpop/table_population.php

*Unleashing the potential of Crowdfunding in the European Union.* (2014). European Commission. Retrieved June 9, 2015, from http://ec.europa.eu/internal_market/finances/docs/crowdfunding/140327-communication_en.pdf

Valančienė, L., & Jegelevičiūtė, S. (2014). Crowdfunding for Creating Value: Stakeholder Approach. *Procedia: Social and Behavioral Sciences, 156*(April), 599–604. doi:10.1016/j.sbspro.2014.11.248

Van Der Schoor, T., & Scholtens, B. (2015). Power to the people: Local community initiatives and the transition to sustainable energy. *Renewable & Sustainable Energy Reviews, 43*, 666–675. doi:10.1016/j.rser.2014.10.089

Van Staveren, I. (2013). Caring Finance Practices. *Journal of Economic Issues, 47*(2), 419–426. doi:10.2753/JEI0021-3624470215

Van Veelen, T. (2015). *Civic crowdfunding: A financial kickstart to urban area development? An analysis of the institutional structure in which civic crowdfunding in urban area development is embedded.* Retrieved June 27, 2015, from http://repository.tudelft.nl/view/ir/uuid%3Abf8caf4f-79a8-4257-ac88-948518f5cc36/

Van Zuylen-Wood, S. (2015). *Bloomberg Businessweek The Struggles of New York City's Taxi King.* Retrieved from www.bloomberg.com/features/2015-taxi-medallion-king/

Vasileiadou, E., Huijben, J. C. C. M., & Raven, R. P. J. M. (2015). Three is a crowd? Exploring the potential of crowdfunding for renewable energy in the Netherlands. *Journal of Cleaner Production.* doi:10.1016/j.jclepro.2015.06.028

Vezzoli, C., Ceschin, F., Diehl, J. C., & Kohtala, C. (2015). New Design Challenges to Widely Implement "Sustainable Product-Service Systems.". *Journal of Cleaner Production*, *97*, 1–12. doi:10.1016/j.jclepro.2015.02.061

Vodafone. (2014). *Corporate Social Responsibility report 2014.* Retrieved from https://www.vodafone.com/content/sustainability/our_vision_and_approach/managing_sustainability/stakeholder_engagement.html

Wadrop, R., Zhang, B., Raghavendra, R., & Gray, M. (2015). Moving mainstream: The European alternative finance benchmarking report. Cambridge Centre for Alternative Finance Report. University of Cambridge.

*We Share Solar*. (n.d.). Retrieved from: http://wesharesolar.com/

Wechsberg, J. (1966). *The Merchant Bankers*. New York: Pocket Books.

Weitzman, M. L. (1986). The share economy conquering stagflation.pdf. *Industrial & Labor Relations Review*, *39*(2), 285–290. doi:10.1177/001979398603900210

Wellman, B. (1992). Which ties provide what kinds of support? *Advances in Group Processes*, *9*, 207–235.

Wellman, B. (2002). Little Boxes, Glocalization, and Networked Individualism. In M. Tanabe, P. van den Besselaar, & T. Ishida (Eds.), *Digital Cities* (pp. 10–25). Berlin: Springer-Verlag.

Wellman, B., Quan-Haase, A., Boase, J., Chen, K. H., Diaz, I., & Miyata, K. (2003). The social affordances of the Internet for networked individualism. *Journal of Computer-Mediated Communication*, *8*(3), 1–43.

Wikipedia. (n.d.). *Shadow Banking*. Retrieved 1 August 2015 from: http://en.wikipedia.org/wiki/ShadowBanking

Wikipedia. (n.d.). *Systemically Important Financial Institutions*. Retrieved 12 September 2015 from http://en.wikipedia.org/wiki/SystemicallyImportantFinancialInstituions

World Bank. (2013). *Crowdfunding's potential for the developing world. Info Dev Growing Innovation Report*. Washington, DC: The World Bank.

*World Development Report*. (n.d.). Washington, DC: World Bank.

Wright, P. K. (2005). Rapid prototyping consumer product. *Communications of the ACM*, *48*(6), 36–41. doi:10.1145/1064830.1064856

Yildiz, Ö. (2014). Financing renewable energy infrastructures via financial citizen participation – The case of Germany. *Renewable Energy*, *68*, 677–685. doi:10.1016/j.renene.2014.02.038

Yoo, J., Choi, S., Choi, M., & Rho, J. (2014). Why people use Twitter: Social conformity and social value perspectives. *Online Information Review*, *38*(2), 265–283. doi:10.1108/OIR-11-2012-0210

Zachary, R., & Mishra, C. S. (2013). Research on Angel Investments: The Intersection of Equity Investments and Entrepreneurship. *Entrepreneurship Research Journal*, *3*(2), 160–170. doi:10.1515/erj-2013-0044

Zhao, J., Wu, J., & Xu, K. (2010). Weak ties: Subtle role of information diffusion in online social networks. *Physical Review*, *82*(1), 1–8. PMID:20866687

Zheng, H., Li, D., Wua, J., & Xu, Y. (2014). The role of multidimensional social capital in crowdfunding: A comparative study in China and US. *Information & Management*, *51*(4), 488–496. doi:10.1016/j.im.2014.03.003

Zhou, Z., Wu, J. P., Zhang, Q., & Xu, S. (2013). Transforming visitors into members in online brand communities: Evidence from China. *Journal of Business Research*, *66*(12), 2438–2443. doi:10.1016/j.jbusres.2013.05.032

# About the Contributors

**Walter Vassallo** (MSc, MBA, PhD) is an economist and entrepreneur with vast professional experience in overall company coordination over multidisciplinary teams. His work is linked to his vision to translate trends and changes into opportunities, to move the way we think in order to act and implement sustainable solutions. Decade experience in policies, research, startups during which he has been involved in developing, acquiring, managing and coordinating innovative projects. He implemented assignments commissioned by private and public entities, including University teaching and mentoring. Currently he is co-founder of MC Shareable, disruptive company based in Monaco. In 2016 he created and launched Letyourboat, a website for people to list, find, and rent accommodations onboard of boats, which mission is to revolutionize the traditional recreational boating charter sector and make it accessible and affordable to everyone. Author of articles, publications and books.

* * *

**Fabio Allegreni** is an entrepreneur and a digital media advisor since 1997, with main focus on projects increasing the company assets value through converged mobile, web, social networks, e-commerce. He has an extensive knowledge of the media industry over Internet, social networks and mobile and his track record includes digital media projects development for large corporations, as well as business development advisory for start-ups, having also launched several tech start-ups each with a successful exit. Since several years, Fabio founded Crowd Advisors a consulting firm specialized in advising start-ups, SMEs and PA on fund raising, with a special focus on crowdfunding and crowd economy. He also publishes Crowdfunding Buzz, the first Italian on line magazine entirely dedicated to crowdfunding.

**Priti Ambani** is the co-founder and COO at The Next Billion, the first impact driven crowdfunding platform for women entrepreneurs enabling an ecosystem that provides women in business with access to finance, markets, skills and mentors. She is a thought leader on the crowd economy and the Principal Strategy Consultant, at CSW2, a global think-tank that helps organisations create value with crowds. Priti is a Professional Engineer and lives in Auckland, New Zealand with her husband and sons. She is a lover of the outdoors, traveling and from-scratch cooking! Follow her on Twitter @envirotarian.

**Yonca Aslanbay** is professor of marketing and the chair at the Communication PhD, Istanbul Bilgi University. She received her Ph.D. degree in 1992 and has been lecturing in different universities since then. Her current teaching agenda and prior publications are in the domains of marketing, consumer behavior and research methodology. Her recent research focuses on the new types of online collectivities and sustainability.

**Birnbaum Bogdan-Iustin** is Policy Assistant at the European Commission, DG for Research and Innovation. Bogdan holds an MA in European Interdisciplinary Studies from the College of Europe in Natolin (2008-09), and an MA in International Relations from the University of Trieste (2006-2008).

**Sebnem Burnaz** is a Professor of Marketing at Istanbul Technical University, Faculty of Management. She holds Ph.D. degree in Management with major in marketing from Bogazici University. Her research interests are in the fields of Marketing, Retailing, Decision Making, and Business Ethics. She published articles which appeared in Advances in International Marketing, Journal of Business Ethics, Sex Roles: A Journal of Research, Qualitative Marketing Research: An International Journal, Journal of Product & Brand Management and Journal of Multi-Criteria Decision Analysis.

**Chiara Candelise** is an experienced energy economist and policy specialist. She is interested in cross disciplinary research in the field and has a forward looking interest into sustainability transitions and large scale changes in socio-economic and technical systems. She is Research Fellow at Bocconi University (Italy) and Imperial College London (UK). Strong track record in delivering high impact publications and research output and contributions to the energy policy debate. In her background work experience as economists at both private and public institutions, including the UK Department for Environment, Food and Rural Affairs (Defra).

**Steven A. Cinelli** has engaged in corporate finance and strategy supporting emerging growth companies throughout many industries. Reared in commercial banking, he expanded his financing activities into private equity, venture capital and mezzanine finance. Mr. Cinelli has been on the forefront of technology applied to finance with his co-founding of OffRoad Capital (1999), the first capital market system to digitize the private placement process. He later founded PRIMARQ, which is intent of improving the housing finance industry in the US and abroad. An active speaker and writer, he has scribed reports, articles and opinion pieces on topics of housing finance, alternative finance, shadow banking and private secondary markets for publications such as the Huffington Post, Forbes, INC Magazine, American Banker, Housing Wire, TheStreet, Equities.com. Mr. Cinelli has lectured at and overseen plenary discussions at Oxford University, Stanford University, Georgetown University and Santa Clara University.

**Melek Demiray** is a Ph.D candidate in Business Administration, Institute of Social Sciences, Istanbul Technical University. She works as a research assistant at Department of Management Engineering in Istanbul Technical University. Her research interest is online communities, social network marketing, e-WOM communication.

**Domenico Rossetti di Valdalbero** is Principal Administrator at the European Commission, DG for Research and Innovation. His specialisation themes are the New paths of economic growth; Global Europe; Socio-ecological transition; and Smart urban futures. He coordinates the Horizon 2020 Societal Challenge on "Inclusive, innovative and reflective societies". Domenico got a PhD in economics. He is the author of more than 100 articles and two books entitled: "The Power of Science" and "Mapping European integration through its Cities". He is Professor at the Belgian royal Academy of Sciences and Secretary General of the Union of European Federalists.

**Tomaso Greco** teaches Sociology of Law at the University of Insubria and has co-founded bookabook, the book crowdfunding in Italy (http://www.bookabook.it). With a degree in Law and a PhD in Philosophy and Sociology of Law, he has devoted his studies to transformations of society, and has been focusing on the continuously evolving relation between rules and new technologies. He is author of two monographs, La crisi del lavoro. Dimensioni, analisi e possibili policies (Job Crisis. Dimensions, analysis and possible policies), Milano, FrancoAngeli, 2013 and Le violenze psicologiche nel mondo del lavoro. Un'analisi sociologico-giuridica del fenomeno mobbing. (Psychological abuse in the world of work. A sociological-juridical analysis of the mobbing phenomenon), Milano, Giuffré, 2009. He collaborates with several (scientific and non-scientific) magazines.

**Payal Jain** possesses a double master's joint degree: Masters in European Public Health sponsored by European Commission from the University of Sheffield, UK (2007) and Jageillonian University, Krakow, Poland (2008). Her undergraduation was in Sociology and Public Administration from the Punjab University, Chandigarh, India. She has worked as a researcher in the public and third sectors in the UK and New Zealand for four years prior to embarking on her doctoral study. Presently, she is pursuing her doctoral studies in social value creation, sponsored by the University of Northampton, based on the implementation of Public Services (Social Value) Act in 2012. She was selected by the British Council to attend an international workshop on Eco Food Supply Chain in Sao Paulo, Brazil in August 2015. The workshop was organised under the Researcher Links programme, funded under Newton Fund. Her interest lies in sociology, sociology of health and illness and applied sociology in various industrial sectors. Her areas of interest include: social entrepreneurship, social value, social policy and social and public sector service delivery.

**Ronald Kleverlaan** is an entrepreneur, author and crowdfunding thought leader. He is one of the most important innovators on crowdfunding in Europe and keynote speaker on crowdfunding trends. He is expert advisor for the European Commission on "Crowd Investing", is co-founder of the European Crowdfunding Network and is member of the advisory board of several leading crowdfunding platforms and initiatives in Europe, Asia and the USA. He is also the founder of the CrowdfundingHub, an international educational network to raise awareness on crowdfunding with local events throughout Europe. Ronald contributed to several books and international publications on crowdfunding and alternative finance and is a well-known speaker on the trends and future of crowdfunding and alternative finance.

**Alessandro M. Lerro** is an Italian attorney who has been advising for more than 25 years in innovation, new technologies and alternative finance, matching a deep expertise in both intellectual property and finance. Chairman of the Italian Equity Crowdfunding Association, Lerro is one of the most well known crowdfunding experts in the world and a leading European lawyer in this area. He is an appreciated international speaker, lecturer and writer in Italian and English about innovation and crowdfunding.

**Emanuele Musa** is an international social entrepreneurship trainer with an engineering and business background as well as proven experience in open-social innovation, crowd-sourcing, business modeling and Lean startup. Has trained over 500 social entrepreneurs through numerous workshops and bootcamps around the world - in Tokyo, Paris, Budapest, Milan, Berlin, Prague, Brussels, Bratislava, Bucharest, Lviv and Rome. Emanuele is cofounder of Babele.co : open-innovation platform that interconnects the key actors supporting the creation of common good in society (from foundations to universities and impact accelerators), in order to open-source innovative business models, spread awareness of the best practices while enabling social entrepreneurs to crowd-source feedback and validation from a wide spectrum of international stakeholders and peer-entrepreneurs. The platform was launched on September of 2013 and gathers together 5300 experts, who give feedback, share ideas and provide solution for 750 social enterprises coming from 99 countries. Previous work experience includes: Program management for Procter & Gamble in Paris, financial analysis at the Fund Manager "AE Global Investment Solutions ltd" in London, strategic analysis at Bombardier Transportation in Brussels, strategy consulting at Advention Business Partners in Paris and management consulting for IBM Global Business Services in Paris.

**Epi Ludvik Nekaj** is the Founder & CEO of Crowdsourcing Week, a global platform that is uniting thought leadership on the crowd economy through global conferences and summits. Recently Epi Ludvik launched CSW2, an action tank to help organizations and innovators embed the power of the crowd in their products and services & connect the dots in the global crowd-powered marketplace. Epi Ludvik is one of the pioneers in the crowdsourcing ad space, an industry expert and thought leader in entertainment advertising. Epi Ludvik earned a BS in Advertising & Marketing from the Fashion Institute of Technology in NYC and has been a serial entrepreneur since graduation. His varied and global experiences, throughout his career, have taken him across the US and Asia, where he honed his skills in advertising and media with high-profile clients and power brands. "Companies and enterprises can engage with their stakeholders on a completely new level -- through social interactions and work together to create value for everyone. Social media is now turning into purpose-driven social productivity."

**Paul Niederer** is one of world's leading authorities in collaborative and equity based crowdfunding, investor aggregation and capital raisings. He is COO of Crowdsourcing Week and CEO of ASSOB International. Paul also consults to a number of organisations and governments internationally on how to structure their funding portals, investor aggregation processes and regulations to successfully manage funding with transparency and governance. Internationally he is known as "Mr Governance" due to the fraud free reign he managed while CEO of ASSOB, the world's oldest equity crowdfunding platform. ASSOB facilitated capital raisings for over 300 Australian companies using techniques tailored by Paul and his team over many years. Paul lives south of Brisbane on Australia's eastern coast and relishes the opportunity to discuss peer to peer finance developments, blockchain evolution and any moves towards better governance in fund raising and the democratisation of capital.

# Index


## A

accredited investors 213-214, 216-219, 222-223, 239
adverse selection 257, 259
all-or-nothing mechanism 130, 133, 147-148
alternative assets 240-241
alternative finance 5, 45, 204, 267

## B

baby boomers 7-8
balance sheet 80, 84, 246, 252
bottom line 54, 57, 301
brand equity 195, 197, 199-201
business plans 47

## C

campaign creators 128-129, 134, 138, 141-142, 145, 149, 181
climate change 21, 23-24, 33, 52, 54, 65, 267
collective intelligence 54-55, 64-65, 68, 73, 144, 291-292
collective wisdom 64, 70
commercial banking 78, 80, 88, 239, 250, 252
commercial banks 78, 84-85, 244, 251-253
community energy 268, 281-283
community funding 224, 235
community share 272
competitive advantage 63
Corporate Social Responsibility (CSR) 26, 58
credit worthiness 87, 200
CRM 187, 239
crowd capital 4-8, 11, 14, 291, 300, 302, 309
crowd economy 1-2, 4-6, 8-12, 17-18, 37, 41, 204
crowd-funder value 288, 293, 299, 302, 309
crowdfunding 4-5, 12-14, 17, 40, 45-48, 52, 83, 85, 90-111, 113-123, 127-130, 133-153, 155-169, 175-183, 185-191, 193-202, 204-222, 224-235, 237-240, 242-245, 247-248, 250-252, 259, 264, 266-272, 274-276, 278-284, 289-292, 295, 298, 300, 302
crowdfunding benefits 175
crowdfunding model 91, 94, 97, 103, 106-108, 120, 162, 164, 166, 205, 208, 268, 271, 276
crowdfunding platforms 40, 48, 52, 95, 103-104, 107-108, 110-111, 116, 122, 130, 140-141, 143, 146, 151-152, 157, 163, 176-178, 191, 194-196, 199, 206, 216, 222, 227, 231-232, 247, 281, 284, 290-291, 293, 296, 298, 300, 302, 309
crowdfunding project 13, 105, 108, 110-111, 231
crowd investing 227, 229
crowd lending 237
crowdsourcing 1-2, 4-6, 10, 12, 14, 17, 37, 41-43, 45, 48-52, 54, 62, 66-68, 71-73, 92-93, 107, 111-112, 120, 122-123, 157-158, 167, 169, 177, 182, 237

## D

donation-based 91-92, 94, 99, 102-104, 106, 108, 111, 161, 163, 166, 204-205, 225, 232
donation based crowdfunding 107
donation crowdfunding 128-129, 205

## E

economic development 23, 113, 123
economic objective 130-131, 133-134, 136, 138, 141, 151
economic system. 25, 34, 155, 158
entrepreneurship 37-41, 52, 58, 60-61, 65, 68, 70, 72, 75, 91, 106, 113, 123, 162, 168, 209, 220, 289, 294-295, 301-302
environmental change 56, 62
environmental quality 54, 57, 60
equity 13-14, 47, 54, 57, 60, 80, 82, 84-85, 97-99, 102, 104, 106, 117, 120, 122, 128, 147, 176, 180, 189, 193-195, 197-202, 204-210, 212-222, 228-231, 237, 240-242, 244, 249-250, 264, 267, 269, 272, 275, 278-281, 289, 298, 301
equity-based 91, 96-99, 102, 104, 106-108, 110-111, 113, 115-117, 119-123, 162-168, 212, 215, 227-228
equity crowdfunding 13-14, 47, 97-99, 104, 120, 128, 147, 180, 193-195, 198-202, 204, 206-207, 209, 212-214, 216-222, 272, 275
exogenous factors 241

## F

FCA 101, 222
Financial Technology (FinTech) 17, 80, 85-86, 88, 264
financing 8, 28, 45, 79, 82-83, 94, 96-97, 100, 102, 113, 120-121, 128-129, 146-147, 158, 161, 193-194, 201, 204, 206, 208, 211-212, 215-216, 219-220, 225, 227, 231, 238, 240, 242-246, 248-252, 255, 259, 266-268, 276, 283-284, 290
finite resources 33, 54-55

## G

global economy 18, 24, 56, 74
global markets 166

## H

horizontal platforms 103, 109-110, 140-143, 150, 152

## I

Identification 122, 138, 155-156, 159, 161-169, 176, 217, 222, 292, 301
individual empowerment 20-21, 34
industrial partnership 199
innovative solutions 28-29, 66, 70
institutional investors 80, 210, 240-241, 251, 283
intellectual property 29, 33, 62, 64, 119, 145, 149, 152, 169, 272
interest rate 47, 82, 84, 95-96, 210, 228, 240-242, 244, 251-252
intrapreneur 37-38
investing 33, 72, 79, 83, 86, 97-98, 166, 176, 182, 187-189, 193-202, 211, 215, 227, 229-230, 239-242, 244, 247, 249, 255, 264, 267-268, 272, 275
investments 29, 79, 82, 86, 98, 102, 113, 119-120, 128, 167, 181, 193, 195, 209, 212-214, 216, 218, 221-223, 238, 240-241, 248-249, 266-268, 270-271, 274-276, 279, 281-283, 294-295, 298
investor network 246, 249, 259

## K

Kotler 157, 188

## L

law 14, 25, 34, 50, 56-58, 79, 102, 127, 132, 138, 168, 205, 208, 213
lean startup 40, 42, 67-68
lending 10, 13, 28, 47, 78-88, 95-96, 99, 102, 106-108, 115-118, 120-121, 129, 165, 189, 204, 210-212, 228, 231, 237, 250-252, 258, 266, 272, 274-276, 278-282, 289, 295, 298
lending-based 91, 95-96, 99, 102, 106-107, 111, 113, 115-117, 120-122, 162, 165, 167, 211-212, 220
localism 268
lofty valuations 264

## M

marketer 197
marketing 2, 4, 10, 21, 55, 58, 81, 122, 131, 136, 146, 150, 162, 178, 181-183, 187-191, 193, 195-198, 200-202, 205, 208, 221, 224, 229-233, 235, 237, 249, 284, 294
market research 43, 95, 224, 227, 230
microcredit 211, 267
monetization strategies 264
motivation 26, 28, 58, 96, 99, 120, 155, 160, 163, 166, 199, 240, 248-249, 291, 302

## N


network system 155
non-bank originators 83, 250, 264
non-profit organizations 93, 99, 106, 114, 121, 225


## O


online communities 40, 42-43, 48, 55, 91, 155-160, 167-169, 294
open platform 149
open source 44, 60, 62-64, 68


## P


participatory democracy 71, 73
pre-selling 94, 178-179, 187, 227
principal risk 242, 246
product development 41, 62-63, 298, 301-302
profit participation 246, 248-249


## Q


quantitative easing 240-241


## R


R&D 29, 33, 69-70, 199, 201, 275
real estate industry 104, 123
real estate investment trust 240, 242
regulation 4, 23, 47, 84, 90, 98, 100-101, 119, 204-206, 209, 212-214, 218-222, 247
renewables 267, 269, 271, 275
reserves 80, 84, 148, 252
residential mortgage 244, 250, 264
reward-based 91, 94, 99, 106-107, 111, 113, 115, 120, 161-164, 168, 204-205, 207, 209, 216, 225-226, 229, 232-233, 237
reward based crowdfunding 272
reward crowdfunding 102, 128-129, 139, 189-190, 193, 195-196, 200-201
risk management 78, 85, 252, 257, 264
Rudyard Kipling 78


## S


SEC 99, 213-214, 216-219, 222, 247
shadow banking 80, 252
shared ownership 268-269, 281, 283
sharing economy 1, 4, 25, 27, 51, 137
social business 59, 68
social capital 37, 166, 168, 288, 290-295, 300, 302
social enterprises 58-59, 65, 67-70, 301
social entrepreneurs 21, 26, 58, 60, 67-70, 288-289, 301
social entrepreneurship 52, 58, 60-61, 65, 68, 70, 75, 289, 295, 301-302
social finance 83, 250, 295, 301
social identity 156, 159-160, 162-165, 288, 291-294, 300, 302, 309
Social Identity Theory 159-160, 288, 291-293, 302
social innovation 20, 28, 30-32, 34, 54, 63, 67-70, 73
social productivity 1, 9-10, 52
social relational capital 309
social reputation 10-11, 48
social value 32, 63, 91, 298, 301, 309
social value creation 298
socio-economic impact 127
software programs 90, 106
solicitation 137, 206, 208, 212, 217-218
stakeholder engagement 60-61, 63, 75
stakeholder participation 63-65, 68
startups 1, 8, 12-13, 17, 41-42, 44, 47, 49, 51, 67-68, 80, 84, 94, 98, 167, 189, 199, 209-210, 213, 220, 228, 271, 275, 283, 289
sustainable development 54-55, 57, 60-61, 71
sustainable economy 20-21
sustainable enterprises 65, 72
symbolic capital 292, 300, 302, 309
Symbolic Convergence Theory 288, 302
syndications 240, 249


## T


technical knowledge 141-142
transaction fees 80, 248
trustful relationships 25, 34


## V


validation 13, 47, 67, 116, 138, 141-144, 164, 175, 180-183, 185, 205
vertical crowdfunding 103-104, 108, 110-111, 140-144, 149, 153
vertical platforms 103, 105, 109, 122, 140-141, 143-144, 150, 152-153


## W


Wall Street Reform 84, 252
worldwide attention 54, 57

www.ingramcontent.com/pod-product-compliance
Lightning Source LLC
LaVergne TN
LVHW060416300926
849162LV00006B/1442